From Sentience to Symbols

READINGS ON CONSCIOUSNESS

Edited by
John Pickering
and
Martin Skinner

HARVESTER WHEATSHEAF

New York London Toronto Sydney Tokyo Singapore

First published 1990 by
Harvester Wheatsheaf
66 Wood Lane End, Hemel Hempstead
Hertfordshire HP2 4RG
A division of
Simon & Schuster International Group

British Library Cataloguing in Publication Data

From sentience to symbols: readings on consciousness.
1. Consciousness
I. Pickering, John, *1946–* II. Skinner, Martin, *1950–*
126

ISBN 0–7450–0877–1
ISBN 0–7450–0803–8 pbk

Printed in Great Britain by
BPCC Wheatons Ltd, Exeter

Contents

Acknowledgements

The editors are grateful to the following copyright holders and publishers for granting permission to reprint material previously published.

CHAPTER 1

Descartes, R., 'Meditations on First Philosophy' and 'The Principles of Philosophy' reprinted from *The Philosophical Works of Descartes*, translated and edited by Haldane, E. S. and Ross, G. T. R., published by Cambridge University Press, 1911; 'Correspondence with Princess Elizabeth (1643)' extracted from *Descartes: Political Writings*, translated by Anscome, E. and Geach, P., published by Nelson, 1954, and reprinted by kind permission of the publishers.

Ryle, G., reprinted from *The Concept of the Mind*, Hutchinson, 1949, and reprinted by kind permission of The Bursar, Hertford College, Oxford University.

Bergson, H., 'Life and Consciousness', The Huxley Lecture, delivered at Birmingham University, May 24th, 1911, and reprinted from *Mind-Energy*, translated by H. Wildon Carr, published by Holt, 1921.

CHAPTER 2

Popper, K. & Eccles, J., extracts from the first three chapters and dialogues of *The Self and its Brain*, Springer, 1977.

Whitehead, A., reprinted from *Science in the Modern World*, Cambridge University Press.

Prigogine, I., excerpt from *Order Out of Chaos*, by Ilya Prigogine and Isabelle Stengers, copyright © 1984 by Ilya Prigogine and Isabelle Stengers. Used by permission of Bantam Books, a division of Bantam, Doubleday, Dell Publishing Group, Inc.

Bohm, D. & Peat, D., reprinted from *Science, Order and Creativity*, Routledge, 1987.

Wald, G., 'Life and Mind in the Universe' in *International Journal of Quantum Chemistry*, 1984, 11:1–15, copyright © John Wiley & Sons, Inc. Reprinted by permission of John Wiley & Sons, Inc. All Rights Reserved.

CHAPTER 3

Huxley, J., extract from *Evolution in Action*, Chatto & Windus, Ltd, 1953. Reprinted by permission of the Peters Fraser & Dunlop Group Ltd.

Griffin, D. R., reprinted from *The Question of Animal Awareness*, Rockefeller University Press, 1981.

Humphrey, N., 'Consciousness: A Just-So Story', first appeared in *New Scientist*, London, the weekly review of science and technology, 1982, vol 95, pg 474.

Popper, K., 'Natural Selection and the Emergence of Mind', reprinted from *Dialectica, Revue Internationale de Philosophie de la Connaissance*, 1978, 32:339–355, by permission of the publisher.

CHAPTER 4

Penfield, W., *The Mystery of the Mind: A Critical Study of Consciousness and the Human Brain*, copyright © 1975 Princeton University Press. Excerpt, 14 pages reprinted with permission of Princeton University Press.

Oakley, D., 'Animal Awareness, Consciousness and Self-image', reprinted from *Brain and Mind*, 1985, Methuen & Co.

Eccles, J., excerpt reprinted with permission of The Free Press, a Division of Macmillan, Inc., from *The Wonder of Being Human: Our Brain and Our Mind* by Sir John Eccles and Daniel N. Robinson. Copyright © 1984 by The Free Press.

Sperry, Roger, *Science and Moral Priority: Merging Mind, Brain, and Human Values* (Praeger Publishers, New York, 1985), abridged from pages 78–103. Copyright © 1983 by Roger Sperry. Abridged and reprinted with permission of the author and Praeger Publishers, an imprint of Greenwood Publishing Group, Inc.

CHAPTER 5

James, W., *Selections from the Principles of Psychology, Part 1*, Macmillan.

Mandler, G., 'Consciousness', Chapter 3 of *Cognitive Psychology*, Lawrence Erlbaum Associates, Inc., 1985, reprinted by permission of the author.

Johnson-Laird, P., 'A Computational Analysis of Consciousness', copyright © P. N. Johnson-Laird, 1988. Reprinted from *Consciousness in Contemporary Science* edited by A. J. Marcel and E. Bisiach (1988), by permission of Oxford University Press.

CHAPTER 6

White, L., 'Four Stages in the Evolution of Minding', reprinted from *Evolution After Darwin*, edited by S. Tax, University of Chicago Press, 1960, copyright © University of Chicago Press.

Sinha, C., 'A Socio-naturalistic Approach to Human Development', reprinted from *Evolution and Developmental Psychology*, edited by Butterworth, G., Rutkowska, J. and Scaife, M., Harvester Wheatsheaf, 1985, copyright © The Harvester Press.

Luria, A., reprinted by permission of the publishers from *Cognitive Development: Its Social and Cultural Foundations*, by A. R. Luria, Cambridge, Mass.: Harvard University Press, copyright © 1976 by the President and Fellows of Harvard College.

CHAPTER 7

Mead, G. H., 'The Mechanism of Social Consciousness', first appeared in *The Journal of Philosophy, Psychology and Scientific Methods*, 1912, 9, 401–406, reprinted in A. Reck (Ed.) *Selected Writings George Herbert Mead* (Ch. 11), University of Chicago Press, 1964.

Meltzer, B. N., *Mead's Social Psychology*, reprinted from *Symbolic Interactionism*, Allyn & Bacon, Boston, 1964, originally from B. N. Meltzer, *Mead's Social Psychology*, (1964), Center for Sociological Research, Western Michigan University.

Berger, Peter L. & Luckmann, Thomas, *The Social Construction of Reality*, Allen Lane The Penguin Press, 1967, copyright © Peter L. Berger and Thomas Luckmann, 1966.

Duval, S. & Wicklund, R., *A Theory of Objective Self Awareness*, Academic Press, 1972, copyright © 1972, by Academic Press Inc.

CHAPTER 8

Lock, A., 'Universals in Human Conception', in *Indigenous Psychologies: The Anthropology of the Self*, edited by Paul Heelas and Andrew Lock, Academic Press, 1981, copyright © 1981, by Academic Press, Inc.

Jaynes, J., 'The Mind of the Iliad', excerpts from *The Origins of Consciousness in the Breakdown of the Bicameral Mind* by Julian Jaynes. Copyright © 1976 by Julian Jaynes. Reprinted by permission of Houghton Mifflin Co.

Logan, R. D., 'Historical Change in Prevailing Sense of the Self', reprinted from *Self and Identity: Psychosocial Perspectives*, edited by K. Yardley and T. Honess, John Wiley and Sons Ltd, 1987, copyright © 1987, John Wiley & Sons Ltd. Reproduced by permission of the John Wiley and Sons Ltd.

Heelas, P., 'Introduction: Indigenous Psychologies', in *Indigenous Psychologies: The Anthropology of the Self*, edited by Paul Heelas and Andrew Lock, Academic Press, 1981, copyright © 1981, by Academic Press, Inc.

CHAPTER 9

Vygotsky, L. S., *Thought and Language*, edited and translated by E. Hanfmann and G. Vakar, M.I.T. Press, 1962, copyright © 1962, by the Massachusetts Institute of Technology.

Whorf, B. L., 'The Relation of Habitual Thought and Behaviour to Language' in *Language Thought and Reality: Selected Writings of Benjamin Lee Whorf*, edited by J. B. Carroll, M.I.T. Press, 1956, copyright © 1956, by the Massachusetts Institute of Technology.

Wittgenstein, L., *Philosophical Investigations*, translated by G. E. M. Anscombe, Oxford, Basil Blackwell, 1963, copyright © Basil Blackwell & Mott, Ltd, 1958.

Fromm, E., 'The Nature of Consciousness Repression and De-repression', from *Psychoanalysis and Zen Buddhism*, Souvenir Press, copyright © 1960, by Erich Fromm 'The Human Situation and Zen Buddhism', renewed 1988 Moshe Bodmor. Reprinted by permission of Harper Collins Publishers and Souvenir Press Ltd.

CHAPTER 10

Freud, S., 'Dissection of the Psychical Personality', acknowledgement is made to Sigmund Freud Copyrights, The Institute of Psychoanalysis, and The Hogarth Press for permission to quote from *The Standard Edition of the Complete Psychological Works of Sigmund Freud*, translated and edited by James Strachey.

Freud, A., 'The Ego as the Seat of Observation', in *The Ego and the Mechanisms of Defence*, The Hogarth Press, 1979, reprinted by permission of Mark Patterson Associates. Copyright © Anna Freud 1966.

Wolf, E., 'Irrationality in a Psychoanalytic Psychology of the Self' in *The Self: Psychological and Philosophical Issues*, Basil Blackwell, 1977, copyright © Basil Blackwell, 1977.

Guntrip, H., *The Turning Point: From Psychology to Object-Relations in Psychoanalytic Theory, Therapy, and the Self* by Harry Guntrip. Copyright © 1971 by Basic Books, Inc. Reprinted by permission of Basic Books, Inc., Publishers, New York.

Craib, I., 'The Unconscious and Language' in *Psychoanalysis and Social Theory*, Harvester Wheatsheaf, 1989, copyright © Ian Craib 1989.

CHAPTER 11

Crook, J. H., 'The Quest for Meaning: Models of Mind and Ego-transcendence', copyright © Oxford University Press 1980. Abridged from *The Evolution of Human Consciousness* by J. H. Crook (1980) by permission of Oxford University Press.

Taylor, E., 'Asian Interpretations: Transcending the Stream of Consciousness' in *The Stream of Consciousness: Scientific Investigations into the Flow of Human Experience*, edited by K. S. Pope and J. L. Singer, Plenum Press, 1978.

Naranjo, C., 'The Domain of Meditation' in *On the Psychology of Meditation*, edited by C. Naranjo and R. Ornstein, Allen & Unwin Ltd, 1972.

Watts, A., 'Empty and Marvellous' and 'Sitting Quietly, Doing Nothing' in *The Way of Zen* by Alan Watts, Thames & Hudson, 1957.

While every attempt has been made where appropriate to trace the copyright holders of the above extracts, the editors and publishers would be pleased to hear from any interested parties.

Preface

This collection of readings has its origins in a course which we began to teach to third year undergraduates about eight years ago. We had both become interested in consciousness, though independently and from different points of view. The readings contributed by John Pickering were originally part of another course on current issues in cognitive science. However, in time it began to seem more appropriate to use them to complement what Martin Skinner had to say in a course devoted entirely to consciousness. For Martin Skinner the impetus came both from a practical interest in meditation and from trying to find ways of teaching social psychology which addressed human self-consciousness.

Our course was planned as an open and eclectic arena within which we could develop our views through discussion with students. As we expanded our own interests we discovered the complementarity between our approaches amounted to a coherent framework. Within this framework this book presents readings which, in increasingly rich combination, relate to the mental life of all sentient beings and which underlie, at different depths, the unique character of human awareness.

In recent years consciousness has resumed a central place in many areas of psychology as well as science and philosophy more generally. We have touched on many of these diverse sources but since the field is very large and because of length constraints, it is inevitable that there will have been omissions and truncations. The constraint on length has also meant that we have often taken material out of context and edited it to make it more self-contained. A line space in the extract indicates where we have deleted parts of the text. We hope we have minimized the inevitable distortion and that the book presents a comprehensible progression of ideas which will encourage readers to look at the original sources.

We are indebted to the authors whose work we have been able to reproduce in this book and to the students whose contributions to the course helped our own understanding. We are particularly grateful to

Farrell Burnett and Sue Cooper for their encouragement and for their work in producing the book.

John Pickering and Martin Skinner
University of Warwick, May 1990.

Introduction

Consciousness remains one of the least understood aspects of human mental life in spite of an enormous and varied body of research and speculation. The selections in this book based on an eclectic undergraduate course, are only a small sample of this body of work and aim to present a range of perspectives on human consciousness and the central role played by symbolic function. The capacity for symbolic thought, for language and for self-awareness is so much more highly developed in human beings than in other species that it amounts to a qualitative break in the evolutionary progression of mental powers. In particular, it appears as if no other organism has anything approaching the human capacity for making itself and its thoughts objects in their own mental life. This reflexive symbolic capacity permits human awareness to become significantly more powerful than the various classes of non-human awareness, ranging from mere sentience to the complex mental worlds of, say, social mammals. We are concerned in this book principally with how it is that the direct awareness of the world which human beings share with other sentient organisms has been transcended to yield the unique human mode of being in the world. Because of the large number of factors involved in the production of this mode of being we have opted to present a broad range of perspectives rather than to go into particular factors in depth.

Should we expect a scientific theory of consciousness to emerge from these various perspectives? (Scientific is being used here in the sense of an account that is explicit and relatively complete.) If we take psychology to be the *science* of mental life, as William James called it, and if we take consciousness to be an object of scientific inquiry like any other, then perhaps the answer should be yes. In fact, we feel the answer is no. This is not just because psychological findings, like any form of scientific knowledge, are essentially provisional. Rather it is because human consciousness depends to a significant extent on a highly developed symbolic capacity. It is possible to show formally that once a symbolic system develops sufficient complexity its productions, its *behaviour* as it were, become quite literally unaccountable, that is, unpredictable and thus non-deterministic. More precisely, the system needs to be capable of self-reference, that is, be able to make symbolic statements about itself; the human symbolic capacity certainly permits self-reference and is thus non-deterministic in this sense. There are necessary limits to predictive knowledge about such systems, especially if, like human consciousness, they are in the process of evolving. Neither their behaviour nor their development can be known to the degree of certainty possible elsewhere in the physical and biological sciences, recent discoveries concerning chaos and indeterminacy notwithstanding. The upshot is that owing to its symbolic basis, the flow of human conscious experience is just what common experience suggests it is — self-determining and not reducible to any finite set of bio-physical causes. A scientific theory of consciousness in this sense faces a barrier of principle.

This aside, there are also practical limits to our understanding of consciousness. The mind-brain system that supports human consciousness is, after all, a uniquely complex mix of physical, biological and socio-cultural systems, integrated within an historical process. The flow of experience generated by the interplay of causes and effects within this process will be correspondingly complex. Describing this flow will require more than just a combination of terms from the natural sciences, and the recognition of this fact is indicated in the search for non-reductionist theories of mind that characterizes contemporary psychology and philosophy.

A further difficulty arises in attempting to deal with consciousness scientifically, in the classic sense of science as a positivist and reductionist enterprise. Scientific knowledge is in this sense assumed to be observer-independent and to exist necessarily in the public domain. However, possibly the most important knowledge available concerning consciousness is *experience* and this satisfies neither criterion. Thus a problem facing a scientific approach to consciousness is how experience, that is knowledge from acquaintance, may be related to objective knowledge of other factors relevant to consciousness. One common solution is to assume that consciousness lies outside the domain of science and that its subjective aspect, human experience, is an effect but not a cause. This solution is not adopted here, not only because of the strong phenomenological evidence for consciousness being an important part of human agency but also because cases for consciousness as a major causal factor in brain function are now beginning to be made. In any case, to take consciousness to be acausal would be to place it outside the natural order and to leave psychology, and science with it, incomplete. Accordingly, this reader has been compiled on the assumption that, for all the difficulties of treating subjectivity within science, a properly complete psychology must bring these two sorts of knowledge into as complete a relationship as possible.

There is a widespread intuition that the human mind is not reducible to any function of knowledge about non-mental things. This intuition may be an insight into our own natures or may represent the fear of being reduced or explained away. However, rather than merely accept intuitions at face value, contemporary psychology generally takes an eclectic approach and deals with whatever aspects of the human mind it can in whatever ways seem productive and appropriate. This reader takes this approach. We assume that relatively broad questions like 'what is consciousness' will be less productive than more focused ones which look at consciousness from specific perspectives. Also, we will need to be wary of assuming that a definitive answer is available at whatever level a question is posed. With respect to consciousness perhaps more than any other psychological issue, the whole is certainly more than the sum of its parts.

Any scientific inquiry, broad or narrow, is vulnerable to implicit assumptions about the type of answer that is expected. Consider as an analogy the scientists of the seventeenth century who asked 'what is heat?'. Heat, they noticed, moves from one place to another, assumes different forms and produces different effects. Now water can move from one place to another and can assume fluid, solid and gaseous forms and has a variety of effects, like wetting, dissolving and promoting growth. They assumed, reasonably enough, that heat was a fluid-like *substance*. Our present understanding is that heat is not so much a *thing* as a *process* — the motion of molecules to be more precise. So in some sense, the original assumption was 'incorrect'. But the 'correct' heat-as-motion model was out of reach for seventeenth-century scientists since the discoveries on which it rests were yet to be made and these discoveries were made while investigating the 'incorrect' heat-as-substance assumption. Clearly, assumptions are a necessary component of inquiry even if they may have misdirected it somewhat at the same time.

Likewise our inquiry into consciousness needs to be focused on particular questions which in turn will need to be motivated by some tentative definition of what sort of a thing it is. However, the overall enterprise could in turn be misdirected if this definition is too ambitious or too restrictive. There is, for example, every reason to think that discoveries yet to be made will be needed before our understanding of the mind comes anywhere near our understanding of heat. Rather than defining what consciousness *is*, we should assume only that using the term creates an arena in which we can examine the enormous and varied literature on the topic. There is certainly no shortage of material. Down the ages has come a steady stream of myth, doctrines and theories on the mind and experience from mystics, theologians, philosophers and scientists roughly in that order. However, diverse as the matters in this arena may seem, their common focus is the universal but extraordinary fact of human awareness.

This reader has been divided into chapters offering a progression of perspectives and signposts to notable features of the arena. Chapters One and Two briefly introduce some contrasting contributions from the philosophy of mind and some recent developments in science. Chapters Three, Four and Five deal respectively with the evolutionary context of consciousness, its neurophysiological basis and some contemporary psychological theory. This biophysical and psychological support for consciousness does not determine its content or nature. Rather, it is a product of social experience and social experience is made possible by the emergence of human symbolic capacity. Chapter Six notes the transition from sentience to symbolic function that has brought the human cultural context into being. It is within this context that the nature of consciousness is largely determined.

The passage from the first to second half of the reader hinges on the fact that human beings can employ symbols in controlling awareness. Whereas much of the first five sections explores how it has come about that we are sufficiently complex to be able to employ symbols, much of the last five is concerned with what happens when symbols are used between and within us. In Chapter Seven we briefly discuss symbols as objects or events standing merely by convention for something else. By 'stand for' we mean they recreate in some recognizable way the experience of the original objects or event which we say they stand for. We could restate this as 'representation' or, better still, '*re*-presentation'. If we can internalize these symbols there can be a new type of control of experience and control of conscious experience. We are principally interested in vocal symbols: words. They stand for the experience of objects and events and if we internalize these symbols we have internalized what they represent. Chapter Seven offers a view on how symbol use might have emerged in the history of the species (the phylogenetic account) and within the life of the individual (the ontogenetic account). Chapter Eight explores the consequences of this view of human consciousness. If it depends on symbols, it depends on a social process since symbols rest on agreements and agreements are a consequence of a social process. Human consciousness, then, seems to depend upon society and not *vice-versa*. This raises the possibility that what people make of their capacity for consciousness might vary from one culture to another and and from time to time throughout history. Chapter Nine looks at symbols explicitly as language and examines the direct relationship between language and consciousness. In Chapter Ten we step momentarily away from language to explore what psychoanalysis has taught us about consciousness through its interest in the unconscious and subsequently in the ego and the self (though even here we shall find ourselves led back to the role of language in the creation of the unconscious). Finally, we step away from our own cultural base to look at approaches to consciousness in Eastern cultures. We find that they parallel the Western approach in significant ways. The parallels are not surprising, because these cultures have employed empirical and practical approaches to consciousness for at least as long as Western European thought has. We hope that Chapter Eleven will show that there is as much to be learned by taking a step back from a topic under investigation as there is from taking a step towards it.

There are many other aspects to the study of consciousness that have had to be omitted. If we believe with Pope that the proper study of man is man, then perhaps we should all be psychologists. If as psychologists we find awareness to be the most intriguing part of mental life then perhaps we should study consciousness. This collection has been assembled for a number of purposes. Among these are: to provide a set of sources which demonstrate the variety of ways by which consciousness may be approached; to show the interrelations of these ways; to act as a guide for students; to emphasize the role of human symbolic function; to stress the dependence of human awareness on the social and cultural matrix of human society; and to show how the study of consciousness relates to current developments in scientific thought. Whether or not these purposes are achieved, we hope that anyone with an interest in the human condition will find something of value here.

1

Consciousness and the Philosophy of Mind

READINGS

René Descartes

Meditations on the First Philosophy *and* The Principles of Philosophy

Gilbert Ryle

The Concept of Mind

Henri Bergson

Life and Consciousness

Philosophy has the pursuit of truth as its end and thought and rational argument as its means. Science has the same end but adds to the means the systematic testing of hypotheses through experiments and observations. Philosophy is thus the complement but also the precursor of science in two senses. In a historical sense science emerged as a distinct discipline from philosophical inquiry into nature as a whole. In a more contemporary sense, it is clear that science's most basic assumptions are philosophical ones.[1]

Scientific understanding rests on accurate observation. The more inaccessible or subtle the phenomenon being observed, the more sophisticated and varied the methods need to be. Scientific understanding of consciousness, perhaps the most complex and diverse phenomenon we know, is thus likely to be correspondingly difficult. Furthermore, special problems follow from the fact that consciousness is not only complex but also intimately bound up with subjective experience. Accordingly this chapter, after a brief historical introduction, offers a small selection of philosophical preliminaries as an appropriate start to this collection of readings.

The philosophy of mind is a relatively recent subdivision of the discipline. Greek thought, with its blend of the rational and the intuitive, had no separate discipline concerned specifically with the nature of the mind. Pythagoras found the perfection of mathematical and geometrical laws so compelling that he concluded that order at all levels of the universe, including human experience, must be governed by such laws. This view of the world and the place of human beings in it was fundamentally religious and Aristotle's philosophy also integrated what would now be distinguished as religion, psychology, aesthetics and science. But Greek science was a science of qualities. All substances were a mixture of the elements earth, air, fire and water (the heavens were made from a fifth element or quintessence more perfect than the other four). The qualities of these elements not only accounted for the properties of matter, but also for how it acted. For example, heavy objects fall because in them the earth element predominates, which makes them heavy, and is attracted (it might even be said it *wishes* to go) to the natural home for such an element, the earth itself. By contrast, flame and heat rise because their predominant element is fire whose natural movement is upward. The overall pattern of world events reflects the nature of the elements from which it is made.

A science of qualities is more concerned with *why* things happen than with *how*. In such a science the flow of action in the soul reflected the particular mix of elements in each person — a theory of personality that was developed by the physician Galen some four centuries after Aristotle and which persisted until well into mediaeval times. Consciousness, being the subjective component of this flow, was a legitimate part of this qualitative science. Plato and Aristotle certainly discussed perception, memory, reasoning and emotion though whether they had an explicit notion of consciousness is open to question.[2] The actions of the mind or soul were treated in terms of how they fitted in with the rest of nature to make or to hinder the making of a harmonious whole. Philosophy aimed to understand this whole and so provide the knowledge that might help human actions and thoughts to be more in concert with the natural order of things.

As Greek thought, particularly that of Aristotle, was developed by Christian and Mohammedan theologians, the place assigned to the human mind in the natural order changed. From being essentially a part of nature, the human mind became a special creation with unique properties, including consciousness, which therefore required special explanation. An extract from *An Oration on the Dignity of Man*, written by Pico della Mirandola in 1486, shows just how special the position of the human mind had become. He concluded his discourse on the Creation with an account of how the Creator gave a unique position and role to human beings, and why:

> Taking man, therefore, this creature of indeterminate image, He set him in the middle of the world and spoke to him thus: 'We have given you, O Adam, no visage proper to yourself, nor any endowment properly your own, in order that whatever place, whatever form, whatever gifts you may, with premeditation, select, these same you may have and possess through your own judgement and decision. The nature of all other creatures is defined and restricted within laws which We have laid down; you, by contrast, impeded by no such restrictions, may, by your own free will, to whose custody we have assigned you, trace for yourself the lineament of your own nature. I have placed you at the centre of the world, so that from that vantage point you may with greater ease glance about you on all that the world contains.[3]

For Pico, human beings had been assigned a specially privileged place which at the same time removed them from the natural order. As we shall see in later chapters, Pico's account has some interesting parallels to accounts of human consciousness offered by contemporary psychology.

The scientific revolution that was well under way in Pico's time naturally influenced the philosophy of mind. As it gained in authority science created an alternative to the religious perspective on mind and consciousness. Descartes, immersed in the new methods and findings of the mid-seventeenth century, introduced a style of philosophy which reflected them. His systematic enquiry into human consciousness, which also appeals to the benign intentions of a Creator, integrated religious doctrine with the powerful insights gained through scientific methods. Descartes' work marks the beginning of the modern phase in the study of the mind and consciousness. The first reading shows something of his style of philosophical inquiry and some of the conclusions to which he came.

It is now some years since I detected how many were the false beliefs that I had from my earliest youth admitted as true, and how doubtful was everything I had since constructed on this basis; and from that time I was convinced that I must once for all seriously undertake to rid myself of all the opinions which I had formerly accepted, and commence to build anew from the foundation, if I wanted to establish any firm and permanent structure in the sciences.

I shall proceed by setting aside all that in which the least doubt could be supposed to exist, just as if I had discovered that it was absolutely false; and I shall ever follow in this road until I have met with something which is certain, or at least, if I can do nothing else, until I have learned for certain that there is nothing in the world that is certain.

I suppose, then, that all the things that I see are false; I persuade myself that nothing has ever existed of all that my fallacious memory represents to me. I consider that I possess no senses; I image that body, figure, extension, movement and place are but the fictions of my mind. What, then, can be esteemed as true? Perhaps nothing at all, unless that there is nothing in the world that is certain.

But how can I know there is not something different from those things that I have just considered, of which one cannot have the slightest doubt? Is there not some God, or some other being by whatever name we call it, who puts these reflections into my mind? That is not necessary, for is it not possible that I am capable of producing them myself? I myself, am I not at least something? But I have already denied that I had senses and body. Yet I hesitate, for what follows from that? Am I so dependent on body and senses that I cannot exist without these? But I was persuaded that there was nothing in all the world, that there was no heaven, no earth, that there were no minds, nor any bodies; was I not then likewise persuaded that I did not exist? Not at all; of a surety I myself did exist since I persuaded myself of something (or merely because I thought of something). But there is some deceiver or other, very powerful and very cunning, who ever employs his ingenuity in deceiving me. Then without doubt I exist also if he deceives me, and let him deceive me as much as he will, he can never cause me to be nothing so long as I think that I am something. So that after having reflected well and carefully examined all things, we must come to the definite conclusion that this proposition: I am, I exist, is necessarily true each time that I pronounce it, or that I mentally conceive it.

But what then am I? A thing which thinks. What is a thing which thinks? It is a thing which doubts, understands, (conceives), affirms, denies, wills, refuses, which also imagines and feels.

Certainly it is no small matter if all these things pertain to my nature. But why should they not so pertain? Am I not that being who now doubts nearly everything, who nevertheless understands certain things, who affirms that one only is true, who denies all the others, who desires to know more, is averse from being deceived . . .

For it is so evident of itself that it is I who doubts, who understands, and who desires, that there is no reason here to add anything to explain it. And I have certainly the power of imagining likewise; for although it may happen (as I formerly supposed) that none of the things which I imagine are true, nevertheless this power of imagining does not cease to be really in use, and it forms part of my thought. Finally, I am the same who feels, that is to say, who perceives certain things, as by the organs of sense, since in truth I see light, I hear noise, I feel heat. But it will be said that these phenomena are false and that I am dreaming. Let it be so; still it is at least quite certain that it seems to me that I see light, that I hear noise and that I feel heat.

I further find in myself faculties employing modes of thinking peculiar to themselves, to wit, the faculties of imagination and feeling, without which I can easily conceive myself clearly and distinctly as a complete being; while, on the other hand, they cannot be so conceived apart from me, that is without an intelligent substance in which they reside for (in the notion we have of these faculties, or, to use the language of the Schools) in their formal concept, some kind of intellection is comprised, from which I infer that they are distinct from me as its modes are from a thing. I observe also in me some other faculties such as that of change of position, the assumption of different figures and such like, which cannot be conceived, any more than can the preceding, apart from some substance to which they are attached, and consequently cannot exist without it; but it is very clear that these faculties, if it be true that they exist, must be attached to some corporeal or extended substance, and not to an intelligent substance, since in the clear and distinct conception of these there is some sort of extension found to be present, but no intellection at all. There is certainly further in me a certain passive faculty of perception, that is, of receiving and recognizing the ideas of sensible things, but this would be useless to me (and I could in no way avail myself of it), if there were not either in me or in some other thing another active faculty capable of forming and producing these ideas. But this active faculty cannot exist in me (inasmuch as I am a thing that thinks) seeing that it does not presuppose thought, and also that those ideas are often produced in me without my contributing in any way to the same, and often even against my will; it is thus necessarily the case that the faculty resides in some substance different from me in which all the reality which is objectively in the ideas that are produced by this faculty is formally or eminently contained, as I remarked before. And this substance is either a body, that is, a corporeal nature in which there is contained formally (and really) all that which is objectively (and by representation) in those ideas, or it is God Himself, or some other creature more noble than body in which that same is contained eminently. But, since God is no deceiver, it is very manifest that He does not communicate to me these ideas immediately and by Himself nor yet by the intervention of some creature in which their reality is not formally, but only eminently, contained. For since He has given me no faculty to recognize that this is the case, but, on the other hand, a very great inclination to believe (that they are sent to me or) that they are conveyed to me by corporeal objects, I do not see how He could be defended from the accusation of deceit if these ideas were produced by causes other than corporeal objects. Hence we must allow that corporeal things exist. However, they are perhaps not exactly what we perceive by the senses, since this comprehension by the senses is in many instances very obscure and confused; but we must at least admit that all things which I conceive in them clearly and distinctly, that is to say, all things which, speaking generally, are comprehended in the object of pure mathematics, are truly to be recognized as external objects.

As to other things, however, which are either particular only, as for example, that the sun is of such and such a figure, etc., or which are less clearly and distinctly conceived, such as light, sound, pain and the like, it is certain that although they are very dubious and uncertain, yet on the sole ground that God is not a deceiver, and that consequently He has not permitted any falsity to exist in my opinion which He has not likewise given me the faculty of correcting, I may assuredly hope to conclude that I have within me the means of arriving at the truth even here. And first of all there is no doubt that in all things which nature teaches me there is some truth contained; for by nature, considered in general, I now understand no other thing than either God himself or else the order and disposition which God has established in created things; and by my nature in particular I understand no other thing than the complexus of all the things which God has given me.

But there is nothing which this nature teaches me more expressly (nor more sensibly) than that I have a body which is adversely affected when I feel pain, which has need of food or drink when I experience the feelings of hunger and thirst, and so on; nor can I doubt there being some truth in all this.

Nature also teaches me by these sensations of pain, hunger, thirst, etc., that I am not only lodged in my body as a pilot in a vessel, but that I am very closely united to it, and so to speak so intermingled with it that I seem to compose with it one whole. For if that were not the case, when my body is hurt, I, who am merely a thinking thing, should not feel pain, for I should perceive this wound by the understanding only, just as the sailor perceives by sight when something is damaged in his vessel; and when my body has need of drink or food, I should clearly understand the fact without being warned of it by confused feelings of hunger and thirst. For all these sensations of hunger, thirst, pain, etc. are in truth none other than certain confused modes of thought which are produced by the union and apparent intermingling of mind and body.

Moreover, nature teaches me that many other bodies exist around mine, of which some are to be avoided, and others sought after. And certainly from the fact that I am sensible of different sorts of colours, sounds, scents, tastes, heat, hardness, etc., I very easily conclude that there are in the bodies from which all these diverse sense-perceptions proceed certain variations which answer to them, although possibly these are not really at all similar to them. And also from the fact that amongst these different sense-perceptions some are very agreeable to me and others disagreeable, it is quite certain that my body (or rather myself in my entirety, inasmuch as I am formed of body and soul) may receive different impressions agreeable and disagreeable from the other bodies which surround it.

. . . I here say, in the first place, that there is a great difference between mind and body, inasmuch as body is by nature always divisible, and the mind is entirely indivisible. For, as a matter of fact, when I consider the mind, that is to say, myself inasmuch as I am only a thinking thing, I cannot distinguish in myself any parts, but apprehend myself to be clearly one and entire; and although the whole mind seems to be united to the whole body, yet if a foot, or an arm, or some other part is separated from my body, I am aware that nothing has been taken away from my mind. And the faculties of willing, feeling, conceiving, etc. cannot be properly speaking said to be its parts, for it is one and the same mind which employs itself in willing and in feeling and understanding. But it is quite otherwise with corporeal or extended objects, for there is not one of these imaginable by me which my mind cannot easily divide into parts, and which consequently I do not recognize as being divisible; this would be sufficient to teach me that the mind or soul of man is entirely different from the body, if I had not already learned it from other sources.

I further notice that the mind does not receive the impressions from all parts of the body immediately, but only from the brain, or perhaps even from one of its smallest parts, to wit, from that in which the common sense is said to reside, which, whenever it is disposed in the same particular way, conveys the same thing to the mind, although meanwhile the other portions of the body may be differently disposed, as is testified by innumerable experiments which it is unnecessary here to recount.

I notice finally that since each of the movements which are in the portion of the brain by which the mind is immediately affected brings about one particular sensation only, we cannot under the circumstances imagine anything more likely than that this movement, amongst all the sensations which it is capable of impressing on it, causes mind to be affected by that one which is best fitted and most generally useful for the conservation of the human body when it is in health. But experience makes us aware that all the feelings with which nature inspires us are such as I have just spoken of; and there is therefore nothing in them which does not give testimony to the power and goodness of the God (who has produced them). Thus, for example, when the nerves which are in the feet are violently or more than usually moved, their movement, passing through the medulla of the spine to the inmost parts of the brain, gives a sign to the mind which makes it feel somewhat, to wit, pain, as though in the foot, by which the mind is excited to do its utmost to remove the cause of the evil as dangerous and hurtful to the foot. It is true that God could have constituted the nature of man in such a way that this same movement in the brain would have conveyed something quite different to the mind; for example, it might have produced consciousness of itself either in so far as it is in the brain, or as it is in the foot, or as it is in some other place between the foot and the brain, or it might finally have produced consciousness of anything else whatsoever; but none of all this would have contributed so well to the conservation of the body. Similarly, when we desire to drink, a certain dryness of the throat is produced which moves its nerves, and by their means the internal portions of the brain; and this movement causes in the mind the sensation of thirst, because in this case there is nothing more useful to us than to become aware that we have need to drink for the conservation of our health; and the same holds good in other instances.

From this it is quite clear that, notwithstanding the supreme goodness of God, the nature of man, inasmuch as it is composed of mind and body, cannot be otherwise than sometimes a source of deception. For if there is any cause which excites, not in the foot but in some part of the nerves which are extended between the foot and the brain, or even in the brain itself, the same movement which usually is produced when the foot is detrimentally affected, pain will be experienced as though it were in the foot, and the sense will thus naturally be deceived; for since the same movement in the brain is capable of causing but one sensation in the mind and this sensation is much more frequently excited by a cause which hurts the foot than by another existing in some other quarter, it is reasonable that it should convey to the mind pain in the foot rather than in any other part of the body. And although the parchedness of the throat does not always proceed, as it usually does, from the fact that drinking is necessary for the health of the body, but sometimes comes from quite a different cause, as is the case with dropsical patients, it is yet much better that it should mislead on this occasion than if, on the other hand, it were always to deceive us when the body is in good health; and so on in similar cases.

It is certain that no thought can exist apart from a thing that thinks; no activity, no accident can be without a substance in which to exist. Moreover, since we do not apprehend the substance itself immediately through itself, but by means only of the fact that it is the subject of certain activities, it is highly rational, and a requirement forced on us by custom, to give diverse names to those substances that we recognize to be the subjects of clearly diverse activities or accidents, and afterwards to inquire whether those diverse names refer to one and the same or to diverse things. But there are *certain* activities, which we call *corporeal*, e.g. magnitude, figure, motion, and all those that cannot be thought of apart from extension in space; and the substance in which they exist is called *body*. It cannot be pretended that the substance that is the subject of figure is different from that which is the subject of spatial motion, etc., since all these activities agree in presupposing extension. Further, there are other activities, which we call *thinking* activities, e.g. understanding, willing, imagining, feeling, etc., which agree in falling under the description of thought, perception, or consciousness. The substance in which they reside we call a *thinking thing* or *the mind*, or any other name we care, provided only we do not confound it with corporeal substance, since thinking activities have no affinity with corporeal activities, and thought, which is the common nature in which the former agree, is totally different from extension, the common term for describing the latter.

But after we have formed two distinct concepts of those two substances, it is easy, from what has been said in the sixth Meditation, to determine whether they are one and the same or distinct.

In philosophizing correctly, there is no need for us to prove the falsity of all those things which we do not admit because we do not know whether they are true. We have merely to take the greatest care not to admit as true what we cannot prove to be true. Thus when I find that I am a thinking substance, and form a clear and distinct concept of that substance, in which there is none of those attributes which belong to the concept of corporeal substance, this is quite sufficient to let me affirm that I, in so far as I know myself, am nothing but a thing which thinks . . .

That each substance has a principal attribute, and that the attribute of the mind is thought, while that of body is extension

But although any one attribute is sufficient to give us a knowledge of substance, there is always one principal property of substance which constitutes its nature and essence, and on which all the others depend. Thus extension in length, breadth and depth, constitutes the nature of corporeal substance; and thought constitutes the nature of thinking substance. For all else that may be attributed to body presupposes extention, and is but a mode of this extended thing; as everything that we find in mind is but so many diverse forms of thinking. Thus, for example, we cannot conceive figure but as an extended thing, nor movement but as in an extended space; so imagination, feeling, and will, only exist in a thinking thing. But, on the other hand, we can conceive extension without figure or action, and thinking without imagination or sensation, and so on with the rest; as is quite clear to anyone who attends to the matter.

How we may have clear and distinct notions of thinking substance, of corporeal substance, and of God

We may thus easily have two clear and distinct notions or ideas, the one of created substance which thinks, the other of corporeal substance, provided we carefully separate all the attributes of thought from those of extension.

Similarly because each one of us is conscious that he thinks, and that in thinking he can shut off from himself all other substance, either thinking or extended, we may conclude that each of us, similarly regarded, is really distinct from every other thinking substance and from every corporeal substance. And even if we suppose that God had united a body to a soul so closely that it was impossible to bring them together more closely, and made a single thing out of the two, they would yet remain really distinct one from the other notwithstanding the union; because however closely God connected them He could not set aside the power which He possessed of separating them, or conserving them one apart from the other, and those things which God can separate, or conceive in separation, are really distinct.

That the soul is united to all the portions of the body conjointly

But in order to understand all these things more perfectly, we must know that the soul is really joined to the whole body, and that we cannot, properly speaking, say that it exists in any one of its parts to the exclusion of the others, because it is one and in some manner indivisible, owing to the disposition of its organs, which are so related to one another that when any one of them is removed, that renders the whole body defective; and because it is of a nature which has no relation to extension, or dimensions, nor other properties of the matter of which the body is composed, but only to the whole conglomerate of its organs, as appears from the fact that we could not in any way conceive of the half or the third of a soul, nor of the space it occupies, and because it does not become smaller owing to the cutting off of some portion of the body, but separates itself from it entirely when the union of its assembled organs is dissolved.

That there is a small gland in the brain in which the soul exercises its functions more particularly than in the other parts

It is likewise necessary to know that although the soul is joined to the whole body, there is yet in that a certain part in which it exercises its functions more particularly than in all the others; and it is usually believed that this part is the brain, or possibly the heart: the brain, because it is with it that the organs of sense are connected, and the heart because it is apparently in it that we experience the passions. But, in examining the matter with care, it seems as though I had clearly ascertained that the part of the body in which the soul exercises its functions immediately is in nowise the heart, nor the whole of the brain, but merely the most inward of all its parts, to wit, a certain very small gland which is situated in the middle of its substance and so suspended above the duct whereby the animal spirits in its anterior cavities have communication with those in the posterior, that the slightest movements which take place in it may alter very greatly the course of these spirits; and reciprocally that the smallest changes which occur in the course of the spirits may do much to change the movements of this gland.

How we know that this gland is the main seat of the soul
The reason which persuades me that the soul cannot have any other seat in all the body than this gland wherein to exercise its functions immediately, is that I reflect that the other parts of our brain are all of them double, just as we have two eyes, two hands, two ears, and finally all the organs of our outside senses are double; and inasmuch as we have but one solitary and simple thought of one particular thing at one and the same moment, it must necessarily be the case that there must somewhere be a place where the two images which come to us by the two eyes, where the two other impressions which proceed from a single object by means of the double organs of the other senses, can unite before arriving at the soul, in order that they may not represent to it two objects instead of one. And it is easy to apprehend how these images or other impressions might unite in this gland by the intermission of the spirits which fill the cavities of the brain; but there is no other place in the body where they can be thus united unless they are so in this gland.

That the seat of the passions is not in the heart
As to the opinion of those who think that the soul receives its passions in the heart, it is not of much consideration, for it is only founded on the fact that the passions cause us to feel some change taking place there; and it is easy to see that this change is not felt in the heart excepting through the medium of a small nerve which descends from the brain towards it, just as pain is felt as in the foot by means of the nerves of the foot, and the stars are perceived as in the heavens by means of their light and of the optic nerves; so that it is not more necessary that our soul should exercise its functions immediately in the heart, in order to feel its passions there, than it is necessary for the soul to be in the heavens in order to see the stars there.

How the soul and the body act on one another
Let us then conceive here that the soul has its principal seat in the little gland which exists in the middle of the brain, from whence it radiates forth through all the remainder of the body by means of the animal spirits, nerves, and even the blood, which, participating in the impressions of the spirits, can carry them by the arteries into all the members. And recollecting what has been said above about the machine of our body, i.e. that the little filaments of our nerves are so distributed in all its parts, that on the occasion of the diverse movements which are there excited by sensible objects, they open in diverse ways the pores of the brain, which causes the animal spirits contained in these cavities to enter in diverse ways into the muscles, by which means they can move the members in all the different ways in which they are capable of being moved; and also

that all the other causes which are capable of moving the spirits in diverse ways suffice to conduct them into diverse muscles; let us here add that the small gland which is the main seat of the soul is so suspended between the cavities which contain the spirits that it can be moved by them in as many different ways as there are sensible diversities in the object, but that it may also be moved in diverse ways by the soul, whose nature is such that it receives in itself as many divine impressions, that is to say, that it possesses as many diverse perceptions as there are divine movements in this gland. Reciprocally, likewise, the machine of the body is so formed that from the simple fact that this gland is diversely moved by the soul, or by such other cause, whatever it is, it thrusts the spirits which surround it towards the pores of the brain, which conduct them by the nerves into the muscles, by which means it causes them to move the limbs.

Princess Elizabeth to Descartes, May 6–16, 1643
. . . I beg of you to tell me how the human soul can determine the movement of the animal spirits in the body so as to perform voluntary acts — being as it is merely a conscious (*pensante*) substance. For the determination of movement seems always to come about from the moving body's being propelled — to depend on the kind of impulse it gets from what sets it in motion, or again, on the nature and shape of this latter thing's surface. Now the first two conditions involve contact, and the third involves that the impelling thing has extension; but you utterly exclude extension from your notion of soul, and contact seems to me incompatible with a thing's being immaterial.

I therefore ask you for a more specific definition of the soul than you give in your metaphysics: a definition of its substance, as distinct from its activity, consciousness (*pensée*). Even if we supposed these to be in fact inseparable — a matter hard to prove in regard to children in their mother's womb and severe fainting-fits — to be inseparable as the divine attributes are: nevertheless we may get a more perfect idea of them by considering them apart.

Descartes to Princess Elizabeth, May 21, 1643
. . . I may truly say that what your Highness is propounding seems to me to be the question people have most right to ask me in view of my published works. For there are two facts about the human soul on which there depends any knowledge we may have as to its nature: first, that it is conscious; secondly, that, being united to a body, it is able to act and suffer along with it. Of the second fact I said almost nothing; my aim was simply to make the first properly understood; for my main object was to prove the distinction of soul and body; and to this end only the first was serviceable, the second might have been prejudicial. But since your Highness sees too clearly for dissimulation to be possible, I will here try to explain how I conceive the union of soul and body and how the soul has the power of moving the body.

My first observation is that there are in us certain primitive notions — the originals, so to say, on the pattern of which we form all other knowledge. These notions are very few in number. First, there are the most general ones, existence, number, duration, etc., which apply to everything we can conceive. As regards body in particular, we have merely the notion of extension and the consequent notions of shape and movement. As regards the soul taken by itself, we have merely the notion of consciousness which comprises the conceptions (*perceptions*) of the intellect and the inclinations of the will. Finally, as regards the soul and body together, we have merely the notion of their union; and on this there depend our notions of the soul's power to move the body, and of the body's power to act on the soul and cause sensations and emotions.

I would also observe that all human knowledge consists just in properly distinguishing these notions and attaching each of them only to the objects that it applies to. If we try to explain some problem by means of a notion that does not apply, we cannot help making mistakes; we are just as wrong if we try to explain one of these notions in terms of another, since, being primitive, each such notion has to be understood in itself. The use of our senses has made us much more familiar with notions of extension, shape, and movement than with others; thus the chief cause of our errors is that ordinarily we try to use these notions to explain matters to which they do not apply; e.g. we try to use our imagination in conceiving the nature of the soul, or to conceive the way the soul moves the body in terms of the way that one body is moved by another body.

In the meditations that your Highness condescended to read, I tried to bring before the mind the notions that apply to the soul taken by itself, and to distinguish them from those that apply to the body taken by itself. Accordingly, the next thing I have to explain is how we are to form the notions that apply to the union of the soul with the body, as opposed to those that apply to the body taken by itself or the mind taken by itself. . . . These simple notions are to be sought only within the soul, which is naturally endowed with all of them, but does not always adequately distinguish between them, or again, does not always attach them to the right objects.

So I think people have hitherto confused the notions of the soul's power to act within the body and the power one body has to act within another; and they have ascribed both powers not to soul, whose nature was so far unknown, but to various qualities of bodies — gravity, heat, etc. These qualities were imagined to be real, i.e. to have an existence distinct from the existence of bodies; consequently, they are imagined to be substances, although they were called qualities. In order to conceive of them, people have used sometimes notions that we have for the purpose of knowing body, and sometimes those that we have for the purpose of knowing the soul, according as they were ascribing to them a material or an immaterial nature. For example, on the supposition that gravity is a real quality, about which we know no more than its power of moving the body in which it occurs towards the centre of the Earth, we find no difficulty in conceiving how it moves the body or how it is united to it; and we do not think of this as taking place by means of real mutual contact between two surfaces; our inner experience shows (*nous expérimentons*) that that

> notion is a specific one. Now I hold that we misuse this notion by applying it to gravity (which, as I hope to show in my *Physics*, is nothing really distinct from body), but that it has been given to us in order that we conceive of the way that the soul moves the body.

Descartes concluded that matter, including the body, and mind were essentially different things. Mind was insubstantial but had the capacity for thought while matter was thoughtless substance. In Cartesian dualism, the human mind and consciousness are separated from the material world on both logical and religious grounds.

As science developed, hitherto unanswerable questions like the control of tides and the nature of heat were recognized as the products of underlying physical causes. There seemed no reason in principle why there should be any limit to this style of explanation where complex things become understandable as the productions of simpler fundamental causes. Now, if all phenomena are reducible to simple underlying causes then why should consciousness be an exception? It began to seem that mind and consciousness would no longer require any special explanation — such as a special act of a Creator or some insubstantial Cartesian mind-stuff. This view, various aspects of which are termed reductionism, positivism or materialism, provides a starting point for much contemporary philosophy of mind.

However, though science raises the possibility that consciousness has no special qualities over and above those of the rest of the material world, experience gives compelling grounds for believing otherwise. Accepting the unique status of conscious experience may not require us to adopt a dualist position but it raises some important questions. The physical world, at least as revealed by physical science, is clearly mindless, so how did mind arise? How does mind relate to the physical world? How can it cause things to happen in the physical world, particularly the body? And so on. Questions like these are, collectively, the mind-body problem.

The care we need to take in asking questions like these is one of the concerns of the next selection which is from Gilbert Ryle's *The Concept of Mind*. Ryle claimed that dualism is a fundamentally flawed position. Here we are not so much interested in deciding whether his claim is right or wrong but in the type of argument he uses to back it up. His is an example of the analytic style which has had a great influence on philosophy in the English-speaking world during this century. Developing out of late nineteenth-century scientific positivism, analytic philosophy is principally concerned to avoid metaphysical talk and thinking.[4] Language must be examined to ensure that it is not only internally, that is logically, consistent but also as consistent in reference as possible. This style of philosophy is a type of linguistic discipline and the excerpts from *Concept of Mind* show how terms like 'mind' and 'consciousness' need to be used with care to avoid dualist conundrums.

The Official Doctrine

There is a doctrine about the nature and place of minds which is so prevalent among theorists and even among laymen that it deserves to be described as the official theory. Most philosophers, psychologists and religious teachers subscribe, with minor reservations, to its main articles and, although they admit certain theoretical difficulties in it, they tend to assume that these can be overcome without serious modifications being made to the architecture of the theory. It will be argued here that the central principles of the doctrine are unsound and conflict with the whole body of what we know about minds when we are not speculating about them.

The official doctrine, which hails chiefly from Descartes, is something like this. With the doubtful exceptions of idiots and infants in arms every human being has both a body and a mind. Some would prefer to say that every human being is both a body and a mind. His body and his mind are ordinarily harnessed together, but after the death of the body his mind may continue to exist and function.

Human bodies are in space and are subject to the mechanical laws which govern all other bodies in space. Bodily processes and states can be inspected by external observers. So a man's bodily life is as much a public affair as are the lives of animals and reptiles and even as the careers of trees, crystals and planets.

But minds are not in space, nor are their operations subject to mechanical laws. The workings of one mind are not witnessable by other observers; its career is private. Only I can take direct cognisance of the states and processes of my own mind. A person therefore lives through two collateral histories, one consisting of what happens in and to his body, the other consisting of what happens in and to his mind. The first is public, the second private. The events in the first history are events in the physical world, those in the second are events in the mental world.

It has been disputed whether a person does or can directly monitor all or only some of the episodes of his own private history; but, according to the official doctrine, of at least some of these episodes he has direct and unchallengeable cognisance. In consciousness, self-consciousness and introspection he is directly and authentically apprised of the present states and operations of his mind. He may have great or small uncertainties about concurrent and adjacent episodes in the physical world, but he can have none about at least part of what is momentarily occupying his mind.

It is customary to express the bifurcation of his two lives and of his two worlds by saying that the things and events which belong to the physical world, including his own body, are external, while the workings of his own mind are internal. This antithesis of outer and inner is of course meant to be construed as a metaphor, since minds, not being in space, could not be described as being spatially inside anything else, or as having things going on spatially inside themselves. But relapses from this good intention are common and theorists are found speculating how stimuli, the physical sources of which are yards or miles outside a person's skin, can generate mental responses inside his skull, or how decisions framed inside his cranium can set going movements of his extremities.

Even when 'inner' and 'outer' are construed as metaphors, the problem how a person's mind and body influence one another is notoriously charged with theoretical difficulties. What the mind wills, the legs, arms and the tongue execute; what affects the ear and the eye has something to do with what the mind perceives; grimaces and smiles betray the mind's moods and bodily castigations lead, it is hoped, to moral improvement. But the actual transactions between the episodes of the private history and those of the public history remain mysterious, since by definition they can belong to neither series. They could not be reported among the happenings described in a person's autobiography of his inner life, but nor could they be reported among those described in someone else's biography of that person's overt career. They can be inspected neither by introspection nor by laboratory experiment. They are theoretical shuttle-cocks which are forever being bandied from the physiologist back to the psychologist and from psychologist back to the physiologist.

Underlying this partly metaphorical representation of the bifurcation of a person's two lives there is a seemingly more profound and philosophical assumption. It is assumed that there are two different kinds of existence or status. What exists or happens may have the status of physical existence, or it may have the status of mental existence. Somewhat as the faces of coins are either heads or tails, or somewhat as living creatures are either male or female, so, it is supposed, some existing is physical existing, other existing is mental existing. It is a necessary feature of what has physical existence that it is in space and time; it is a necessary feature of what has mental existence that it is in time but not in space. What has physical existence is composed of matter, or else is a function of matter; what has mental existence consists of consciousness, or else is a function of consciousness. . . .

The Absurdity of the Official Doctrine

Such in outline is the official theory. I shall often speak of it, with deliberate abusiveness, as 'the dogma of the Ghost in the Machine'. I hope to prove that it is entirely false, and false not in detail but in principle. It is not merely an assemblage of particular mistakes. It is one big mistake and a mistake of a special kind. It is, namely, a category-mistake. It represents the facts of mental life as if they belonged to one logical type or category (or range of types or categories), when they actually belong to another. The dogma is therefore a philosopher's myth. In attempting to explode the myth I shall probably be taken to be denying well-known facts about the mental life of human beings, and my plea that I aim at doing nothing more than rectify the logic of mental-conduct concepts will probably be disallowed as mere subterfuge.

I must first indicate what is meant by the phrase 'Category-mistake'. This I do in a series of illustrations.

A foreigner visiting Oxford or Cambridge for the first time is shown a number of college, libraries, playing fields, museums, scientific departments and administrative offices. He then asks 'But where is the University? I have seen where the members of the Colleges live, where the Registrar works, where the scientists experiment and the rest. But I have not yet seen the University in which reside and work the members of your University'. It has then to be explained to him that the University is not another collateral institution, some ulterior counterpart to the colleges, laboratories and offices which he has seen. The University is just the way in which all that he has already seen is organized. When they are seen and when their co-ordination is understood, the University has been seen. His mistake lay in his innocent assumption that it was correct to speak of Christ Church, the Bodleian Library, the Ashmolean Museum *and* the University, to speak, that is, as if 'the University' stood for an extra member of the class of which these other units are members. He was mistakenly allocating the University to the same category as that to which the other institutions belong.

The same mistake would be made by a child witnessing the march-past of a division who, having had pointed out to him such and such battalions, batteries, squadrons, etc., asked when the division was going to appear. He would be supposing that a division was a counterpart to the units already seen, partly similar to them and partly unlike them. He would be shown his mistake by being told that in watching the battalions, batteries and squadrons marching past he had been watching the division marching past. The march-past was not a parade of battalions, batteries, squadrons *and* a division; it was a parade of the battalions, batteries and squadrons *of* a division. . . .

My destructive purpose is to show that a family of radical category-mistakes is the source of the double-life theory. The representation of a person as a ghost mysteriously ensconced in a machine derives from this argument. Because, as is true, a person's thinking, feeling and purposive doing cannot be described solely in the idioms of physics, chemistry and physiology, therefore they must be described in counterpart idioms. As the human body is a complex organized unit, so the human mind must be another complex organized unit, though one made of a different sort of stuff and with a different sort of structure. Or, again, as the human body, like any other parcel of matter, is a field of causes and effects, so the mind must be another field of causes and effects, though not (Heaven be praised) mechanical causes and effects.

The Origin of the Category-mistake

One of the chief intellectual origins of what I have yet to prove to be the Cartesian category-mistake seems to be this. When Galileo showed that his methods of scientific discovery were competent to prove a mechanical theory which should cover every occupant of space, Descartes found in himself two conflicting motives. As a man of scientific genius he could not but endorse the claims of mechanics, yet as a religious and moral man he could not accept, as Hobbes accepted, the discouraging rider to those claims, namely that human nature differs only in degree of complexity from clockwork. The mental could not be just a variety of the mechanical.

He and the subsequent philosophers naturally but erroneously availed themselves of the following escape-routes. Since mental-conduct words are not to be construed as signifying the occurrence of mechanical processes, they must be construed as signifying the occurrence of non-mechanical processes; since mechanical laws explain movements in space as the effects of other movements in space, other laws must explain some of the non-spatial workings of minds as the effects of other non-spatial workings of minds. The difference between the human behaviours which we describe as intelligent and those which we describe as unintelligent must be difference in their causation; so, while some movements of human tongues and limbs are the effects of mechanical causes, others must be the effects of non-mechanical causes, i.e. some issue from movements of particles of matter, others from workings of the mind.

The differences between the physical and the mental were thus represented as differences inside the common framework of the categories of 'thing', 'stuff', 'attribute', 'state', 'process', 'change', 'cause' and 'effect'. Minds are things, but different sorts of things from bodies; mental processes are causes and effects, but different sorts of causes and effects from bodily movements. And so on. Somewhat as the foreigner expected the University to be an extra edifice, rather like a college but also considerably different, so the repudiators of mechanism represented minds as extra centres of causal processes, rather like machines but also considerably different from them. Their theory was a para-mechanical hypothesis.

That this assumption was at the heart of the doctrine is shown by the fact that there was from the beginning felt to be a major theoretical difficulty in explaining how minds can influence and be influenced by bodies. How can a mental process, such as willing, cause spatial movements like the movements of the tongue? How can a physical change in the optic nerve have among its effects a mind's perception of a flash of light? . . .

When two terms belong to the same category, it is proper to construct conjunctive propositions embodying them. Thus a purchaser may say that he bought a left-hand glove and a right-hand glove, but not that he bought a left-hand glove, a right-hand glove and a pair of gloves. 'She came home in a flood of tears and a sedan-chair' is a well-known joke based on the absurdity of conjoining terms of different types. It would have been equally ridiculous to construct the disjunction 'She came home either in a flood of tears or else in a sedan-chair'. Now the dogma of the Ghost in

> the Machine does just this. It maintains that there exist both bodies and minds; that there occur physical processes and mental processes; that there are mechanical causes of corporeal movements and mental causes of corporeal movements. I shall argue that these and other analogous conjunctions are absurd; but it must be noticed, the argument will not show that either of the illegitimately conjoined propositions is absurd in itself. I am not, for example, denying that there occur mental processes. Doing long division is a mental process and so is making a joke. But I am saying that the phrase 'there occur mental processes' does not mean the same sort of thing as 'there occur physical processes', and, therefore, that it makes no sense to conjoin or disjoin the two.
>
> If my argument is successful, there will follow some interesting consequences. First, the hallowed contrast between Mind and Matter will be dissipated, but dissipated not by either the equally hallowed absorptions of Mind by Matter, or of Matter by Mind, but in quite a different way. For the seeming contrast of the two will be shown to be as illegitimate as would be the contrast of 'she came home in a flood of tears' and 'she came home in a sedan-chair'. The belief that there is a polar opposition between Mind and Matter is the belief that they are terms of the same logical type.
>
> It will also follow that both Idealism and Materialism are answers to an improper question. The 'reduction' of the material world to mental states and processes, as well as the 'reduction' of mental states and processes to physical states and processes, presuppose the legitimacy of the disjunction 'Either there exist minds or there exist bodies (but not both)'. It would be like saying, 'Either she bought a left-hand and a right-hand glove or she bought a pair of gloves (but not both)'.
>
> It is perfectly proper to say, in one logical tone of voice, that there exist minds and to say, in another logical tone of voice, that there exist bodies. But these expressions do not indicate two different species of existence, for 'existence' is not a generic word like 'coloured' or 'sexed'. They indicate two different senses of 'exist', somewhat as 'rising' has different senses in 'the tide is rising', 'hopes are rising', and 'the average age of death is rising'. A man would be thought to be making a poor joke who said that three things are now rising, namely the tide, hopes and the average age of death. It would be just as good or bad a joke to say that there exist prime numbers and Wednesdays and public opinions and navies; or that there exist both minds and bodies.

Notice that Ryle does not deny that mental life, exists nor does he prohibit or dismiss the use of mentalistic language. What he does do is suggest that our use of such language needs to be critically examined. As any treatment of consciousness will almost certainly use mentalistic language we will do well to bear his suggestions in mind. Careful attention to the way we talk about mental life is one of the enduring contributions of the analytic philosophy of which Ryle's work is a classic example. This contribution has had wide influence on the discipline generally as can be seen from the many good introductions to the philosophy of mind that are now available.[5] As science and philosophy have progressed, positions taken on the mind-body problem have increased in number and subtlety. No position is going to be final. Instead we can select among positions according to the emphasis we wish to place on different aspects of mental-physical interaction. The last selection emphasizes what is possibly the most problematic aspect of consciousness.

In very different ways the following article reflects a single enigmatic issue, which, as a recent philosopher puts it: '. . . makes the mind-body problem intractable'.[6] This issue is subjectivity. Unlike any other topic in the natural sciences, consciousness is *essentially* experience. Consciousness is a flow of mental states whose moment-to-moment coherence and continuity depends upon one thing — that this flow is the possession of a single psychological subject. This fundamental fact, so simple but so powerful, has to be addressed in any treatment of consciousness.

Phenomenology is a style of philosophy and psychology which takes subjective experience as the primary fact to be addressed.[7] While Bergson is not strictly a phenomenologist, the following extract is included here as a lively example of a philosopher taking consciousness-as-experience seriously. The contrast with Ryle shows how large the gulf between analytic and phenomenological philosophy was during the early and middle parts of the twentieth century. Just prior to this period, around the turn of the century, Bergson was extremely influential not only in philosophy but also in art, ethics and biology. His fundamental contribution was to suggest, in clear distinction to classical mechanics, that the natural sense of a real direction to time, especially in connection with living processes, should be taken as a fundamental aspect of nature.[8] He is perhaps better remembered for suggesting that an animating principle, the *élan vital*, guides evolution. This romantic-seeming notion is primarily responsible for the marginalized position of his ideas, at least in the English-speaking philosophical world.

When a lecture is dedicated to the memory of a distinguished man of science, one cannot but feel some constraint in the choice of subject. It must be a subject that would have specially interested the person honoured. I feel no embarrassment on this account in regard to the great name of Huxley; the difficulty would be to find any problem to which his mind would have been indifferent, one of the greatest minds the England of the Nineteenth Century produced. And yet it seems to me that if one subject more than another would have appealed with particular force to the mind of a naturalist who was also a philosopher, it is the threefold problem of consciousness, of life and of their relation. For my part, I know no problem more fundamental in its importance, and it is this which I have chosen.

In dealing with this problem we cannot reckon much on the support of systems of philosophy. The problems men have most deeply at heart, those which distress the human mind with anxious and passionate insistence, are not always the problems which hold the place of importance in the speculations of the metaphysicians. Whence are we? What are we? Whither tend we? These are the vital questions, which immediately present themselves when we give ourselves up to philosophical reflexion without regard to philosophical systems. But, between us and these problems, systematic philosophy interposes other problems. 'Before seeking the solution of a problem,' it says, 'must we not first know how to seek it? Study the mechanism of thinking, then discuss the nature of knowledge and criticize the faculty of criticizing: when you have assured yourself of the value of the instrument, you will know how to use it.' That moment, alas! will never come. I see only one means of knowing how far I can go: that is by going.

The first line or direction which I invite you to follow is this. When we speak of mind we mean, above everything else, consciousness. What is consciousness? There is no need to define so familiar a thing, something which is continually present in every one's experience. I will not give a definition, for that would be less clear than the thing itself; I will characterize consciousness by its most obvious feature: it means, before everything else, memory. Memory may lack amplitude; it may embrace but a feeble part of the past; it may retain only what is just happening; but memory is there, or there is no consciousness. A consciousness unable to conserve its past, forgetting itself unceasingly, would be a consciousness perishing and having to be reborn at each moment: and what is this but unconsciousness? When Leibniz said of matter that it is 'a momentary mind,' did he not declare it, whether he would or no, insensible? All consciousness, then, is memory,— conservation and accumulation of the past in the present.

But all consciousness is also anticipation of the future. Consider the direction of your mind at any moment you like to choose; you will find that it is occupied with what now is, but always and especially with regard to what is about to be. Attention is expectation, and there is no consciousness without a certain attention to life. The future is there; it calls up, or rather, it draws us to it; its uninterrupted traction makes us advance along the route of time and requires us also to be continually acting. All action is an encroachment on the future.

To retain what no longer is, to anticipate what as yet is not,— these are the primary functions of consciousness.

Consciousness is then, as it were, the hyphen which joins what has been to what will be, the bridge which spans the past and the future. But what purpose does the bridge serve? What is consciousness called on to do?

In order to reply to the question, let us inquire what beings are conscious and how far in nature the domain of consciousness extends. But let us not insist that the evidence shall be complete, precise and mathematical; if we do, we shall get nothing. To know with scientific certainty that a particular being is conscious, we should have to enter into it, coincide with it, be it. It is literally impossible for you to prove, either by experience or by reasoning, that I, who am speaking to you at this moment, am a conscious being. I may be an ingeniously constructed natural automaton, going, coming, discoursing; the very words I am speaking to affirm that I am conscious may be being pronounced unconsciously. Yet you will agree that though it is not impossible that I am an unconscious automaton, it is very improbable. Between us there is an evident external resemblance; and from that external resemblance you conclude by analogy there is an internal likeness. Reasoning by analogy never gives more than a probability; yet there are numerous cases in which that probability is so high that it amounts to practical certainty. Let us then follow the thread of the analogy and inquire how far consciousness extends, and where it stops.

It is sometimes said that, in ourselves, consciousness is directly connected with a brain, and that we must therefore attribute consciousness to living beings which have a brain and deny it to those which have none. But it is easy to see the fallacy of such an argument. It would be just as though we should say that because in ourselves digestion is directly connected with a stomach, therefore only living beings with a stomach can digest. We should be entirely wrong, for it is not necessary to have a stomach, nor even to have special organs, in order to digest. An amoeba digests, although it is an almost undifferentiated protoplasmic mass. What is true is that in proportion to the complexity and perfection of an organism there is a division of labour; special organs are assigned special functions; and the faculty of digesting is localized in the stomach, or rather in a general digestive apparatus, which works better because confined to that one function alone. In like manner, consciousness in man is unquestionably connected with the brain: but it by no means follows that a brain is indispensable to consciousness. The lower we go in the animal series, the more the nervous centres are simplified and separate from one another, and at last they disappear altogether, merged in the general mass of an organism with hardly any differentiation. If then, at the top of the scale of living beings, consciousness is attached to very complicated nervous centres, must we not suppose that it accompanies the nervous system down its whole descent, and that when at last the nerve stuff is merged in the yet undifferentiated living matter, consciousness is still there, diffused, confused, but not reduced to nothing? Theoretically, then, everything living might be conscious. *In principle*, consciousness is co-extensive with life. Now, is it so *in fact*? Does not consciousness, occasionally, fall asleep or slumber? This is probable, and here is a second line of facts which lead to this conclusion.

In the living being which we know best, it is by means of the brain that consciousness works. Let us then cast a glance at the human brain and see how it functions. The brain is part of a nervous system which includes, together with the brain proper, the spinal cord, the nerves, etc. In the spinal cord there are mechanisms set up, each of which contains, ready to start, a definite complicated action which the body can carry out at will, just as the rolls of perforated paper which are used in the pianola mark out beforehand the tunes which the instrument will play. Each of these mechanisms can be set working directly by an external cause: the body, then, at once responds to the stimulus received by executing a number of interco-ordinated movements. But in some cases the stimulus, instead of obtaining immediately a more or less complicated reaction from the body by addressing itself directly to the spinal cord, mounts first to the brain, then redescends and calls the mechanism of the spinal cord into play after having made the brain intervene. Why is this indirect path taken? What purpose is served by the intervention of the brain? We may easily guess, if we consider the general structure of the nervous system. The brain is in a general relation to all the mechanisms in the spinal cord and not only to some particular one among them; also it receives every kind of stimulus, not only certain special kinds. It is therefore a crossway, where the nervous impulse arriving by any sensory path can be directed into any motor path. Or, if you prefer, it is a commutator, which allows the current received from one point of the organism to be switched in the direction of any motor contrivance. When the stimulus, then, instead of following the direct path, goes off to the brain, it is evidently in order that it may set in action a motor mechanism which has been chosen, instead of one which is automatic. The spinal cord contains a great number of ready-formed responses to the question which the circumstances address to it; the intervention of the brain secures that the most appropriate among them shall be given. The brain is an organ of choice.

Now, the further we descend the scale of the animal series, the less and less definite we find the separation becoming between the functions of the spinal cord and those of the brain. The faculty of choosing, at first localized in the brain, extends gradually to the spinal cord, which then, probably, constructs somewhat fewer mechanisms and also mounts them with less precision. At last, when we come to the nervous system which is rudimentary, still more when distinct nervous elements have disappeared altogether, automatism and choice are fused into one. The reaction is now so simple that it appears almost mechanical; it still hesitates and gropes, however, as though it would be voluntary. The amoeba, for instance, when in presence of a substance which can be made food, pushes out towards it filaments able to seize and enfold foreign bodies. These pseudopodia are real organs and therefore mechanisms; but they are only temporary organs created for the particular purpose, and it seems they still show the rudiments of choice. From top to bottom, therefore, of the scale of animal life we see being exercised, though the form is ever vaguer as we descend, the faculty of choice, that is, the responding to a definite stimulus by movements more or less unforeseen. This then is what we find along the second line of facts. It re-enforces the conclusion we had come to before; for if, as we said, consciousness retains the past and anticipates the future, it is probably because it is called on to make a choice. In order to choose, we must know what we can do and remember the consequences, advantageous or injurious, of what we have already done; we must foresee and we must remember. And now we are going to see that our first conclusion, re-enforced by this new line of facts, supplies an intelligible answer to the question before us: are all living beings conscious, or does consciousness cover a part only of the domain of life?

If consciousness mean choice and if its role be to decide, it is unlikely that we shall meet it in organisms which do not move spontaneously, and which have no decision to take. Strictly speaking, there is no living being which appears completely incapable of spontaneous movement. Even in the vegetable world, where the organism is generally fixed to the soil, the faculty of movement is dormant rather than absent; it awakens when it can be of use. I believe all living beings, plants and animals, possess it in right; but many of them have renounced it in fact,— some animals, especially those which have become parasitic on other organisms and have no need of moving about to find their nourishment, and the vast majority of plants: has it not been said that plants are earth-parasites? It appears to me therefore extremely likely that consciousness, originally immanent in all that lives, is dormant where there is no longer spontaneous movement, and awakens when life tends to free activity. We can verify the law in ourselves. What happens when one of our actions ceases to be spontaneous and becomes automatic? Consciousness departs from it. In learning an exercise, for example, we begin by being conscious of each of the movements we execute. Why? Because we originate the action, because it is the result of a decision and implies a choice. Then, gradually, as the movements become more and more linked together and more and more determine one another mechanically, dispensing us from the need of choosing and deciding, the consciousness of them diminishes and disappears. On the other hand, when is it that our consciousness attains its greatest liveliness? Is it not at those moments of inward crisis when we hesitate between two, or it may be several, different courses to take, when we feel that our future will be what we make it? The variations in the intensity of our consciousness seem then to correspond to the more or less considerable sum of choice or, as I would say, to the amount of creation, which our conduct requires. Everything leads us to believe that it is thus with consciousness in general. If consciousness means memory and anticipation, it is because consciousness is synonymous with choice.

Let us then imagine living matter in its elementary form, such as it may have been when it first appeared: a simple mass of protoplasmic jelly like the amoeba, which can undergo change of form at will, and is therefore vaguely conscious. Now, for it to grow and evolve, there are two ways open. It may take the path towards movement and action,— movement growing ever more effective, action growing freer and freer. The path towards movement involves risk and adventure, but also it involves consciousness, with its growing degrees of intensity and depth. It may take the other path, it may abandon the faculty of acting and choosing, the potentiality of which it carries within it, may accommodate itself to obtain from the spot where it is all it requires for its support, instead of going abroad to seek it. Existence is then assured to it, a tranquil, unenterprising existence, but this existence is also torpor, the first effect of immobility: the torpor soon becomes fixed; this is unconsciousness. These are the two paths which lie open before the evolution of life. Living matter finds itself committed partly to the one path, partly to the other. Speaking generally, the first path may be said to mark the direction of the animal world (we have to qualify it, because many animal species renounce movement and with it probably consciousness also); the second may be said to mark the direction of the vegetable world (again it has to be qualified, for mobility, and therefore probably consciousness also, may occasionally be awakened in plants).

When, now, we reflect on this bias or tendency of life at its entry into the world, we see it bringing something which encroaches on inert matter. The world left to itself obeys fatalistic laws. In determinate conditions matter behaves in a determinate way. Nothing it does is unforeseeable. Were our science complete and our calculating power infinite, we should be able to predict everything which will come to pass in the inorganic material universe, in its mass and in its elements, as we predict an eclipse of the sun or moon. Matter is inertia, geometry, necessity. But with life there appears free, predictable, movement. The living being chooses or tends to choose. Its role is to create. In a world where everything else is determined, a zone of indetermination surrounds it. To create the future requires preparatory action in the present, to prepare what will be is to utilize what has been: life therefore is employed from its start in conserving the past and anticipating the future in a duration in which past, present and future tread one on another, forming an indivisible continuity. Such memory, such anticipation, are consciousness itself. This is why, in right if not in fact, consciousness is co-extensive with life.

Consciousness and matter appear to us, then, as radically different forms of existence, even as antagonistic forms, which have to find a *modus vivendi*. Matter is necessity, consciousness is freedom; but though diametrically opposed to one another, life has found the way of reconciling them. This is precisely what life is,— freedom inserting itself within necessity, turning it to its profit. Life would be an impossibility were the determinism of matter so absolute as to admit no relaxation. Suppose, however, that at particular moments and at particular points matter shows a certain elasticity, then and there will be the opportunity for consciousness to instal itself. It will have to humble itself at first; yet, once installed, it will dilate, it will spread from its point of entry and not rest till it has conquered the whole, for time is at its disposal, and the slightest quantity of indetermination, by continually adding to itself, will make up as much freedom as you like.

All I wish to say is that . . . consciousness appears as a force seeking to insert itself in matter in order to get possession of it and turn it to its profit.

On the one hand, there is matter, subject to necessity, devoid of memory, or at least with no more than suffices to form the bridge between two of its moments, each of which can be deduced from its antecedent, each of which adds nothing to what the world already contains. On the other hand, there is consciousness, memory with freedom, continuity of creation in a duration in which there is real growth;— a duration which is drawn out, wherein the past is preserved indivisible; a duration which grows like a plant, but like the plant of a fairy tale transforms its leaves and flowers from moment to moment. We may surmise that these two realities, matter and consciousness, are derived from a common source. If, as I have tried to show in a previous work (*Creative Evolution*), matter is the inverse of consciousness, if consciousness is action unceasingly creating and enriching itself, whilst matter is action continually unmaking itself or using itself up, then neither matter nor consciousness can be explained apart from one another. I will not return to this theme now, I will merely say that I see in the whole evolution of life on our planet a crossing of matter by a creative consciousness, and effort to set free, by force of ingenuity and invention, something which in the animal still remains imprisoned and is only finally released when we reach man.

We need not go into the details of the scientific investigations which since Lamarck and Darwin have come more and more to confirm the idea of an evolution of species, that is, of the generation of species from one another, the organized forms from the simpler. We can hardly refuse to accept a hypothesis which has the threefold support of comparative anatomy, of embryology and of paleontology. Science has shown, moreover, along the whole evolution of life, the various consequences attending upon the fact that living beings must be adapted to the conditions of the environment. Yet this necessity would seem to explain the arrest of life in various definite forms, rather than the movement which carries the organization ever higher. A very inferior organism is as well adapted as ours to the conditions of existence, judged by its success in maintaining its life: why, then, does life which has succeeded in adapting itself go on complicating itself, and complicating itself more and more dangerously? Some living forms to be met with today have come down unchanged from remotest palaeozoic times; they have persisted, unchanged, throughout the ages. Life then might have stopped at some one definite form. Why did it not stop wherever it was possible? Why has it gone on? Why,— unless it be that there is an impulse driving it to take ever greater and greater risks towards its goal of an ever higher and higher efficiency?

Things have happened just as though an immense current of consciousness, interpenetrated with potentialities of every kind, had traversed matter to draw it towards organization and make it, notwithstanding that it is necessity itself, an instrument of freedom. But consciousness has had a narrow escape from being itself ensnared. Matter, enfolding it, bends it to its own automatism, lulls it to sleep in its own unconsciousness. On certain lines of evolution, those of the vegetable world in particular, automatism and unconsciousness are the rule: the freedom immanent in evolution is shown even here, no doubt, in the creation of unforeseen forms which are veritably works of art; but, once created, the individual has no choice. On other lines, consciousness succeeds in freeing itself sufficiently for the individual to acquire feeling, and therewith a certain latitude of choice; but the necessities of existence restrict the power of choosing to a simple aid of the need to live. So, from the lowest to the highest rung of the ladder of life, freedom is riveted in a chain which at most it succeeds in stretching. With man alone a sudden bound is made; the chain is broken. The human brain closely resembles the animal brain, but it has, over and above, a special factor which furnishes the means of opposing to every contracted habit another habit, and to every automatism an antagonistic automatism. Freedom, coming to itself whilst necessity is at grips with itself, brings back matter to the condition of being a mere instrument. It is as though it had divided in order to rule.

That the united efforts of physics and chemistry to manufacture matter resembling living matter may one day be successful is by no means improbable, for life proceeds by insinuating, and the force which drew matter away from pure mechanism could not have taken hold of matter had it not first itself adopted that mechanism. In such wise, the points of the railway coincide at first with the lines from which they will shunt the train. In other words, life must have installed itself in a matter which had already acquired some of the characters of life without the work of life. But matter left to itself would have stopped there; and the work of our laboratories will probably go no further. We shall reproduce, that is to say, some characters of living matter; we shall not obtain the push in virtue of which it reproduces itself and, in the meaning of transformism, evolves. Now, reproduction and evolution are life itself. Both are the manifestation of an inward impulse, of the twofold need of increasing in number and wealth by multiplication in space and complication in time, of two instincts which make their appearance with life and later become the two great motives in human activity, love and ambition. Visibly there is a force working, seeking to free itself from trammels and also to surpass itself, to give first all it has and then something more than it has. What else is mind? How can we distinguish the force of mind, if it exist, from other forces save in this, that it has the faculty of drawing from itself more than it contains? Yet we must take into account the obstacles of every kind that such a force will meet on its way. The evolution of life, from its early origins up to man, presents to us the image of a current of consciousness flowing against matter, determined to force for itself a subterranean passage, making tentative attempts to the right and to the left, pushing more or less ahead, for the most part encountering rock and

> breaking itself against it, and yet, in one direction at least, succeeding in piercing its way through and emerging into the light. That direction is the line of evolution which ends in man.
>
> To conclude . . . life, as I imagine it, is still a life of striving, a need of invention, a creative evolution . . . I admit that this is no more than a hypothesis. We were just now in the region of the probable, this is the region of the simply possible. Let us confess our ignorance, but let us not resign ourselves to the belief that we can never know. . . . Recollect what has happened in regard to another beyond, that of ultra-planetary space. Auguste Comte declared the chemical composition of the heavenly bodies to be for ever unknowable by us. A few years later the spectroscope was invented, and today we know, better than if we had gone there, what the stars are made of.

Bergson presents consciousness as an organizing force 'flowing against' the tendency of matter to be inert and automatic. Though this vitalistic thread of Bergson's thought has been taken more as a matter of literary imagination than philosophy, we can take from it that consciousness is essentially bound up with purposiveness, choice and awareness.

Bergson reminds us that consciousness is fundamentally an act of knowing by a psychological subject. Though this irreducibly subjective character of consciousness makes any physicalist account of it incomplete, the readings in the next chapter are from scientists and philosophers who explore how subjectivity can be put into some relationship with the findings and methods of science. This question arises naturally, since while the physical and mental worlds are qualitatively distinct they are obviously intimately connected at the same time. Exploring the correspondence between knowledge of the physical world and the world of experience seems a natural next step.

Notes

1. See Chapter One of C. Waddington, *Tools for Thought*, 1977, Jonathan Cape.
2. Hardie, W., 'Concepts of Consciousness in Aristotle', *Mind*, 85:388–411. For a more recent work on Aristotle's view on the mind, see M. Wedin, *Mind and Imagination in Aristotle*, 1988, Yale University Press.
3. *Pico della Mirandola: Oration on the Dignity of Man*, A. Caponigri, trs., 1956, Gateway Editions. A more extensive discussion of this work and its

context can be found Chapter Seven of E. Harth, *Windows on the Mind*, 1985, Pelican.
4. See O. Neurath, 'Unified Science and Psychology' in B. McGuiness, *Unified Science*, 1987, Reidel.
5. Other good introductions are: G. Vesey, ed., *Body and Mind*, 1965, Allen & Unwin; J. Fodor, 'The Mind Body Problem', *Scientific American*, January, 1981; and P. Churchland, *Matter and Consciousness*, 1984, MIT Press.
6. Nagel, T., 'What's It Like To Be a Bat?', *The Philosophical Review*, October, 1974. This is reprinted in D. Dennett and D. Hofstadter, eds., *The Mind's I*, 1981, Harvester Press.
7. Husserl and Heidegger are perhaps the most frequently cited phenomenological philosophers. Husserl tried to show that, with sufficiently close examination of experience, it was possible to reveal mental processes which would be as true of all human minds as the experiments of scientists reveal laws which are true of all material nature. Heidegger developed a philosophy in which the flow of thought and feeling was taken as primary reality. For an introduction to phenomenology, see R. Grossberg, *Phenomenology and Existentialism*, 1984, Routledge & Kegan Paul.
8. A good survey of this issue can be found on pp. 89–93 of I. Prigogine and I. Stengers, *Order out of Chaos*, 1985, Fontana.

2

Consciousness and the Physical World

READINGS

Karl Popper and John Eccles
The Self and its Brain

Alfred Whitehead
Science and the Modern World

Ilya Prigogine and Isabelle Stengers
Order out of Chaos

David Bohm and F. David Peat
Science, Order and Creativity

George Wald
Life and Mind in the Universe

This chapter looks at how consciousness, the subjective mental world, may be related to the physical world, as objectively described by science. The first reading is by Karl Popper, a philosopher of science, from a book he has written with John Eccles, a neuropsychologist. He is prepared to consider consciousness as real and causal. Popper suggests all phenomena fall into three groups which he calls worlds one, two and three. World one is the physical realm of matter and energy itself. World two is the realm of conscious experience; this realm is subjective, but indissoluably linked to world one. World three, the realm of cultural products like language and theories, is objective but is the product of world two. He discusses mind and consciousness against the backdrop of this ontological system and reminds us that consciousness has an evolutionary and cultural context.

Whether or not biology is reducible to physics, it appears that all physical and chemical laws are binding for living things — plants and animals, and even viruses. Living things are material bodies. Like all material bodies, they are processes; and like some other material bodies (clouds, for example) they are open systems of molecules: systems that exchange some of their constituent parts with their environment. They belong to the *universe of physical entities*, or states of physical things, or physical states.

The entities of the physical world — processes, forces, fields of forces — interact among one another, and therefore with material bodies. Thus we conjecture them to be real (in the sense discussed in section 4 above) even though their reality remains conjectural.

Besides the physical objects and states, I conjecture that there are *mental states*, and that these states are real since they interact with our bodies.

It is sometimes said that it is the task of the solution of the brain-mind problem to make the interaction between such different things as physical states or events and mental states or events understandable.

I agree that the main task of science is to further our understanding. But I also think that complete understanding, just like complete knowledge, is unlikely ever to be achieved. Moreover, understanding can be deceptive: we had, for centuries, what appeared to be a perfect understanding of the working of clockwork mechanisms in which the cogs of the cogwheels push each other along. But this turned out to be a very superficial understanding, and the push given by a physical body to another had to be explained by the repulsion between the negatively charged electron shells of their atoms. However, this explanation and this understanding are also superficial, as is shown by the facts of adhesion and cohesion. Thus final understanding is not easy, not even in what seems the most elementary part of physical science. And when we move to the interaction between light and matter, then we get into a region of knowledge which left one of the greatest pioneers in this field, Niels Bohr, baffled; so much so that he said that in quantum theory we had to renounce the hope of understanding our subject. However, though it seems that the ideal of *complete* understanding has to be renounced, a detailed description may lead to some *partial* understanding.

Thus an understanding such as we once mistakenly believed we possessed in the case of mechanical push is not available even in physics. And we can hardly expect it in the case of brain-mind interaction, although a more detailed knowledge of the working of the brain may give us that partial understanding which, it seems, is realizable in science.

In this section, I have talked of physical states and of mental states. I think, however, that the problems with which we are dealing can be made considerably clearer if we introduce a *tripartite* division. First, there is the physical world — the universe of physical entities — to which I referred at the beginning of this section; this I will call 'World 1'. Second, there is the world of mental states, including states of consciousness and psychological dispositions and unconscious states; this I will call 'World 2'. But there is also a *third* such world, the world of the contents of thought, and, indeed, of the products of the human mind; this I will call 'World 3', and it will be discussed in the next few sections.

I think that some increase of understanding can be obtained by studying the role of World 3.

By World 3 I mean the world of the products of the human mind, such as stories, explanatory myths, tools, scientific theories (whether true or false), scientific problems, social institutions, and works of art. World 3 objects are of our own making, although they are not always the result of planned production by individual men.

Many World 3 objects exist in the form of material bodies, and belong in a sense to both World 1 and World 3. Examples of sculptures, paintings, and books, whether devoted to a scientific subject or to literature. A book is a physical object, and it therefore belongs to World 1; but what makes it a significant product of the human mind is its *content*: that which remains invariant in the various copies and editions. And this content belongs to World 3.

One of my main theses is that World 3 objects can be real not only in their World 1 materializations or embodiments, but also in their World 3 aspects. As World 3 objects, they may induce men to produce other World 3 objects and, thereby, to act on World 1; and interaction with World 1 — even indirect interaction — I regard as a decisive argument for calling a thing real.

Many World 3 objects like books or new synthetic medicines or computers or aircraft are embodied in World 1 objects: they are material artefacts, they belong to both World 3 and World 1. Most works of art are like this. Some World 3 objects exist only in encoded form, as musical scores (perhaps never performed) or as gramophone records. Others — poems, perhaps, and theories — may also exist as World 2 objects, as memories, presumably also encoded as memory traces in certain human brains (World 1) and perishing with them.

The main reason why I consider the existence of unembodied World 3 objects so important is this. If unembodied World 3 objects exist, then it cannot be a true doctrine that our grasp or understanding of a World 3 object always depends upon our sensual contact with its material embodiment; for example, upon our reading a statement of a theory in a book. As against this doctrine I assert that the most characteristic way of grasping World 3 objects is by a method which depends little, if at all, upon their embodiment or upon the use of our senses. My thesis is that the human mind grasps World 3 objects, if not always directly, then by an indirect method (which will be discussed); a method which is independent of their embodiment, and which, in the case of those World 3 objects (such as books) that belong also to World 1, abstracts from the fact that they are embodied.

Thus we learn, not by direct vision or contemplation, but by practice, by active participation, how to make World 3 objects, how to understand them, and how to 'see' them. This includes the 'sensing' of open problems, even of problems not yet formulated. It may incite us to think, to examine the existing theories; to discover a vaguely suspected problem; and to produce theories which we hope will solve it. In this process, published theories — embodied theories — may play a role. But the not yet explored logical relations between existing theories may also play a role. Both these theories and their logical relations are World 3 objects, and in general it makes no difference, neither to their character as World 3 objects nor to our World 2 grasp of them, whether or not these objects are embodied. Thus a not yet discovered and not yet embodied logical problem situation may prove decisive for our thought processes, and may lead to actions with repercussions in the physical World 1, for example to a publication. (An example would be the search for, and the discovery of, a suspected new proof of a mathematical theorem.)

In this way World 3 objects, including logical possibilities so far not fully explored, may act on World 2; that is to say, on our minds, on us. And we in turn may act on World 1.

This process may of course be described without mentioning what I call World 3. Thus we may say that, incited by their knowledge about World 1, certain physicists (Szilard, Fermi, Einstein) suspected the physical possibility of making a nuclear bomb, and that these World 2 thoughts brought about the realization of their conjecture. Descriptions such as this are perfectly in order. But they hide the fact that by 'their knowledge about World 1', are meant *theories* which can be objectively investigated, from a logical as well as an empirical point of view, and that these are World 3 objects rather than World 2 objects (though they can be grasped and therefore have World 2 correlates); similarly by the words 'suspected the physical possibility', conjectures about *physical theories* are meant — again World 3 objects, to be investigated logically. It is perfectly true that the physicist is primarily interested in World 1. But in order to learn more about World 1 he must theorize; and this means that he must use World 3 objects as his tools. This forces him to take an interest — a secondary interest, may be — in his tools, in the World 3 objects. And only by investigating them, and working out their logical consequences, can he do 'applied science'; that is, make use of his World 3 products as tools, in order to change World 1.

Thus even unembodied World 3 objects may be regarded as real, and not only the papers and books in which our physical theories are published, or the material instruments which are based on these publications.

It is one of the central conjectures of *The Self and Its Brain* that the consideration of World 3 [see selection 4 above] can throw some new light on the mind-body problem. I will briefly state three arguments.

The first argument is as follows.

(1) World 3 objects are abstract (even more abstract than physical forces), but none the less real; for they are powerful tools for changing World 1. (I do not wish to imply that this is the only reason for calling them real, or that they are nothing but tools.)

(2) World 3 objects have an effect on World 1 only through human intervention, the intervention of their makers; more especially, through being grasped, which is a World 2 process, a mental process, or more precisely, a process in which World 2 and World 3 interact.

(3) We therefore have to admit that both World 3 objects and the processes of World 2 are real — even though we may not like this admission, out of deference, say, to the great tradition of materialism.

I think that this is an acceptable argument — though, of course, it is open to someone to deny any one of its assumptions. He may deny that theories are abstract, or deny that they have an effect on World 1, or claim that abstract theories can directly affect the physical world. (I think, of course, that he would have a difficult time in defending any of these views.)

The second argument partly depends upon the first. If we admit the interaction of the three worlds, and thus their reality, then the interaction between Worlds 2 and 3, which we can to some extent understand, can perhaps help us a little towards a better understanding of the interaction between Worlds 1 and 2, a problem that is part of the mind-body problem.

For one kind of interaction between Worlds 2 and 3 ('grasping') can be interpreted as a making of World 3 objects and as a matching of them by critical selection; and something similar seems to be true for the visual perception of a World 1 object. This suggests that we should look upon World 2 as active — as productive and critical (making and matching). But we have reason to think that some unconscious neurophysiological processes achieve precisely this. This makes it perhaps a little easier to 'understand' that conscious processes may act along similar lines: it is, up to a point, 'understandable' that conscious processes perform tasks similar to those performed by nervous processes.

A third argument bearing on the mind-body problem is connected with the status of human language.

The capacity to learn a language — and even a strong need to learn a language — are, it appears, part of the genetic make-up of man. By contrast, the actual learning of a particular language, though influenced by unconscious inborn needs and motives, is not a gene regulated process and therefore not a natural process, but a cultural process, a World 3 regulated process. Thus language learning is a process in which genetically based dispositions, evolved by natural selection, somewhat overlap and interact with a conscious process of exploration and learning, based on cultural evolution. This supports the idea of an interaction between World 3 and World 1; and in view of our earlier arguments, it supports the existence of World 2.

Several eminent biologists have discussed the relationship between genetic evolution and cultural evolution. Cultural evolution, we may say, continues genetic evolution by other means: by means of World 3 objects.

It is often stressed that man is a tool-making animal, and rightly so. If by tools material physical bodies are meant, it is, however, of considerable interest to notice that none of the human tools is genetically determined, not even the stick. The only tool that seems to have a genetic basis is language. Language is non-material, and appears in the most varied physical shapes — that is to say, in the form of very different systems of physical sounds.

There are behaviourists who do not wish to speak of 'language', but only of the 'speakers' of one or the other particular language. Yet there is more to it than that. All normal men speak; and speech is of the utmost importance for them; so much so that even a deaf, dumb and blind little girl like Helen Keller acquired with enthusiasm, and speedily, a substitute for speech through which she obtained a real mastery of the English language and of literature. Physically, her language was vastly different from spoken English; but it had a one-to-one correspondence with written or printed English. There can be no doubt that she would have acquired any other language in place of English. Her urgent though unconscious need was for language — language in the abstract.

As shown by their numbers and their differences, the various languages are manmade: they are cultural World 3 objects, though they are made possible by capabilities, needs, and aims which have become genetically entrenched. Every normal child acquires a language through much active work, pleasurable and perhaps also painful. The intellectual achievement that goes with it is tremendous. This effort has, of course, a strong feedback effect on the child's personality, on his relations to other persons, and on his relations to his material environment.

Thus we can say that the child is, partly, the product of his achievement. He is himself, to some extent, a World 3 product. Just as the child's mastery and consciousness of his material environment are extended by his newly acquired ability to speak, so also is his consciousness of himself. The self, the personality, emerges in interaction with the other selves and with the artefacts and other objects of his environment. All this is deeply affected by the acquisition of speech; especially when the child becomes conscious of his name, and when he learns to name the various parts of his body; and, most important, when he learns to use personal pronouns.

Becoming a fully human being depends on a maturation process in which the acquisition of speech play an enormous part. One learns not only to perceive, and to interpret one's perceptions, but also to be a person, and to be a self. I regard the view that our perceptions are 'given' to us as a mistake: they are 'made' by us, they are the result of active work. Similarly I regard it as a mistake to overlook the fact that the famous Cartesian argument 'I think, therefore I am' presupposes language, and the ability to use the pronoun (to say nothing of the formulation of the highly sophisticated problem which this argument is supposed to settle).

I propose that the evolution of consciousness, and of conscious intelligent effort, and later that of language and of reasoning — and of World 3 — should be considered teleologically, as we consider the evolution of bodily organs: as serving certain purposes, and as having evolved under certain selection pressures. [Compare section III of the previous selection.]

The problem can be put as follows. Much of our purposeful behaviour (and presumably of the purposeful behaviour of animals) happens without the intervention of consciousness. What, then, are the biological achievements that are helped by consciousness?

I suggest as a first reply: the solution of *problems of a non-routine kind*. Problems that can be solved by routine do not need consciousness. This may explain why intelligent speech (or still better, writing) is such a good example of a conscious achievement (of course, it has its unconscious roots). As has been often stressed, it is one of the characteristics of human language that we constantly produce new *sentences* — sentences never before formulated — and understand them. As opposed to this major achievement, we constantly make use of *words* (and, of course, of phonemes) which are used routinely, again and again, though in a most varied context. A fluent speaker produces most of these words unconsciously, without paying attention to them, except where the choice of the best word may create a problem — a new problem, not solved by routine. '. . . new situations and the new responses they prompt are kept in the light of consciousness', Erwin Schrödinger writes; 'old and well practised ones are no longer so [kept].'

A closely related idea concerning the function of a consciousness is the following. Consciousness is needed to select, critically, new expectations or theories — at least on a certain level of abstraction. If any one expectation or theory is invariably successful, under certain conditions, its application will after a time turn into a matter of routine, and become unconscious. But an unexpected event will attract attention, and thus consciousness. We may be unconscious of the ticking of a clock, but 'hear' that it has stopped ticking.

We cannot know, of course, how far animals are conscious. But novelty can excite their attention; or more precisely, it can excite behaviour which, because of its similarity to human behaviour, many observers will describe as 'attention', and interpret as conscious.

But the role of consciousness is perhaps clearest where an aim or purpose (perhaps even an unconscious or instinctive aim or purpose) can be achieved by *alternative means*, and when two or more means are tried out, after deliberation. It is the case of making a new decision. (Of course, the classical case is Köhler's chimpanzee Sultan who fitted a bamboo stick into another, after many attempts to solve the problem of obtaining fruit out of his reach: a detour strategy in problem solving.) A similar situation is the choice of a non-routine programme, or of a new aim, such as the decision whether or not to accept an invitation to lecture, in addition to much work in hand. The acceptance letter, and the entry into the engagement calendar, are World 3 objects, anchoring our action programme; and the general principles we may have developed for accepting or rejecting such invitations are also programmes, also belonging to World 3, though perhaps on a higher hierarchical level.

From the biological point of view it is, especially in the case of the higher animals, the individual organism that is fighting for its existence; that is relaxing; that is acquiring new experiences and skills; that is suffering; and that is ultimately dying. In the case of the higher animals it is the central nervous system which 'INTEGRATES' (to use Sherrington's phrase) all the activities of the individual animal (and, if I may say so, all its 'passivities' which will include *some* 'reflexes'). Sherrington's famous idea of 'the integrative action of the nervous system' is perhaps best illustrated by the innumerable nervous actions which have to co-operate in order to keep a man standing quietly upright, at rest.

A great many of these integrative actions are automatic and unconscious. But some are not. To these belong, especially, the selection of means to certain (often unconscious) ends, that is to say, the making of decisions, the selection of programmes.

Decision making or programming is clearly a biologically important function of whatever the entity is that rules, or controls, the behaviour of animals or men. It is essentially an integrative action, in Sherrington's sense: it relates the behaviour at different instants of time to expectations; or in other words, it relates present behaviour to impending or future behaviour. And it directs *attention*, by selecting what are relevant objects, and what is to be ignored.

As a wild conjecture I suggest that it is out of four biological functions that consciousness emerges: pain, pleasure, expectation and attention. Perhaps attention emerges out of primitive experiences of pain and pleasure. But attention is, as a phenomenon, almost identical with consciousness: even pain may sometimes disappear if attention is distracted and focused elsewhere.

The question arises: how far can we explain the individual unity of our consciousness, or our selfhood, by an appeal to the biological situation? I mean by an appeal to the fact that we are animals, animals in whom the instinct for individual survival has developed, as well as, of course, an instinct for racial survival.

Konrad Lorenz writes of the sea urchin that its 'non-centralized nervous system . . . makes it impossible for such animals to inhibit completely one of a number of potentially possible ways of behaviour, and thus to 'decide' in favour of an alternative way. But such a decision (as shown so convincingly by Erich von Holst in the case of the earth worm) is the most fundamental and the most important achievement of a brainlike central nervous organ.' In order to achieve this, the relevant situation must be signalled to the central organ in an adequate manner (that is to say, both in a realistic manner and, by suppressing the irrelevant aspects of the situation, in an idealizing manner). Thus a unified centre must inhibit some of the possible ways of behaviour and only allow one single way at a time to proceed: a way, Lorenz says, 'which in the situation just existing can contribute to survival. . . . The greater the number of possible ways of behaviour, the higher the achievement which is required from the central organ.'

Thus (1) the individual organism — the animal — is a unit; (2) each of the various ways of behaving — the items of the behavioural repertoire — is a unit, the whole repertoire forming a set of mutually exclusive alternatives; (3) the central organ of control must act as a unit (or rather, it will be more successful if it does).

Together these three points, (1), (2), and (3), make even of the animal an active, problem solving *agent*: the animal is always actively attempting to control its environment, in either a positive sense, or, when it is 'passive', in a negative sense. In the latter case it is undergoing or suffering the actions of an (often hostile) environment that is largely beyond its control. Yet even if it is merely contemplating, it is actively contemplating: it is never merely the sum of its impressions, or of its experiences. Our mind (and, I venture to suggest, even the animal mind) are never a mere 'stream of consciousness', a stream of experiences. Rather, our active attention is focused at every moment on just the relevant aspects of the situation, selected and abstracted by our perceiving apparatus, into which a selection programme is incorporated; a programme which is adjusted to our available repertoire of behavioural responses.

I do not think that what I have said here or in the preceding sections clears up any mystery; but I do think that we need not regard as mysterious either the individuality, or the unity, or the uniqueness of the self, or our personal identity; at any rate not as more mysterious than the existence of consciousness, and ultimately that of life, and of individualized organisms. The emergence of full consciousness, capable of self-reflection, which seems to be linked to the human brain and to the descriptive function of language, is indeed one of the greatest of miracles. But if we look at the long evolution of individuation and of individuality, at the evolution of a central nervous system, and at the uniqueness of individuals (due partly to genetic uniqueness and partly to the uniqueness of their experience), then the fact that consciousness and intelligence and unity are linked to the biological individual organism (rather than, say, to the germ plasm) does not seem so surprising. For it is in the individual organism that the germ plasm — the genome, the programme for life — has to stand up to tests.

Popper's views create a basis for dialogue between scientists and others concerned with consciousness. However, Popper rejects any subjectivist interpretation of quantum physics and holds to the view of a physical world whose parts are separable and have independent meaning.[1]

While the author of the next reading might accept that it is possible to analyse physical reality into parts, he also holds that it is a fundamental mistake to think that these parts have independent meaning. Alfred Whitehead is chiefly known as the co-author, with Bertrand Russell, of *Principia Mathematica*, an influential work on the logical foundations of mathematics. But in his subsequent work he concentrated on developing what has been variously labeled an Organicist or Process philosophy. Very informally, the distinctive critical claim in this philosophy is that modern Western science may be missing the wood for the trees. For Whitehead, the causal structure of reality is not to be found by looking into smaller and smaller details. The world, at physical, biological and psychological levels of organization is essentially a process, active and generative. To hold this process still and look deeply into isolated parts of it can produce only a limited view.

Although the influence of Whitehead's later philosophy has not yet been very great, some of his work is offered here for two reasons. First, he represents a contemporary form of organicism — a world view characteristic of pre-socratic philosophy which is beginning to re-emerge in various forms.[2] Second, he has been a significant influence on a particular group of scientists with greater than usual interest in consciousness.

When you are criticising the philosophy of an epoch, do not chiefly direct your attention to those intellectual positions which its exponents feel it necessary explicitly to defend. There will be some fundamental assumptions which adherents of all the variant systems within the epoch unconsciously presuppose. Such assumptions appear so obvious that people do not know what they are assuming because no other way of putting things has ever occurred to them. With these assumptions a certain limited number of types of philosophic systems are possible, and this group of systems constitutes the philosophy of the epoch.

One such assumption underlies the whole philosophy of nature during the modern period. It is embodied in the conception which is supposed to express the most concrete aspect of nature. The Ionian philosophers asked, What is nature made of? The answer is couched in terms of stuff, or matter, or material,— the particular name chosen is indifferent — which has the property of simple location in space and time, or, if you adopt the more modern ideas, in space-time. What I mean by matter, or material, is anything which has this property of *simple location*. By simple location I mean one major characteristic which refers equally both to space and to time, and other minor characteristics which are diverse as between space and time.

The characteristic common both to space and time is that material can be said to be *here* in space and *here* in time, or *here* in space-time, in a perfectly definite sense which does not require for its explanation any reference to other regions of space-time.

The answer, therefore, which the seventeenth century gave to the ancient question of the Ionian thinkers, 'What is the world made of?' was that the world is a succession of instantaneous configurations of matter,— or of material, if you wish to include stuff more subtle than ordinary matter, the ether for example.

We cannot wonder that science rested content with this assumption as to the fundamental elements of nature. The great forces of nature, such as gravitation, were entirely determined by the configurations of masses. Thus the configurations determined their own changes, so that the circle of scientific thought was completely closed. This is the famous mechanistic theory of nature, which has reigned supreme ever since the seventeenth century. It is the orthodox creed of physical science. Furthermore, the creed justified itself by the pragmatic test. It worked. Physicists took no more interest in philosophy. They emphasized the anti-rationalism of the Historical Revolt. But the difficulties of this theory of materialistic mechanism very soon became apparent. The history of thought in the eighteenth and nineteenth centuries is governed by the fact that the world had got hold of a general idea which it could neither live with nor live without.

This simple location of instantaneous material configurations is what Bergson has protested against, so far as it concerns time and so far as it is taken to be the fundamental fact of concrete nature. He calls it a distortion of nature due to the intellectual 'spatialisation' of things. I agree with Bergson in his protest: but I do not agree that such distortion is a vice necessary to the intellectual apprehension of nature. I shall endeavour to show that this spatialisation is the expression of more concrete facts under the guise of very abstract logical constructions. There is an error; but it is merely the accidental error of mistaking the abstract for the concrete. It is an example of what I will call the 'Fallacy of Misplaced Concreteness.' This fallacy is the occasion of great confusion in philosophy.

It is not necessary for the intellect to fall into the trap, though in this example there has been a very general tendency to do so.

It is at once evident that the concept of simple location is going to make great difficulties for induction. For, if in the location of configurations of matter throughout a stretch of time there is no inherent reference to any other times, past or future, it immediately follows that nature within any period does not refer to nature at any other period. Accordingly, induction is not based on anything which can be observed as inherent in nature.

There is another presupposition of thought which must be put beside the theory of simple location. I mean the two correlative categories of Substance and quality.

I hold that substance and quality afford another instance of the fallacy of misplaced concreteness. Let us consider how the notions of substance and quality arise. We observe an object as an entity with certain characteristics. Furthermore, each individual entity is apprehended through its characteristics. For example, we observe a body; there is something about it which we note. Perhaps, it is hard, and blue, and round, and noisy. We observe something which possesses these qualities: apart from these qualities we do not observe anything at all. Accordingly, the entity is the substratum, or substance, of which we predicate qualities. Some of the qualities are essential, so that apart from them the entity would not be itself; while other qualities are accidental and changeable. In respect to material bodies, the qualities of having a quantitative mass, and of simple location somewhere, were held by John Locke at the close of the seventeenth century to be essential qualities. Of course, the location was changeable, and the unchangeability of mass was merely an experimental fact except for some extremists.

So far, so good. But when we pass to blueness and noisiness a new situation has to be faced. In the first place, the body may not be always blue, or noisy.

But . . . in the second place, the seventeenth century exposed a real difficulty. The great physicists elaborated transmission theories of light and sound, based upon their materialistic views of nature. There were two hypotheses as to light: either it was transmitted by the vibratory waves of a materialistic ether, or — according to Newton — it was transmitted by the motion of incredibly small corpuscles of some subtle matter. We all know that the wave theory of Huyghens held the field during the nineteenth century, and at present physicists are endeavouring to explain some obscure circumstances attending radiation by a combination of both theories. But whatever theory you choose, there is no light or colour as a fact in external nature. There is merely motion of material. Again, when the light enters your eyes and falls on the retina, there is merely motion of material. Then your nerves are affected and your brain is affected, and again this is merely motion of material. The same line of argument holds for sound, substituting waves in the air for waves in the ether, and ears for eyes.

We then ask in what sense are blueness and noisiness qualities of the body. By analogous reasoning, we also ask in what sense is its scent a quality of the rose.

Galileo considered this question, and at once pointed out that, apart from eyes, ears, or noses, there would be no colours, sounds, or smells. Descartes and Locke elaborated a theory of primary and secondary qualities. For example, Descartes in his 'Sixth Meditation' says: 'And indeed, as I perceive different sorts of colours, sounds, odours, tastes, heat, hardness, etc., I safely conclude that there are in the bodies from which the diverse perceptions of the senses proceed, certain varieties corresponding to them, although, perhaps, not in reality like them; . . .'

Also in his *Principles of Philosophy*, he says: 'That by our senses we know nothing of external objects beyond their figure [or situation], magnitude, and motion.'

Locke, writing with a knowledge of Newtonian dynamics, places mass among the primary qualities of bodies. In short, he elaborates a theory of primary and secondary qualities in accordance with the state of physical science at the close of the seventeenth century. The primary qualities are the essential qualities of substances whose spatio-temporal relationships constitute nature. The orderliness of these relationships constitutes the order of nature. The occurrences of nature are in some way apprehended by minds, which are associated with living bodies. Primarily, the mental apprehension is aroused by the occurrences in certain parts of the correlated body, the occurrences in the brain, for instance. But the mind in apprehending also experiences sensations which, properly speaking, are qualities of the mind alone. These sensations are projected by the mind so as to clothe appropriate bodies in external nature. Thus the bodies are perceived as with qualities which in reality do not belong to them, qualities which in fact are purely the offspring of the mind. Thus nature gets credit which should in truth be reserved for ourselves: the rose for its scent: the nightingale for his song: and the sun for his radiance. The poets are entirely mistaken. They should address their lyrics to themselves, and should turn them into odes of self-congratulation on the excellency of the human mind. Nature is a dull affair, soundless, scentless, colourless; merely the hurrying of material, endlessly, meaninglessly.

However you disguise it, this the practical outcome of the characteristic scientific philosophy which closed the seventeenth century.

In the first place, we must note its astounding efficiency as a system of concepts for the organisation of scientific research. In this respect, it is fully worthy of the genius of the century which produced it. It has held its own as the guiding principle of scientific studies ever since. It is still reigning. Every university in the world organises itself in accordance with it. No alternative system of organising the pursuit of scientific truth has been suggested. It is not only reigning, but it is without a rival.

And yet — it is quite unbelievable. This conception of the universe is surely framed in terms of high abstractions, and the paradox only arises because we have mistaken our abstraction for concrete realities.

The seventeenth century had finally produced a scheme of scientific thought framed by mathematicians, for the use of mathematicians. The great characteristic of the mathematical mind is its capacity for dealing with abstractions; and for eliciting from them clear-cut demonstrative trains of reasoning, entirely satisfactory so long as it is those abstractions which you want to think about. The enormous success of the scientific abstractions, yielding on the one hand *matter* with its *simple location* in space and time, on the other hand *mind*, perceiving, suffering, reasoning, but not interfering, has foisted onto philosophy the task of accepting them as the most concrete rendering of fact.

Thereby, modern philosophy has been ruined. It has oscillated in a complex manner between three extremes. There are the dualists, who accept matter and mind as on equal basis, and the two varieties of monists, those who put mind inside matter, and those who put matter inside mind. But this juggling with abstractions can never overcome the inherent confusion introduced by the ascription of *misplaced concreteness* to the scientific scheme of the seventeenth century.

. . . I have traced the evolution, during the seventeenth century, of the scheme of scientific ideas which has dominated thought ever since. It involves a fundamental duality, with *material* on the one hand, and on the other hand *mind*. In between there lie the concepts of life, organism, function, instantaneous reality, interaction, order of nature, which collectively form the Achilles heel of the whole system.

I also express my conviction that if we desired to obtain a more fundamental expression of the concrete character of natural fact, the element in this scheme which we should first criticise is the concept of *simple location*. In view therefore of the importance which this idea will assume in these lectures, I will repeat the meaning which I have attached to this phrase. To say that a bit of matter has *simple location* means that, in expressing its spatio-temporal relations, it is adequate to state that it is where it is, in a definite finite region of space, and throughout a definite finite duration of time, apart from any essential reference of the relations of that bit of matter to other regions of space and to other durations of time. Again, this concept of simple location is independent of the controversy between the absolutist and the relativist views of space or of time. So long as any theory of space, or of time, can give a meaning, either absolute or relative, to the idea of a definite region of space, and of a definite duration of time, the idea of simple location has a perfectly definite meaning. This idea is the very foundation of the seventeenth century scheme of nature. Apart from it, the scheme is incapable of expression. I shall argue that among the primary elements of nature as apprehended in our immediate experience, there is no element whatever which possesses this character of simple location. It does not follow, however, that the science of the seventeenth century was simply wrong.

I hold that by a process of constructive abstraction we can arrive at abstractions which are the simply-located bits of material, and at other abstractions which are the minds included in the scientific scheme. Accordingly, the real error is an example of what I have termed: The Fallacy of Misplaced Concreteness.

The point before us is that this scientific field of thought is now, in the twentieth century, too narrow for the concrete facts which are before it for analysis. This is true even in physics, and is more especially urgent in the biological sciences. Thus, in order to understand the difficulties of modern scientific thought and also its reactions on the modern world, we should have in our minds some conception of a wider field of abstraction, a more concrete analysis, which shall stand nearer to the complete concreteness of our intuitive experience. Such an analysis should find in itself a niche for the concepts of matter and spirit, as abstractions in terms of which much of our physical experience can be interpreted. It is in the search for this wider basis for scientific thought that Berkeley is so important. He launched his criticism shortly after the schools of Newton and Locke had completed their work, and laid his finger exactly on the weak spots which they had left. I do not propose to consider either the subjective idealism which have been derived from him, or the schools of development which trace their descent from Hume and Kant respectively. My point will be that — whatever the final metaphysics you may adopt — there is another line of development embedded in Berkeley, pointing to the analysis which we are in search of. Berkeley overlooked it, partly by reason of the over intellectualism of philosophers, and partly by his haste to have recourse to an idealism with its objectivity grounded in the mind of God. You will remember that I have already stated that the key of the problem lies in the notion of simple location. Berkeley, in effect, criticises this notion. He also raises the question, What do we mean by things being realised in the world of nature?

In Sections 23 and 24 of his *Principles of Human Knowledge*, Berkeley gives his answer to this latter question. I will quote some detached sentences from those Sections:

"23. But, say you, surely there is nothing easier than for me to imagine trees, for instance, in a park, or books existing in a closet, and nobody by to perceive them. I answer, you may so, there is no difficulty in it; but what is all this, I beseech you, more than framing in your mind certain ideas which you call books and trees, and at the same time omitting to frame the idea of any one that may perceive them? . . ."

"When we do our utmost to conceive the existence of external bodies, we are all the while only contemplating our own ideas. But the mind *taking no notice of itself*, is deluded to think it can and does conceive bodies existing unthought of or without the mind, though at the same time they are apprehended by or exist in itself. . . ."

"24. It is very obvious, upon the least inquiry into our thoughts, to know whether it be possible for us to understand what is meant by the *absolute existence of sensible objects in themselves, or without the mind.*

To me it is evident those words mark out either a direct contradiction, or else nothing at all. . . ."

Again there is a very remarkable passage in Section 10, of the fourth Dialogue of Berkeley's *Alciphron*. I have already quoted it, at greater length, in my *Principles of Natural Knowledge*:

"*Euphranor*. Tell me, Alciphron, can you discern the doors, window and battlements of that same castle?

Alciphron. I cannot. At this distance it seems only a small round tower.

Euph. But I, who have been at it, know that it is no small round tower, but a large square building with battlements and turrets, which it seems you do not see.

Alc. What will you infer from thence?

Euph. I would infer that the very object which you strictly and properly perceive by sight is not that thing which is several miles distant.

Alc. Why so?

Euph. Because a little round object is one thing, and a great square object is another. Is it not so? . . ."

Some analogous examples concerning a planet and a cloud are then cited in the dialogue, and this passage finally concludes with:

"*Euphranor*. Is it not plain, therefore, that neither the castle, the planet, nor the cloud, *which you see here*, are those real ones which you suppose exist at a distance?"

It is made explicit in the first passage, already quoted, that Berkeley himself adopts an extreme idealistic interpretation. For him mind is the only absolute reality, and the unity of nature is the unity of ideas in the mind of God. Personally, I think that Berkeley's solution of the metaphysical problem raises difficulties not less than those which he points out as arising from a realistic interpretation of the scientific scheme. There is, however, another possible line of thought, which enables us to adopt anyhow an attitude of provisional realism, and to widen the scientific scheme in a way which is useful for science itself.

I recur to the passage from Francis Bacon's *Natural History*, already quoted in the previous lecture:

"It is certain that all bodies whatsoever, though they have no sense, yet they have perception: . . . and whether the body be alterant or altered, evermore a perception precedeth operation; for else all bodies would be alike one to another. . . ."

Also in the previous lecture I construed *perception* (as used by Bacon) as meaning *taking account of* the essential character of the thing perceived, and I construed *sense* as meaning *cognition*. We certainly do take account of things of which at the time we have no explicit cognition. We can even have a cognitive memory of the taking account, without having had a contemporaneous cognition. Also, as Bacon points out by his statement, ". . . for else all bodies would be alike one to another," it is evidently some element of the essential character which we take account of, namely something on which diversity is founded and not mere bare logical diversity.

The word '*perceive*' is, in our common usage, shot through and through with the notion of cognitive apprehension. So is the word '*apprehension*', even with the adjective *cognitive* omitted. I will use the word '*prehension*' for *uncognitive apprehension*: by this I mean *apprehension* which may or may not be cognitive. Now take Euphranor's last remark:

"Is it not plain, therefore, that neither the castle, the planet, nor the cloud, *which you see here*, are those real ones which you suppose exist at distance?" Accordingly, there is a prehension, *here* in this place, of things which have a reference to *other* places.

Now go back to Berkeley's sentences, quoted from his *Principles of Human Knowledge*. He contends that what constitutes the realisation of natural entities is the being perceived within the unity of mind.

We can substitute the concept, that the realisation is a gathering of things into the unity of a prehension; and that what is thereby realised is the prehension, and not the things. This unity of a prehension defines itself as a *here* and a *now*, and the things so gathered into the grasped unity have essential reference to other places and other times. For Berkeley's *mind*, I substitute a process of prehensive unification.

There are more entities involved in nature than the mere sense-objects, so far considered. But, allowing for the necessity of revision consequent on a more complete point of view, we can frame our answer to Berkeley's question as to the character of the reality to be assigned to nature. He states it to be the reality of ideas in mind. A complete metaphysic which has attained to some notion of mind, and to some notion of ideas, may perhaps ultimately adopt that view.

We can be content with a provisional realism in which nature is conceived as a complex of prehensive unifications. Space and time exhibit the general scheme of interlocked relations of these prehensions. You cannot tear any one of them out of its context. Yet each one of them within its context has all the reality that attaches to the whole complex. Conversely, the totality has the same reality as each prehension; for each prehension unifies the modalities to be ascribed, from its standpoint, to every part of the whole. A prehension is a process of unifying. Accordingly, nature is a process of expansive development, necessarily transitional from prehension to prehension. What is achieved is thereby passed beyond, but it is also retained as having aspects of itself present to prehensions which lie beyond it.

Thus nature is a structure of evolving processes. The reality is the process. It is nonsense to ask if the colour red is real. The colour red is ingredient in the process of realisation. The realities of nature are the prehensions in nature, that is to say, the events in nature.

Now that we have cleared space and time from the taint of simple location, we may partially abandon the awkward term prehension. This term was introduced to signify the essential unity of an event, namely, the event as one entity, and not as a mere assemblage of parts or of ingredients. It is necessary to understand that space-time is nothing else than a system of pulling together of assemblages into unities. But the word *event* just means one of these spatio-temporal unities. Accordingly, it may be used instead of the term 'prehension' as meaning the thing prehended.

An event has contemporaries. This means that an event mirrors within itself the modes of its contemporaries as a display of immediate achievement. An event has a past. This means that an event mirrors within itself the modes of its predecessors, as memories which are fused into its own content. An event has a future. This means that an event mirrors within itself such aspects as the future throws back onto the present, or, in other words, as the present has determined concerning the future. Thus an event has anticipation:

> 'The prophetic soul
> Of the wide world dreaming on things to come.' [cvii]

These conclusions are essential for any form of realism. For there is in the world for our cognisance, memory of the past, immediacy of realisation, and indication of things to come.

In this sketch of an analysis more concrete than that of the scientific scheme of thought, I have started from our own psychological field, as it stands for our cognition. I take it for what it claims to be: the self-knowledge of our bodily event. I mean the total event, and not the inspection of the details of the body. This self-knowledge discloses a prehensive unification of modal presences of entities beyond itself. I generalise by the use of the principle that this total bodily event is on the same level as all other events, except for an unusual complexity and stability of inherent pattern. The strength of the theory of materialistic mechanism has been the demand, that no arbitrary breaks be introduced into nature, to eke out the collapse of an explanation. I accept this principle. But if you start from the immediate facts of our psychological experience, as surely as empiricist should begin, you are at once led to the organic conception of nature of which the description has been commenced here.

It is the defect of the eighteenth century scientific scheme that it provides none of the elements which compose the immediate psychological experiences of mankind. Nor does it provide any elementary trace of the organic unity of a whole, from which the organic unities of electrons, protons, molecules, and living bodies can emerge. According to that scheme, there is no reason in the nature of things why portions of material should have any physical relations to each other. Let us grant that we cannot hope to be able to discern the laws of nature to be necessary. But we can hope to see that it is necessary that there should be an order of nature. The concept of the order of nature is bound up with the concept of nature as the locus of organisms in process of development.

We quickly find that the Western peoples exhibit on a colossal scale a peculiarity which is popularly supposed to be more especially characteristic of the Chinese. Surprise is often expressed that a Chinaman can be of two religions, a Confucian for some occasions and a Buddhist for other occasions. Whether this is true of China I do not know; nor do I know whether, if true, these two attitudes are really inconsistent. But there can be no doubt that an analogous fact is true of the West, and that the two attitudes involved are inconsistent. A scientific realism, based on mechanism, is conjoined with an unwavering belief in the world of men and of the higher animals as being composed of self-determining organisms. This radical inconsistency at the basis of modern thought accounts for much that is half-hearted and wavering in our civilisation.

Wordsworth was not bothered by any intellectual antagonism. What moved him was a moral repulsion. He felt that something had been left out, and that what had been left out comprised everything that was most important. Tennyson is the mouthpiece of the attempts of the waning romantic movement in the second quarter of the nineteenth century to come to terms with science. By this time the two elements in modern thought had disclosed their fundamental divergence by their jarring interpretations of the course of nature and the life of man.

Tennyson goes to the heart of the difficulty. It is the problem of mechanism which appalls him,

> ' "The stars," she whispers, "blindly run." '

This line states starkly the whole philosophy problem implicit in the poem. Each molecule blindly runs. The human body is a collection of molecules. Therefore, the human body blindly runs, and therefore there can be no individual responsibility for the actions of the body. If you once accept that the molecule is definitely determined to be what it is, independently of any determination by reason of the total organism of the body, and if you further admit that the blind run is settled by the general mechanical laws, here can be no escape from this conclusion. But mental experiences are derivative from the actions of the body, including of course its internal behaviour. Accordingly, the sole function of the mind is to have at least some of its experiences settled for it, and to add such others as may be open to it independently of the body's motions, internal and external.

There are then two possible theories as to the mind. You can either deny that it can supply for itself any experiences other than those provided for it by the body, or you can admit them.

If you refuse to admit the additional experiences, then all individual moral responsibility is swept away. If you do admit them, then a human being may be responsible for the state of his mind though he has no responsibility for the actions of his body. The enfeeblement of thought in the modern world is illustrated by the way in which this plain issue is avoided in Tennyson's poem. There is something kept in the background, a skeleton in the cupboard.

I have stated the arguments concisely, because in truth the issue is a very simple one. Prolonged discussion is merely a source of confusion. The question as to the metaphysical status of molecules does not come in. The statement that they are mere formulae has no bearing on the argument. For presumably the formulae mean something. If they mean nothing, the whole mechanical doctrine is likewise without meaning, and the question drops. But if the formulae mean anything, the argument applies to exactly what they do mean. The traditional way of evading the difficulty — other than the simple way of ignoring it — is to have recourse to some form of what is now termed 'vitalism.' This doctrine is really a compromise. It allows a free run to mechanism throughout the whole of inanimate nature, and holds that the mechanism is partially mitigated within living bodies. I feel that this theory is an unsatisfactory compromise. The gap between living and dead matter is too vague and problematical to bear the weight of such an arbitrary assumption, which involves an essential dualism somewhere.

The doctrine which I am maintaining is that the whole concept of materialism only applies to very abstract entities, the products of logical discernment. The concrete enduring entities are organisms, so that the plan of the *whole* influences the very characters of the various subordinate organisms which enter into it. In the case of an animal, the mental states enter into the plan of the total organism and thus modify the plans of the successive subordinate organisms until the ultimate smallest organisms, such as electrons, are reached. Thus an electron within a living body is different from an electron outside it, by reason of the plan of the body. The electron blindly runs either within or without the body; but it runs within the body in accordance with its character within the body; that is to say, in accordance with the general plan of the body, and this plan includes the mental state. But the principle of modification is perfectly general throughout nature, and represents no property peculiar to living bodies. This doctrine involves the abandonment of the traditional scientific materialism, and the substitution of an alternative doctrine of organism.

I would term the theory of *organic mechanism*. In this theory, the molecules may blindly run in accordance with the general laws, but the molecules differ in their intrinsic characters according to the general organic plans of the situations in which they find themselves.

Science is taking on a new aspect which is neither purely physical, nor purely biological. It is becoming the study of organisms. Biology is the study of the larger organisms; whereas physics is the study of the smaller organisms. There is another difference between the two divisions of science. The organisms of biology include as ingredients the smaller organisms of physics; but there is at present no evidence that the smaller of the physical organisms can be analysed into component organisms. It may be so. But anyhow we are faced with the question as to whether there are not primary organisms which are incapable of further analysis. It seems very unlikely that there should be any infinite regress in nature. Accordingly, a theory of science which discards materialism must answer the question as to the character of these primary entities. There can be only one answer on this basis. We must start with the event as the ultimate unit of natural occurrence. An event has to do with all that there is, and in particular with all other events. This interfusion of events is effected by the aspects of those eternal objects, such as colours, sounds, scents, geometrical characters, which are required for nature and are not emergent from it. Such an eternal object will be an ingredient of one event under the guise, or aspect, of qualifying another event. There is a reciprocity of aspects, and there are patterns of aspects. Each event corresponds to two such patterns; namely, the pattern of aspects of other events which it grasps into its own unity, and the pattern of its aspects which other events severally grasp into their unities. Accordingly, a non-materialistic philosophy of nature will identify a primary organism as being the emergence of some particular pattern as grasped in the unity of a real event.

Thus in the process of analysing the character of nature in itself, we find that the emergence of organisms depends on a selective activity which is akin to purpose. The point is that the enduring organisms are now the outcome of evolution; and that, beyond these organisms, there is nothing else that endures. On the materialistic theory, there is material — such as matter or electricity — which endures. On the organic theory, the only endurances are structures of activity, and the structures are evolved.

Only if you take *material* to be fundamental, this property of endurance is an arbitrary fact at the base of the order of nature; but if you take *organism* to be fundamental, this property is the result of evolution.

Accordingly the key to the mechanism of evolution is the necessity for the evolution of a favourable environment, conjointly with the evolution of any specific type of enduring organisms of great permanence. Any physical object which by its influence deteriorates its environment, commits suicide.

There are thus two sides to the machinery involved in the development of nature. On one side, there is a given environment with organisms adapting themselves to it. The scientific materialism of the epoch in question emphasised this aspect. From this point of view, there is a given amount of material, and only a limited number of organisms can take advantage of it. The givenness of the environment dominates everything. Accordingly, the last words of science appeared to be the Struggle for Existence, and Natural Selection. Darwin's own writings are for all time a model of refusal to go beyond the direct evidence, and of careful retention of every possible hypothesis. But those virtues were not so conspicuous in his followers, and still less in his camp-followers. The imagination of European sociologists and publicists was stained by exclusive attention to this aspect of conflicting interests. The idea prevailed that there was a peculiar strong-minded realism in discarding ethical considerations in the determination of the conduct of commercial and national interests.

> The other side of the evolutionary machinery, the neglected side, is expressed by the word *creativeness*. The organisms can create their own environment. For this purpose, the single organism is almost helpless. The adequate forces require societies of co-operating organisms. But with such co-operation and in proportion to the effort put forward, the environment has a plasticity which alters the whole ethical aspect of evolution.
>
> In the immediate past, and at present, a muddled state of mind is prevalent. The increased plasticity of the environment for mankind, resulting from the advances in scientific technology, is being construed in terms of habits of thought which find their justification in the theory of a fixed environment.
>
> The riddle of the universe is not so simple. There is the aspect of permanence in which a given type of attainment is endlessly repeated for its own sake; and there is the aspect of transition to other things,— it may be of higher worth, and it may be of lower worth. Also there are its aspects of struggle and of friendly help. But romantic ruthlessness is no nearer to real politics, than is romantic self-abnegation.

Whitehead's views appear similar in some ways to those of Bergson, particularly in an emphasis on process. In fact they reached their respective positions by very different routes and there are fundamental differences between them. But, as the next reading shows, for one scientist interested in living processes, they have had a common influence.

Ilya Prigogine was awarded the Nobel Prize in 1977 for work on thermodynamics that has turned out to have significance when considering biological and possibly psychological levels of organization. Before looking at Prigogine a brief word on thermodynamics is in order.

The discipline now known as thermodynamics arose in the early nineteenth century and, as its name shows, is concerned with the movement of heat. The machines of the industrial age had made this a practical matter since their efficiency was crucially dependent upon how quickly heat could be moved from one place to another. Thermodynamics rapidly became more than an applied matter as its deeper implications were recognized. What they showed was that heat, and hence work, could only be obtained at the cost of destroying organization. For example, once burned, the energy and structure of a piece of coal was distributed more thinly within the environment and it became harder to get any more useful work from it. All processes, natural or contrived, inevitably have this character. It is this which makes perpetual motion machines impossible and which makes it so much easier to get work from high density energy sources like oil than from low density ones like the wind. In the universe, as in life, there's no such thing as a free lunch.

Even though thermodynamics shows that the universe as a whole is running down and becoming disordered there is obviously growth, evolution and the development of new levels of complexity within the biosphere. Prigogine's work has helped to show how this comes about and, though controversial, has stimulated a lively debate on the nature of biological and other types of organization. His popular book, *Order out of Chaos*, is an informal account of the thermodynamics of irreversibility, i.e. of a directed sense of time, which is absent from classical mechanics but which is so fundamentally attached to consciousness. What follows is not the technical details of Prigogine's thermodynamics but rather his exploration of Whitehead and Bergson and some of his speculations on the wider implications of his own work.

At the end of the nineteenth century, when Bergson undertook his search for an acceptable alternative to the science of his time, he turned to intuition as a form of speculative knowledge, but he presented it as quite different from that of the Romantics. He explicitly stated that intuition is unable to produce a system but produces only results that are always partial and nongeneralizable, results to be formulated with great caution.

Science and intuitive metaphysics 'are or can become equally precise and definite. They both bear upon reality itself. But each one of them retains only half of it so that one could see in them, if one wished, two subdivisions of science or two departments of metaphysics, if they did not mark divergent directions of the activity of thought.'

The definition of these two divergent directions may also be considered as the historical consequence of scientific evolution. For Bergson, it is no longer a question of finding scientific alternatives to the physics of his time. In his view, chemistry and biology had definitely chosen mechanics as their model. In Bergson's view, science is a whole and must therefore be judged as a whole. And this is what he does when he presents science as the product of a practical intelligence whose aim is to dominate matter and that develops by abstraction and generalization the intellectual categories needed to achieve this domination. Science is the product of our vital need to exploit the world, and its concepts are determined by the necessity of manipulating objects, of making predictions, and of achieving reproducible actions. This is why rational mechanics represents the very essence of science, its actual embodiment. The other sciences are more vague, awkward manifestations of an approach that is all the more successful the more inert and disorganized the terrain it explores.

For Bergson all the limitations of scientific rationality can be reduced to a single and decisive one: it is incapable of understanding *duration* since it reduces time to a sequence of instantaneous states linked by a deterministic law.

'Time is invention, or it is nothing at all.' Nature is change, the continual elaboration of the new, a totality being created in an essentially open process of development without any preestablished model. 'Life progresses and endures in time.' The only part of this progressions that intelligence can grasp is what it succeeds in fixing in the form of manipulable and calculable elements and in referring to a time seen as sheer juxtaposition of instants.

> Therefore, physics 'is limited to coupling simultaneities between the events that make up this time and the positions of the mobile *T* on its trajectory. It detaches these events from the whole, which at every moment puts on a new form and which communicates to them something of its novelty. It considers them in the abstract, such as they would be outside of the living whole, that is to say, in a time unrolled in space. It retains only the events or systems of events that can be thus isolated without being made to undergo too profound a deformation, because only these lend themselves to the application of its method. Our physics dates from the day when it was known how to isolate such systems.'

When it comes to understanding duration itself, science is powerless. What is needed is intuition, a 'direct vision of the mind by the mind.' 'Pure change, real duration, is something spiritual. Intuition is what attains the spirit, duration, pure change.'

Can we say Bergson has failed in the same way that the post-Kantian philosophy of nature failed? He has failed insofar as the metaphysics based on intuition he wished to create has not materialized. He has not failed in that, unlike Hegel, he had the good fortune to pass judgment upon science that was, on the whole, firmly established — that is, classical science at its apotheosis, and thus identified problems which are indeed still our problems. But, like the post-Kantian critics, he identified the science of his time with science in general. He thus attributed to science *de jure* limitations that were only *de facto*. As a consequence he tried to define once and for all a *statu quo* for the respective domains of science and other intellectual activities. Thus the only perspective remaining open for him was to introduce a way in which antagonistic approaches could at best merely coexist.

In conclusion, even if the way in which Bergson sums up the achievement of classical science is still to some extent acceptable, we can no longer accept it as a statement of the eternal limits of the scientific enterprise. We conceive of it more as a program that is beginning to be implemented by the metamorphosis science is now undergoing. In particular, we know that time linked with motion does not exhaust the meaning of time in physics. Thus the limitations Bergson criticized are beginning to be overcome, not by abandoning the scientific approach or abstract thinking but by perceiving the limitations of the concepts of classical dynamics and by discovering new formulations valid in more general situations.

As we have emphasized, the element common to Kant, Hegel, and Bergson is the search for an approach to reality that is different from the approach of classical science. This is also the fundamental aim of Whitehead's philosophy, which is resolutely pre-Kantian. In his most important book, *Process and Reality*, he puts us back in touch with the great philosophies of the Classical Age and their quest for rigorous conceptual experimentation.

Whitehead sought to understand human experience as a process belonging to nature, as physical existence. This challenge led him, on the one hand, to reject the philosophic tradition that defined subjective experience in terms of consciousness, thought, and sense perception, and, on the other, to conceive of all *physical* existence in terms of enjoyment, feeling, urge, appetite, and yearing — that is, to cross swords with what he calls 'scientific materialism,' born in the seventeenth century. Like Bergson, Whitehead was thus led to point out the basic inadequacies of the theoretical scheme developed by seventeenth-century science:

> The seventeenth century had finally produced a scheme of scientific thought framed by mathematicians, for the use of mathematicians. The great characteristic of the mathematical mind is its capacity for dealing with abstractions; and for eliciting from them clear-cut demonstrative trains of reasoning, entirely satisfactory so long as it is those abstractions which you want to think about. The enormous success of the scientific abstractions, yielding on the one hand matter with its simple location in space and time, on the other hand mind, perceiving, suffering, reasoning, but not interfering, has foisted on to philosophy the task of accepting them as the most concrete rendering of fact.
>
> Thereby, modern philosophy has been ruined. It has oscillated in a complex manner between three extremes. There are the dualists, who accept matter and mind as on equal basis, and the two varieties of monists, those who put mind inside matter, and those who put matter inside mind. But this juggling with abstractions can never overcome the inherent confusion introduced by the ascription of misplaced concreteness to the scientific scheme of the seventeenth century.

However, Whitehead considered this to be only a temporary situation. Science is not doomed to remain a prisoner of confusion.

We have already raised the question of whether it is possible to formulate a philosophy of nature that is not directed against science. Whitehead's cosmology is the most ambitious attempt to do so. Whitehead saw no basic contradiction between science and philosophy. His purpose was to define the conceptual field within which the problem of human experience and physical processes could be dealt with consistently and to determine the conditions under which the problem could be solved. What had to be done was to formulate the principles necessary to characterize all forms of existence, from that of stones to that of man. It is precisely this universality that, in Whitehead's opinion, defines his enterprise as 'philosophy.' While each scientific theory selects and abstracts from the world's complexity a peculiar set of relations, philosophy cannot favor any particular region of human experience. Through conceptual experimentation it must construct a consistency that can accommodate all dimensions of experience, whether they belong to physics, physiology, psychology, biology, ethics, etc.

Whitehead understood perhaps more sharply than anyone else that the creative evolution of nature could never be conceived if the elements composing it were defined as permanent, individual entities that maintained their identity throughout all changes and interactions. But he also understood that to make all permanence illusory, to deny being in the name of becoming, to reject entities in favor of a continuous and ever-changing flux meant falling once again into the trap always lying in wait for philosophy — to 'indulge in brilliant feats of explaining away.'

Thus for Whitehead the task of philosophy was to reconcile permanence and change, to conceive of things as processes, to demonstrate that becoming forms entities, individual identities that are born and die. It is beyond the scope of this book to give a detailed presentation of Whitehead's system. Let us only emphasize that he demonstrated the connection between a philosophy of *relation* — no element of nature is a permanent support for changing relations; each receives its identity from its relations with others — and a philosophy of *innovating becoming*. In the process of its genesis, each existent unifies the multiplicity of the world, since it adds to this multiplicity an extra set of relations. At the creation of each new entity 'the many become one and are increased by one.'

In the conclusion of this book, we shall again encounter Whitehead's question of permanence and change, this time as it is raised in physics; we shall speak of entities formed by their irreversible interaction with the world. Today physics has discovered the need to assert both the distinction and interdependence between units and relations. It now recognizes that, for an interaction to be real, the 'nature' of the related things must derive from these relations, while at the same time the relations must derive from the 'nature' of the things. This is a forerunner of 'self-consistent' descriptions as expressed, for instance, by the 'bootstrap' philosophy in elementary-particle physics, which asserts the universal connectedness of all particles. However, when Whitehead wrote *Process and Reality*, the situation of physics was quite different, and Whitehead's philosophy found an echo only in biology.

Whitehead's case as well as Bergson's convince us that only an opening, a widening of science can end the dichotomy between science and philosophy. This widening of science is possible only if we revise our conception of time. To deny time — that is, to reduce it to a mere deployment of a reversible law — is to abandon the possibility of defining a conception of nature coherent with the hypothesis that nature produced living beings, particularly man. It dooms us to choosing between an antiscientific philosophy and an alienating science.

Science certainly involves manipulating nature, but it is also an attempt to understand it, to dig deeper into questions that have been asked generation after generation. One of these questions runs like a leitmotiv, almost as an obsession, through this book, as it does through the history of science and philosophy. This is the question of the relation between being and becoming, between permanence and change.

One of the main sources of fascination in modern science was precisely the feeling that it had discovered eternal laws at the core of nature's transformations and thus had exorcised time and becoming. This discovery of an order in nature produced the feeling of intellectual security described by French sociologist Lévy-Bruhl:

> Our feeling of intellectual security is so deeply anchored in us that we even do not see how it could be shaken. Even if we suppose that we could observe some phenomenon seemingly quite mysterious, we still would remain persuaded that our ignorance is only provisional, that this phenomenon must satisfy the general laws of causality, and that the reasons for which it has appeared will be determined sooner or later. *Nature around us is order and reason, exactly as is the human mind*. Our everyday activity implies a perfect confidence in the universality of the laws of nature.

This feeling of confidence in the 'reason' of nature has been shattered, partly as the result of the tumultuous growth of science in our time. As we stated in the Preface, our vision of nature is undergoing a radical change toward the multiple, the temporal, and the complex.

We were seeking general, all-embracing schemes that could be expressed in terms of eternal laws, but we have found time, events, evolving particles. We were also searching for symmetry, and here also we were surprised, since we discovered symmetry-breaking processes on all levels, from elementary particles up to biology and ecology. We have described in this book the clash between dynamics, with temporal symmetry it implies, and the second law of thermodynamics, with its directed time.

A new unity is emerging: irreversibility is a source of order at all levels. Irreversibility is the mechanism that brings order out of chaos. How could such a radical transformation of our views on nature occur in the relatively short time span of the past few decades? We believe that it shows the important role intellectual construction plays in our concept of reality. This was very well expressed by Bohr, when he said to Werner Heisenberg on the occasion of a visit at Kronberg Castle:

> Isn't it strange how this castle changes as soon as one imagines that Hamlet lived here? As scientists we believe that a castle consists only of stone, and admire the way the architect put them together. The stones, the green roof with its patina, the wood carvings in the church, constitute the whole castle. None of this should be changed by the fact that Hamlet lived here, and yet it is changed completely. Suddenly the walls and the ramparts speak a different language. . . . Yet all we really know about Hamlet is that his name appears in a thirteenth-century chronicle. . . . But everyone knows the questions Shakespeare had him ask, the human depths he was made to reveal, and so he too had to have a place on earth, here in Kronberg.

The question of the meaning of reality was the central subject of a fascinating dialogue between Einstein and Tagore. Einstein emphasized that science had to be independent of the existence of any observer. This led him to deny the reality of time as irreversibility, as evolution. On the contrary, Tagore maintained that even if absolute truth could exist, it would be inaccessible to the human mind. Curiously enough, the present evolution of science is running in the direction stated by the great Indian poet. Whatever we call reality, it is revealed to us only through the active construction in which we participate.

The world of dynamics, be it classical or quantum, is a reversible world. As we have emphasized in Chapter VIII, no evolution can be ascribed to this world; the 'information' expressed in terms of dynamical units remains constant. It is therefore of great importance that the existence of an evolutionary paradigm can now be established in physics — not only on the level of macroscopic description but also on all levels. Of course, there are conditions: as we have seen, a minimum complexity is necessary. But the immense importance of irreversible processes shows that this requirement is satisfied for most systems of interest. Remarkably, the perception of oriented time increases as the level of biological organization increases and probably reaches its culminating point in human consciousness.

How general is this evolutionary paradigm? It includes isolated systems that evolve to disorder and open systems that evolve to higher and higher forms of complexity. It is not surprising that the entropy metaphor has tempted a number of writers dealing with social or economic problems. On the human level irreversibility is a more fundamental concept, which is for us inseparable from the meaning of our existence. Still it is essential that in this perspective we no longer see the internal feeling of irreversibility as a subjective impression that alienates us from the outside world, but as marking our participation in a world dominated by an evolutionary paradigm.

Exploring the implications and the coherence of those fundamental concepts, which appear both scientific and philosophical, may be risky, but it can be very fruitful in the dialogue between science and philosophy. Let us illustrate this with some brief references to Whitehead.

In Whitehead's philosophy being is inseparable from becoming. Whitehead wrote: 'The elucidation of the meaning of the sentence "everything flows" is one of metaphysics' main tasks.' Physics and metaphysics are indeed coming together today in a conception of the world in which process, becoming, is taken as a primary constituent of physical existence and where existing entities can interact and therefore also be born and die.

We agree completely with Herman Weyl:

> Scientists would be wrong to ignore the fact that theoretical construction is not the only approach to the phenomena of life; another way, that of understanding from within (interpretation), is open to us. . . . Of myself, of my own acts of perception, thought, volition, feeling and doing, I have a direct knowledge entirely different from the theoretical knowledge that represents the 'parallel' cerebral processes in symbols. This inner awareness of myself is the basis for the understanding of my fellowmen whom I meet and acknowledge as beings of my own kind, with whom I communicate sometimes so intimately as to share joy and sorrow with them.

Until recently, however, there was a striking contrast. The external universe appeared to be an automaton following deterministic causal laws, in contrast with the spontaneous activity and irreversibility we experience. The two worlds are now drawing closer together. Is this a loss for the natural sciences?

Classical science aimed at a 'transparent' view of the physical universe. In each case you would be able to identify a cause and an effect. Whenever a stochastic description becomes necessary, this is no longer so. We can no longer speak of causality in each individual experiment; we can only speak about statistical causality. This has, in fact, been the case ever since the advent of quantum mechanics, but it has been greatly amplified by recent developments in which randomness and probability play an essential role, even in classical dynamics or chemistry. Therefore, the modern trend as compared to the classical one leads to a kind of 'opacity' as compared to the transparency of classical thought.

Is this a defeat for the human mind? This is a difficult question. As scientists, we have no choice; we cannot describe for you the world as we would like to see it, but only as we are able to see it through the combined impact of experimental results and new theoretical concepts.

It is hard to avoid the impression that the distinction between what exists in time, what is irreversible, and, on the other hand, what is outside of time, what is eternal, is at the origin of human symbolic activity. Perhaps this is especially so in artistic activity. Indeed, one aspect of the transformation of a natural object, a stone, to an object of art is closely related to our impact on matter. Artistic activity breaks the temporal symmetry of the object. It leaves a mark that translates our temporal dissymmetry into the temporal dissymmetry of the object. Out of the reversible, nearly cyclic noise level in which we live arises music that is both stochastic and time-oriented.

It is quite remarkable that we are at a moment both of profound change in the scientific concept of nature and of the structure of human society as a result of the demographic explosion. As a result, there is a need for new relations between man and nature and between man and man. We can no longer accept the old a priori distinction between scientific and ethical values. This was possible at a time when the external world and our internal world appeared to conflict, to be nearly orthogonal. Today we know that time is a construction and therefore carries an ethical responsibility.

The ideas to which we have devoted much space in this book — the ideas of instability, of fluctuation — diffuse into the social sciences. We know now that societies are immensely complex systems involving a potentially enormous number of bifurcations exemplified by the variety of cultures that have evolved in the relatively short span of human history. We know that such systems are highly sensitive to fluctuations. This leads both to hope and a threat: hope, since even small fluctuations may grow and change the overall structure. As a result, individual activity is not doomed to insignificance. On the other hand, this is also a threat, since in our universe the security of stable, permanent rules seems gone forever. We are living in a dangerous and uncertain world that inspires no blind confidence, but perhaps only the same feeling of qualified hope that some Talmudic texts appear to have attributed to the God of Genesis:

> Twenty-six attempts preceded the present genesis, all of which were destined to fail. The world of man has arisen out of the chaotic heart of the preceding debris; he too is exposed to the risk of failure, and the return to nothing. 'Let's hope it works' [*Halway Sheyaamod*] exclaimed God as he created the World, and this hope, which has accompanied all the subsequent history of the world and mankind, has emphasized right from the outset that this history is branded with the mark of radical uncertainty.

Prigogine shows how the sense of a direction to time, so conspicuously absent from classical mechanics but so central to consciousness, is also a characteristic of natural physical processes once certain conditions are met. While Prigogine does not address consciousness directly he notes that his work makes it possible to bring our knowledge of physical reality into relationship with the world of experience.

This is also the concern of David Bohm, co-author of the next reading, who also acknowledges the influence of Whitehead.[3] Bohm is perhaps best known for important contributions to the development of quantum physics and for work on its metaphysical implications. His later writings show the influence of contacts with such figures as Krishnamurti and the Dalai Lama, but he is cautious on parallels between science and mysticism. Bohm's principal metaphysical proposal is what he calls the implicate order.[4] Under this view, all manifest objects or events, whether physical, biological or psychological, are taken as arising from hidden levels of organization in the world. The explicit, perceptible structure of the world is enfolded within more subtle, implicit levels of order, a frequent analogy for this enfoldment being an image contained in a hologram.

There is not the space to deal with the implicate order in the detail it requires but readers should note that it provides a unique framework for describing the interaction of mental and physical levels of organization. In the following extract Bohm and co-author David Peat develop the idea of implicate order by suggesting that all forms of organization whether inanimate, living or experiential, originate in a generative level of order within the universe. This level of order they identify with active conscious intelligence.

Awareness and attention bring about a movement of content from the more manifest physical levels toward the subtler levels of the generative order. The response to this is a movement in the other direction, an unfoldment of the creative action of intelligence. This originates, ultimately, in the depths of the generative order.

All of this emphasizes the universality of creative intelligence, which has ultimately the same origin in every area of life. In this connection it is not appropriate to think of experience as being something which exists on its own, and which is from time to time somewhat modified by the perceptions, thoughts, and actions that come out of creative intelligence. Rather, every aspect of such experience, whether physical or mental, emotional or intellectual, can be profoundly affected by creative intelligence, wherever this is able to act. For through its action everything may take on a new meaning.

It should be emphasized in this connection that intellect, emotion, and will cannot actually be separated, except for the purpose of analysis in thought.

The inseparable nature of emotion, intellect, and will is in harmony with what is known about the general structure of the brain. For example, a very thick bundle of nerves connects the frontal lobes, which have an intellectual function, to the base of the brain, which is more associated with the emotions and from which the whole organism is bathed in chemicals that affect all parts profoundly. Recent knowledge goes much further toward revealing a similar but far more extensive and pervasive interconnection in various structures and processes that must be involved in the actual operation of thought, feeling, and will. The activity of each individual nerve cell is directly connected, via synaptic links which involve the activity of various neurochemicals, to some thousand other nerves. In any neural network, therefore, the number of interconnections is astronomical. In turn, the strength of each interconnection is influenced by neurochemicals, such as the neurotransmitters, as well as by the actual electrical activity within the network. The system is of an almost unanalyzable complexity and subtlety in the operation of its various processes, for individual nerve impulses are mediated by an enormous range of chemical and electrical responses, some of them local and others global, some general and others highly specific in nature. An extreme reductionist view may suggest that the nerve impulses are 'processing data' relating to movement, senses, and the intellect, while the neurochemical bath would be close to a basis of experiencing an emotional response. But in view of the enormous complexity of the brain, such an image is clearly too crude a simplification. Rather, the insights of neurochemistry and the nature of nerve networks indicate very strongly that there can be no fundamental separation at this level between thought, feeling, and will.

The subtle mental side to these processes indicates that there is even less reason for making such a separation. For it is possible to sense and experience directly an intimate connection between thought, feeling, and will and show that there is no point at which one of them ends and the others begin. Moreover, creative intelligence can profoundly affect the whole meaning of these functions, as well as the entire way in which they proceed at the level of physics and chemistry. Indeed, it has already been suggested why a large part of this content probably cannot be understood in terms of the current laws of physics and chemistry and requires a level of explanation that goes beyond the superimplicate order.

Ultimately the origin of all this lies in the creative intelligence, which is beyond anything that can be discussed in the manifest physical side. This intelligence is universal and acts in every area of mental operation.

In its depths, such intelligence can involve no separation between knowing, feeling, and will. Thus, one way in which intelligence becomes manifest is by organizing the categories, orders, and structures of the intellect in new ways. It may orchestrate feelings in an ever-changing movement, like that which is experienced in music. Such a movement goes beyond the sort of succession of fixed patterns of feelings that can be identified in fairly well defined forms, such as pain, pleasure, fear, anger, desire, and hatred. Orchestrated movement of feelings may perhaps be what is meant most deeply by words such as *love, beauty, vitality*, and so on. But where these feelings emerge from the whole of the generative order, they must evidently have meanings that are not adequately signified by what is commonly conveyed by these words.

In everyday consciousness, however, the mind is absorbed largely in the tacit infrastructure of ideas and dispositions to feel and act, which are mainly mechanical in their operation. In a metaphorical sense, at least, this activity of the mind could be said to be 'programmed.' But it should be clear that these programs, while both useful and necessary, are limited, since something more and something different, creative intelligence, is always in principle available.

Those with a distaste for metaphysics often question what use can be made of it. Useless or not, questions of ultimate meaning arise naturally in the human mind and metaphysics is always implicit in science though the current fashion is to underplay it. Occasionally, scientists will share their metaphysical concerns with their colleagues, though this usually only happens once they have a solid reputation of conventional scientific work behind them.

Bohm is something of an exception since his metaphysical interest has generally been prominent in his work from an early stage. George Wald, a biochemist and winner of the Nobel Prize in 1967 for his work on the neurophysiology of vision, is perhaps more typical. In the last reading of this chapter, he presents something of a valediction to two major problems which confront him at the end of a long and distinguished scientific career. These are the problems of consciousness and cosmology.

His own speculations on the links between them have come, as he puts it, as '. . . a shock at first to my scientific sensibilities . . .'.

The extract from Wald's paper omits parts of the first section. These presented evidence from particle physics, chemistry and cosmology which supports what is known as the strong anthropic principle. This, simply put, holds that the physical universe appears to be uniquely suited to support life when it might have been otherwise.[5] In the second part of the paper Wald takes as full a measure of the problem of consciousness as he can and concludes, somewhat paradoxically, that while it is an issue for science it also lies beyond science. What is of interest is that Wald links the anthropic principle to the fact of consciousness by proposing that consciousness is, in some sense, the *purpose* of the physical universe.

I have come to the end of my scientific life facing two great problems. Both are rooted in science; and I approach them as only a scientist would. Yet both I believe to be in essence unassimilable as science. That is scarcely to be wondered at, since one involves cosmology, the other consciousness.

I will begin with the cosmology.

We know that we live in a historical universe, one in which not only living organisms but stars and galaxies are born, mature, grow old, and die. There is good reason to believe that it is a universe permeated with life, in which life arises — given enough time — wherever the conditions exist that make it possible.

How many such places are there? I like Arthur Eddington's old formula: 10^{11} stars make a galazy; 10^{11}1 galaxies make a universe. Our own galaxy, the Milky Way, contains about 10^{11} (one-hundred billion) stars; and the lowest estimate I have ever seen of the fraction of them that might possess a planet that could support life is 1%. That means one billion such places in our home galaxy; and with about one billion such galaxies within reach of our telescopes, the lowest estimate we might put to the number of places in the already observed universe that could support life would be of the order of one billion billion — 10^{18}.

And now, to the first problem: If any one of a considerable number of physical properties of the universe we know were other than it is — some of those properties basic, others seeming trivial, almost accidental — life, that now appears so prevalent, would become impossible, here or anywhere.

The nub . . . is that our universe possesses a remarkably detailed constellation of properties, as it happens just that constellation that breeds life. It takes no great intelligence or imagination to conceive of other universes, indeed any number of them, each of which might be stable and workable, yet lifeless.

How did it happen that, with what seem to be so many other options, our universe came out just as it did? From our own self-centred point of view, that is the best way to make a universe: But what I want to know is, how did the universe find that out?

It may be objected that the question would not arise if we were not here to ask it. Yet here we are, and strangely insistent on asking that kind of question.

Perhaps that indeed is the answer: That this is a life-breeding universe, precisely in order eventually to bring forth creatures that ask and attempt to answer such questions; so that through them the universe can come not only to be, but to be known; indeed can come to know itself. We are plunged at once into the deep ambiguities that exist between the concepts of *being* and *being known*; of which more below.*

The problem of consciousness was hardly to be avoided by one who has spent most of his scientific life working on mechanisms of vision. That is by now a very active field, with thousands of workers. We have learned a lot, and expect to learn much more; yet none of it touches or even points however tentatively in the direction of *what it means to see*.

When one examines the visual systems of human beings and frogs, they come out much alike. The retinas contain both rods and cones, the visual pigments are closely similar in chemistry and behavior, the neural connections in the retina and with the brain are comparable. But I know that *I see*. I have good reason to believe that other persons see. Does a frog see? It reacts to light; so does a photocell-activated garage door. But does it *see*, is it *aware* of visual images, does it *know* that it is responding?

There is nothing whatever that I can do as a scientist to answer that question. That is a problem of consciousness. It is altogether impervious to scientific approach. As I worked on visual systems — it would have been the same for any other sensory mode, let alone more subtle or complex manifestations of mental activity — this realization lay always in the background. Now for me it is in the foreground. I think that it involves a permanent condition: that it never will become possible to identify physically the presence *or absence* of consciousness, much less its content.†

---

* The properties that make this a life-breeding universe give it also a more general character. For if it were deprived of almost any of the properties that make life possible, it would become also a less variegated, less mutable, less innovative universe. The breeding of life is one special manifestation of that more general *creative* quality.

† In this article I have used the terms *mind* and *consciousness* more loosely than I like, at times almost interchangeably. Others have done the same. When I ask whether a frog — or a garage door — is conscious, I mean is it *aware*, as we humans are aware. I speak also of consciousness — or better, mind — as a timeless and pervasive property, a complementary aspect of all reality. I hope that in each case the context will make my meaning clear.

The only unequivocally sure thing is what goes on in my own consciousness. Everything else that I think I know involves some degree of inference. As already said, I feel sure that other persons are conscious. It helps that they tell me so, and display other evidences of consciousness in speech and writing, art and technology. I believe that other mammals are conscious; and birds — why else would they sing? But frogs — and fishes? Those animals at least respond reasonably to light and some visual images. But I have worked also on the electrical responses to light of the numerous and anatomically magnificent eyes of scallops, without finding any indication that these animals see, beyond reacting to a passing shadow. And with some marine worms with great bulging eyes whose electrical responses we measured, we could find no behavioral responses to light at all.

The important point is that any assumption regarding the presence or absence of awareness in any nonhuman animal remains just that: an unsupported assumption. Matters are no different with inorganic devices. Does the photoelectrically activated garage door resent having to open when a car's headlights shine on it? I think not. Does a computer that has just beaten a human opponent at chess feel elated? I believe not. But there is nothing I can do to shore up those assumptions either. (It will be objected that the chess-playing computer was programmed by a human chess player; but was not also its human opponent?)

Consciousness is not part of that universe of space and time, of observable and measurable quantitites, that is amenable to scientific investigation. For a scientist, it would be a relief to dismiss it as unreal or irrelevant. I have heard distinguished scientists do both. In a discussion with the physicist P. W. Bridgman some years ago, he spoke of consciousness as 'just a way of talking.' His thesis was that only terms that can be defined operationally have meaning; and there are no operations that define consciousness. In the same discussion the psychologist B. F. Skinner dismissed consciousness as irrelevant to science, since confined to a private world, not accessible to others.

Unfortunately for such attitudes, consciousness is not just an epiphenomenon, a strange concomitant of our neural activity that we project onto physical reality. On the contrary, all that we know, including all our science, is in our consciousness. It is part, not of the superstructure, but of the foundations. No consciousness, no science, public or private. Perhaps, indeed, no consciousness, no reality — of which more later.

Though consciousness is the essential condition for all science, science cannot deal with it. It does not lie as an unassimilable element within science, but just the opposite: science is the highly digestible element within consciousness, which includes science as a limited but highly definable territory within the much wider reality of whose existence we are conscious.

Yet within its own sphere, science is limitless. There it expands endlessly, every answer breeding new questions, facing in all directions an endlessly expanding horizon. It is the endless quest — for what? To bring more and more of the material world into our consciousness; to recognize, hence to become aware of, more and more subtle aspects of physical reality. That recognition is itself a kind of creation. Physical reality is not just 'out there,' waiting to be discovered. Its discovery in a sense creates it.

In my picture of these relationships, science stretches like a shoreless ocean, within the boundless space that is consciousness.

Consciousness itself lies outside the parameters of space and time that would make it accessible to science. That realization carries an enormous consequence: consciousness cannot be located. But more: *It has no location*.

Some years ago I talked about this with Wilder Penfield, the great Canadian neurosurgeon. In the course of his therapeutic activities he had unprecedented opportunities to explore the exposed brains of conscious patients, and hoped in this way to discover the seat of human consciousness. I asked him, 'Why do you think consciousness is in the brain?' He chuckled, and said, 'Well, I'll keep on trying.' When we met again a few years later, he said, 'I'll tell you one thing: it's not in the cerebral cortex!'

Shortly afterward came the exciting announcement that the so-called reticular formation in the brain stem of mammals contains an arousal center, a center that stimulates awareness. The dilemma involved in all such observations is that one cannot know whether one is dealing with a source, or with part of the machinery of reception and transmission. It is as though, finding that the removal of a transistor from a television set stopped the transmission, one concluded it to be the source of the program.

How could one possibly locate a phenomenon that one has no means of identifying — neither its presence nor absence — nor any known parameters of space, time, energy exchange, by which to characterize merely its occurrence, let alone its content? The very idea of a location of consciousness is absurd. Just as with Heisenberg's uncertainty principle, we have more to deal with here than technical inadequacy, with a perhaps temporary lack of means of observation and measurement. What we confront is an intrinsic condition of reality. It is not only that we are unable to locate consciousness: *It has no location*.

So that is the problem of mind — consciousness — a vast, unchartable domain that includes all science, yet that science cannot deal with, has no way of approaching; not even to identify its presence *or absence*; that offers nothing to measure, and nothing to locate, since *it has no location*.

A few years ago it occurred to me — albeit with some shock to my scientific sensibilities — that my two problems, that of a life-breeding universe, and that of consciousness that can neither be identified nor located, might be brought together. That would be with the thought that mind, rather than being as most biologists suppose, a late development in the evolution of organisms, had existed always: that this is a life-breeding universe because the constant presence of mind had made it so.

As I say, that came initially as a shocking idea, a bit of intellectual play, elating as a game is elating. But very quickly I realized that I was in excellent company. Not alone are rudiments of this kind of thoght deeply embedded in millenia-old Eastern philosophies; it is stated explicitly or strongly implied in the writings of quite recent and some present physicists.

Perhaps being a biologist made it seem at first so strange to me. Biologists tend to be embarrassed by consciousness. As an attribute of some living organisms, they feel that they should know about it, and should indeed be in a position to straighten out physicists about it — none of which has come their way. Hence the discomfort, the avoidance, or worse: talking ancillary mechanisms that dodge the essential problem. This is so plain as to have drawn comment from the physicist Eugene Wigner.

Whereas physicists live day in and out with the problem of consciousness. Early in this century it became evident to all physicists that the observer is an intrinsic component of every physical observation. Physical reality is what physicists recognize to be real. One cannot separate the recognition of existence from existence. As Erwin Schrödinger put it: 'The world is a construct of our sensations, perceptions, memories. It is convenient to regard it as existing objectively on its own. But it certainly does not become manifest by its mere existence'.

A simple example of the intervention of mind in physical observation: Everyone by now is aware that radiation — light, indeed all elementary particles — exhibits two sets of what are essentially incompatible properties, those of waves and particles. This is the prime example of a widespread class of relationships that Neils Bohr brought together in his Principle of Complementarity. Enter consciousness: the physicist, setting up an experiment, decides beforehand which of those sets of properties he will encounter. If he does a wave experiment, he gets a wave answer; from a particle experiment he gets a particle answer.

Some years ago I began to entertain the thought that through the development of such science-making creatures as we, a universe wanting not only to be but to be known, can come to know itself: 'Out of the ninety two natural elements, four — C, N, O, and H — have unique properties that make life possible. Suppose, as one easily can, that they were less unique, more like the other elements. One wouldn't have to go far in that direction to make life impossible.

'Let me talk a little frank nonsense about this, make of it what you will. It would be a poor thing to be an atom in a universe without physicists. And physicists are made of atoms. A physicist is the atom's way of knowing about atoms'.‡

Of course, implicit in such talk is the recognition that a universe in which mind can eventually achieve such overt expression as in science, art and technology must be at its core, from its inception, in some sense

---

‡ In this article, as in my thinking, I tend to speak of as *physics* any aspect of science that goes beyond simple description and classification. In that sense such aspects of biology become the physics of living organisms.

a knowing universe; that it must in some sense possess mind as its pervasive and enduring attribute. The stuff of this universe, then, is ultimately mind-stuff. What we recognize as the material universe, the universe of space and time and elementary particles and energies, is then an avatar, the materialization of primal mind. In that sense there is no waiting for consciousness to arise. It is there always. What we wait for in the evolution of life is only the culminating event, the emergence of creatures that in their self-awareness can *articulate* consciousness, can give it a voice and being also *social* creatures, can embody it in culture, in technology, art and science.

It is primarily physicists who in recent times have expressed most clearly and forthrightly this pervasive relationship between mind and matter, and indeed at times the primacy of mind. So Eddington (1928): 'the stuff of the world is mind-stuff . . . The mind-stuff is not spread in space and time. . . . Recognizing that the physical world is entirely abstract and without "actuality" apart from its linkage to consciousness, we restore consciousness to the fundamental position . . .' von Weizsäcker (1971) states as 'a new and, I feel, intelligible interpretation of quantum theory' what he calls his 'Identity Hypothesis: Consciousness and matter are different aspects of the same reality.' I prefer most of all Wolfgang Pauli's formulation (1952): 'To us . . . the only acceptable point of view appears to be the one that recognizes *both* sides of reality — the quantitative and the qualitative, the physical and the psychical — as compatible with each other, and can embrace them simultaneously . . . It would be most satisfactory of all if *physis* and *psyche* (i.e., matter and mind) could be seen as complementary aspects of the same reality.'

What this kind of thought means essentially is that one has no more basis for considering the existence of matter without its complementary aspect of mind, than for asking that elementary particles not also be waves.

As for this seeming — at least until one gets used to it — a strange viewpoint for a scientist, I should like to say two things: first, that a fair number of major scientists have come to this conclusion. Perhaps all that is needed is to think deeply about it. And also, as in so many other instances, what is wanted is not so much an acceptable concept as an acceptable rhetoric.

If I say, with Eddington 'To put the conclusion crudely — the stuff of the world is mind-stuff,' that has a metaphysical ring. But if I say that ultimate reality is expressed in the solutions of the equations of quantum mechanics, quantum electrodynamics, and quantum field theory — that sounds like good, modern physics. Yet what are those equations, indeed what is mathematics, but mind-stuff? — virtually the ultimate in mind-stuff and for that reason deeply mysterious.

I think that we now possess, at least in outline, all that is needed to shape a credible view of the plan of this universe, and of the place in it of life and mind.

That view begins with a sense of the deep interpenetration of the concepts, *to be*, and *to be known* — existence, and its recognition.

Let us entertain the supposition that a universe that to be, needs to be known, to that end has taken on a design that breeds and fosters life; so that life might eventually, here and there, evolve scientists who could cast back upon the history that produced them, and could begin to understand it; through whose knowing, the universe could achieve increasingly the reality of becoming known, of coming to know itself.

Recently, to my surprise and that of some of the physicists most involved, this kind of thinking has been given the rather pretentious name: the anthropic principle. It states essentially that the universe possesses the properties it does in order eventually to produce physicists. Apart from meanderings in this direction, it emerged explicitly in a short paper by Dicke at Princeton, pointing out, for example, that the Hubble age of the universe is not wholly arbitrary, since about that much time was needed for physicists to appear. I had enjoyed hearing this idea expressed some time earlier in the silly question, Why is the world 5 — or 10 or 20 — billion years old? — and the answer: Because it took that long to find that out!

But all of this, provocative as it is, does not yet express what I think is the crux of the matter.

A universe that through breeding life evolves eventually science-, art-, and technology-making creatures, presumably in many places, enters thereby a new phase in its evolution, that now includes means for the independent evolution of consciousness.

For such creatures found societies and establish cultures. They invent languages, writing, institutions for accumulating, storing and propagating information, speculation and belief.

Those creatures were evolved anatomically and physiologically through natural selection, a process that involves three components: a ceaseless outpouring of inherited variations, advantageous and disadvantageous; a mechanism of inheritance; and a competition for survival, in the course of which those organisms and properties that function better are retained, and those that function less well are discarded.

Cultures — on our planet, human cultures — exhibit all these elements. They too display endless variations, both advantageous and disadvantageous to survival; mechanisms of *cultural* inheritance, mainly if not entirely Lamarckian (nongenic), since they involve almost entirely acquired (i.e., learned) characters; and continuous competition, an unceasing interplay of absorption and rejection, domination and subjection.

So is launched an independent evolution of consciousness, superimposed upon and parallel with the continuing evolution of anatomy and physiology. It takes its place as an intrinsic development in cosmic evolution. That universe in which, from its inception, matter and mind have been the complementary aspects of reality, now comes to develop, and regularly, in many places, complementary systems of evolution by natural selection, physical and mental.

> I think that is the substantive outcome of our argument. It accords to humankind and its like elsewhere in the universe a great place in cosmic evolution. It gives our species here a transcendent worth and dignity, among the many kinds of creatures that inhabit our planet. It tells us our place in the universe: it is to know and create, and to try to understand, as we alone can do under our sun.

To speak, as Wald does, of the universe as having a purpose is to close the circle with the organicist view of early Greek times. Most scientists would find such language empty and point to how much there is to be achieved and known without indulging in metaphysics. However, most science does not deal with consciousness.

What might be called a neo-organicism has been emphasized in this chapter on the grounds that the demise of simple reductive materialism has left something of a metaphysical vacuum. With reductionism has gone the unnecessary mutual exclusivity of dualism and its alternatives. When the material world was firmly identified as non-mind-like, then it was inevitable that mind and consciousness were seen as non-material. But they were discussed as if they had causal powers. This, as Ryle showed, encourages the mistake that because it is possible to talk about something as if it were a certain type of thing, it *is* that thing.

What will take the place of simple reductive materialism is not yet clear. However, the readings in this chapter suggest that it is now possible to accept that mind and consciousness have causal powers without having to propose non-material bearers of those powers. The physical world is not as inimical to consciousness as was previously thought. Accordingly, we can return consciousness to something like an organicist relationship to the physical world, modified to reflect contemporary physical knowledge. However, the nature of mind and consciousness is not absolute or fixed; it has a history and is the product of evolutionary processes with many different levels and stages. The next chapter begins the psychological treatment of consciousness by placing it in an evolutionary setting.

Notes

1. Popper, K., *Quantum Theory and the Schism in Physics*, 1982, Allen & Unwin.
2. Bergman, M., *The Reenchantment of the World*, 1981, Cornell University

Press. See also K. Wilber, 'Physics, Mysticism and the New Holographic Paradigm' in *The Holographic Paradigm and Other Paradoxes*, 1982, Routledge & Kegan Paul.

3. See D. Bohm, *Unfolding Meaning*, 1987, Routledge; also *Wholeness and the Implicate Order*, 1980, Routledge & Kegan Paul.
4. Bohm's *Wholeness and the Implicate Order* carries the most formalized statement of his ideas but a more accessible account is in Chapter Four of *Science, Order and Creativity* which he wrote with Peat.
5. Greenstein, G., *The Symbiotic Universe*, 1988, William Morrow.

|| 3 ||

An Evolutionary Perspective

READINGS

Julian Huxley
Evolution in Action

D. R. Griffin
The Question of Animal Awareness

Nicholas Humphrey
Consciousness: a Just-So Story

Karl Popper
Natural Selection and the Emergence of Mind

Discussions of consciousness occasionally treat it as if it were absolute, a propery of mind that is either totally present or totally absent. However, it is clear that whatever consciousness is, this is one of the things it is not. First, consciousness is not absolute and unvarying but graded and heterogenous. When considered over various species and over different psychological functions it is present in different varieties and to different degrees. Human beings have a highly developed form, but other grades and types of consciousness exist in other species. Second, the nature of consciousness is subject to change, over both long and short time scales. Short time scales may be taken to cover momentary and developmental changes, while longer time scales imply evolutionary changes in patterns of awareness.

There are, in any case, many different things to which the label 'consciousness' gets attached. It can mean the registering of a sensation, it can refer to a level or degree of psychological activation, it can mean our sense of individuality and so on. This list does not exhaust the meanings and no one definition will be final or complete. For the present, we will use consciousness as a generalized term for these and other aspects of awareness and hope to refine this definition as we proceed.

The longer term gradations and changes in consciousness, how they emerged and how they relate to each other is the subject of this chapter. However, just as there was no single sense of the term consciousness, there is also more than one sense of the term evolution. The reading from Popper included in the previous chapter pointed out that different evolutionary processes occur at different levels of organization. The following extract from Julian Huxley develops this point.

> All phenomena have a historical aspect. From the condensation of nebulae to the development of the infant in the womb, from the formation of the earth as a planet to the making of a political decision, they are all processes in time; and they are all interrelated as partial processes within the single universal process of reality. All reality, in fact, *is* evolution, in the perfectly proper sense that it is a one-way process in time; unitary; continuous; irreversible; self-transforming; and generating variety and novelty during its transformations. I am quite aware that many people object to the use of the term evolution for anything but the transformations of living substance. But I think this is undesirably narrow. Some term is undoubtedly needed for the comprehensive process in all its aspects, and no other convenient designation exists at present save that of evolution.

The overall process of evolution in this comprehensive sense comprises three main phases. Although there is continuity between them, they are very distinct in their main features, and represent three sectors of reality, in which the general process of evolution operates in three quite different ways. We may call these three phases the inorganic or, if you like, cosmological; the organic or biological; and the human or psycho-social. The three sectors of the universal process differ radically in their extent, both in space and time, in the methods and mechanisms by which their self-transformations operate, in their rates of change, in the results which they produce, and in the levels of organization which they attain. They also differ in their time-relations. The second phase is only possible on the basis of the first, the third on the basis of the second; so that, although all three are in operation today, their origins succeeded each other in time. There was a critical point to be surmounted before the second could arise out of the first, or the third out of the second.

The inorganic sector I must deal with extremely briefly. For further details I must refer my readers to standard works like those of Sir James Jeans or Sir Arthur Eddington, or the more recent picture so vividly sketched by Fred Hoyle in his little book *The Nature of the Universe*. The chief points which have a bearing on my theme of evolution seem to me to be these. This sector of reality comprises all the purely physico-chemical aspects of the universe throughout the whole of space, intergalactic as well as interstellar, all the galaxies, all the stars and stellar nebulae. The diameter of that part of it visible with the new 200-inch telescope is nearly a thousand million light-years; and there is a celestial region of unknown size beyond the range of any telescope that we may ever be able to construct. There are over a hundred million visible galaxies; and each of these contains anything from a hundred to ten thousand million stars. Obviously, then, the inorganic sector is by far the largest in spatial extent. It is also the largest in temporal extent: astronomers put the age of our own galaxy at up to five thousand million years — probably rather less — and most of them think the universe as a whole is of about the same age, though some believe it is considerably older.

But the mechanism of its transformation is of the simplest kind — physical, and very occasionally chemical, interaction. The degree of organization to be found in it is correspondingly simple: most of this vast sector consists of nothing but radiations, subatomic particles, and atoms; only here and there in it is matter able to attain the molecular level, and nowhere are its molecules at all large or complicated. Very few of them contain more than half a dozen atoms, as opposed to the many thousands of atoms in the more complex organic molecules found in living substance. Many of the results are extremely large — stars and galaxies; but their organization is of a very low order: the simple spirals of the galaxies, the concentric arrangement of layers in the stars. In the tiny fraction that has turned into planetary systems, the level of organization is higher, but only a little higher. Nowhere in all its vast extent is there any trace of purpose, or even of prospective significance. It is impelled from behind by blind physical forces, a gigantic and chaotic jazz dance of particles and radiations, in which the only overall tendency we have so far been able to detect is that summarized in the Second Law of Thermo-dynamics — the tendency to run down.

By contrast, the spatial extension of the biological sector is very much restricted. Living substance could not come into being except in that small minority of stars which have produced planetary systems. Within them, it is restricted to that small minority of their planets which are of the right size and in the right stage of their history for complicated self-copying organic molecules to be produced; and in them again to an infinitesimal surface shell. The number of such potential homes of life in our own galaxy is put by a few astronomers as high as a hundred thousand, but by most at only a few thousand or even a few hundreds. Whatever the truth turns out to be, the biological sector, considered spatially as the area occupied by life, cannot at the very outside constitute more than a million-million-millionth part of the extent of the visible universe, and probably much less. And of course the only spot of which we have actual knowledge is our own planet, with the possibility of Mars in addition. On the earth the extension of the biological sector in time appears to be about two thousand million years.

On the other hand, the level of organization reached is almost infinitely greater than in the preceding sector. The proteins, the most essential chemical constituents of living substance, have molecules with tens or even hundreds of thousands of atoms, all arranged in patterns characteristic for each kind of protein. Each single tiny cell has a highly complex organization of its own, with a nucleus, chromosomes, and genes, and other cell-organs, and is built out of a number of different kinds of proteins and other types of chemical units, mostly large and complex. But that is only the beginning, for large higher mammals such as men and whales may have in their bodies over a hundred million million or even over a thousand million million cells of many different types, and organized in the most elaborate patterns. As Professor J. Z. Young has set forth in his recent book, *Doubt and Certainty in Science*, the number of cells in our 'thinking parts' alone — the cerebral cortex of our brain — is about seven times the total human population of the world, and their organization is of a scarcely conceivable complexity.

Evolutionary transformation in this sector is brought about by the wholly new method of natural selection, which was not available during the thousands of millions of years before the emergence of living substance. This new method is responsible for the much higher level of organization which evolution here produces, as well as the greater variety of organization. It is also responsible for the much faster tempo of change: quite large changes in biological organization take only a few tens of millions of years; and really major ones, much more radical than any which can have occurred during the entire inorganic phase, only a hundred million or so.

At first sight the biological sector seems full of purpose. Organisms are built as if purposefully designed, and work as if in purposeful pursuit of a conscious aim. But the truth lies in those two words 'as if'. As the genius of Darwin showed, the purpose is only an apparent one. However, this at least implies prospective significance. Natural selection operates in relation to the future — the future survival of the individual and the species. And its products, in the shape of actual animals and plants, are correspondingly oriented towards the future, in their structure, their mode of working, and their behaviour. A few of the later products of evolution, notably the higher mammals, do show true purpose, in the sense of the awareness of a goal. But the purpose is confined to individuals and their actions. It does not enter into the basic machinery of the evolutionary process, although it helps the realization of its results. Evolution in the biological phase is still impelled from behind; but the process is now structured so as to be directed forwards.

The human phase of evolution, what I have called the psycho-social sector, is again enormously more limited in spatial extent. On this earth it is restricted to one among over a million species of organisms; elsewhere anything that could be called a psycho-social sector assuredly cannot have been attained in more than a very small fraction — perhaps a hundredth, perhaps only a ten-thousandth — of the planets habitable by some kind of life. It is still more limited in its temporal extent: its existence on this earth, from its first dim dawn to the present, occupies only one-half of one-tenth per cent of the history of life as a whole; and it has only operated at anything like full swing for perhaps a tenth of that tiny fraction of time.

Once again, a new main method of transformation has become available in this sector — the method of cumulative experience combined with conscious purpose. This has produced a new kind of result, in the shape of transmissible cultures; the main unit of evolution in the human phase is not the biological species, but the stream of culture, and genetic advance has taken a back seat as compared with changes in the transmissible techniques of cultural advance — arts and skills, moral codes and religious beliefs, and above all knowledge and ideas. It has also meant not only a more rapid tempo, but a new kind of tempo — an acceleration instead of a more or less steady average rate over long periods. In the long prologue of human evolution, each major change demanded something of the order of a hundred thousand years; immediately after the end of the Ice Age, something like a thousand years; during much of recorded history, the time-unit of major change was around a century; while recently it has been reduced to a decade or so. And again correlated with this increased tempo of change, we find an enormous increase in the variety of the results produced and in the levels of organization attained. In a way most important, purpose has now entered the process of transformation itself; both the mechanisms of psycho-social evolution and its products have a truly purposeful component, and evolution in this sector is pulled on consciously from in front as well as being impelled blindly from behind.

Organisms differ from man's machines in being able to construct themselves. In constructing itself, every organism goes through a process of individual development — what is technically called its ontogeny.

You, like me and every other human being, were once a microscopic spherical ovum, then in turn a double sheet of undifferentiated cells, an embryo with enormous outgrowths enabling you to obtain food and oxygen parasitically from your mother, a creature with an unjointed rod — what biologists call the notochord — in place of jointed backbone; you once had gill-clefts like a fish, you once had a tail, and once were covered with dense hair like a monkey; you were once a helpless infant which had to learn to distinguish objects and to talk; you underwent the transformation of your body and mind that we call puberty; you learned a job. You are in fact a self-transforming process.

Organisms can not only construct themselves, they can also reproduce themselves. One of the most important advances of nineteenth-century biology was the discovery of the physical basis of reproduction. The answer was simple — reproduction depends on continuity of substance. New individuals develop from portions of the living substance of other individuals. The original individual may simply split into two; or it may detach a portion of its substance to serve as a basis for the new individual's development. Even in very large organisms, the detached portions may be only microscopic single cells, as in the spores of plants. In sexual reproduction, two such detached cells, the sperm and the ovum, fuse to form one. But in every case, there is a continuity of living substance, a reproductive stream of life flowing down the generations.

Life has two aspects, a material and a mental. Its mental aspect increases in importance during evolutionary time. Later animal deployments have reached a higher level of mental organization than earlier ones: the higher animals have a larger mental components in their make-up. This fact leads to an important conclusion — that mind is not a pale epiphenomenon, not a mere 'ghost in the machine', to use Professor Ryle's phrase, but an *operative* part of life's mechanism. For no evolutionary trend can be maintained except by natural selection, and natural selection can only work on what is biologically useful to its possessors.

Mental activity is intensified and mental organization improved during evolution: like bodily organization, it is improved in different ways in relation to different needs.

For a biologist, much the easiest way is to think of mind and matter as two aspects of a single, underlying reality — shall we call it world substance, the stuff out of which the world is made. At any rate, this fits more of the facts and leads to fewer contradictions than any other view. In this view, mental activities are among the inevitable properties of world substance when this is organized in the form of the particular kind of biological machinery we find in a brain. The electrical properties of living substance provides us with a useful analogy. We now know that all activities in the body are accompanied by electrical changes — but changes so minute that they were only detected when special instruments were invented during the nineteenth century. All living substance, indeed all substance, inorganic as well as organic, has electrical properties, and all its properties have an electrical aspect. But these minute electrical changes can be intensified and utilized for biological ends. In nervous tissue they are utilized to transmit messages within the body: and a few fish, like the torpedo and the electric eel, possess organs for intensifying them to such a degree that they can be used to give dangerous electric shocks. I find myself driven to assume that the analogy with mind holds good — in other words that all living substance has mental, or we had better say mind-like, properties; but that these are, for the most part, far below the level of detection. They could only be utilized for biological ends when organs were evolved capable of intensifying them. It is by means of them that mind emerges as an operative factor in evolution.

The evolutionary approach brings out another important point about mind. Granted that natural selection is the only effective agency for producing change in biological evolution, a high degree of mental activity and mental organization could only have come into being if it was of biological advantage to its possessors. This at one stroke overthrows all theories of materialism, for they deny the effective reality of mind, or reduce it to a mere fly on the material wheel.

Verbal language was perhaps the greatest technical invention of living substance. It enables human beings to communicate and share with each other, and in so doing it automatically gives rise to the second major uniqueness of man — a common pool of experience for a group. This is not a pool in the sense of a static water tank. It is something which can grow and develop. The pooled experience is organized, and its organization changes and evolves with time.

Nothing of the sort exists in any other organism. It provides a new kind of environment for life to inhabit. It needs a name of its own: following Père Teilhard de Chardin, the French palaeontologist and philosopher, I shall call it the *nöosphere*, the world of mind. As fish swim in the sea and birds fly through the air, so we think and feel our way through this collective mental world. Our life is a voyage of exploration through its vast and varied landscape; as with all other kinds of exploration, hard work and passion and discipline are needed for success. Each one of us can only explore a limited area in any detail, but we can arrive at an idea of the whole, just as we can have an idea of the earth as a globe without physically journeying all over its surface. Only by exploring it and utilizing its resources can a man achieve the dual task of building a self and transcending the self that he has built. It is a world of possibilities, not merely of actualities.

The great complexity of human mental organization gives it an enormous range and depth of new consequential possibilities. And evolution in the human phase is essentially the adventurous and stormy story of the emergence of ever more of these possibilities into actuality.

Purely biological progress, in fact, has come to an end, but human progress is just beginning. There is a radical difference between them, which is correlated with the equally radical difference between any kind of animal life and any kind of human life. We begin by minimizing the difference between animals and ourselves by unconsciously projecting our own qualities into them: this is the way of children and of primitive peoples. Though early scientific thinkers, like Descartes, tried to make the difference absolute, later applications of the method of scientific analysis to man have, until quite recently, tended to reduce it again. This is partly because we have often been guilty of the fallacy of mistaking origins for explanations — what we may call the 'nothing but' fallacy: if sexual impulse is at the base of love, then love is to be regarded as nothing but sex; if it can be shown that man originated from an animal, then in all essentials he is nothing but an animal. This, I repeat, is a dangerous fallacy.

We have tended to misunderstand the nature of the difference between ourselves and animals. We have a way of thinking that if there is continuity in time there must be continuity in quality. A little reflection would show that this is not the case. When we boil water there is a continuity of substance between water as a liquid and water as steam; but there is a critical point at which the substance H_2O changes its properties. This emergence of new properties is even more obvious when the process involves change in organization, as in all cases when chemical elements combine to produce a chemical compound.

The critical point in the evolution of man — the change of state when wholly new properties emerged in evolving life — was when he acquired the use of verbal concepts and could organize his experience in a common pool. It was this which made human life different from that of all other organisms; and we can now begin to grasp the nature and profundity of the difference. The development of animals is always closed; their evolution is always sooner or later restricted. Man's individual development, on the other hand, is potentially open. It continues throughout his life, and it can take place in all sorts of directions; while in animals there is only one normal pattern to be realized. The same sort of thing holds for man as a type — his pooled experience can be indefinitely added to, and it can be organized in an indefinite number of different ways. Animal types have limited possibilities, and sooner or later exhaust them: man has an unlimited field of possibilities, and he can never realize all of them. He has developed a new method of evolution: the transmission of organized experience by way of tradition, which supplements and largely overrides the automatic process of natural selection as the agency of change in the human phase.

This puts mind, in all its aspects, into the business of evolution. Thus, under this new dispensation, beliefs are inevitably brought into being; and once they have been brought into being, they become tools of living. And the same is true of ideals or purposes or scientific theories or religious systems — they are among the emergent properties of the new, human type of organization. They cannot help coming into existence, and then they cannot help becoming operative factors for further change. Thus, once life had become organized in human form it was impelled forward, not merely by the blind forces of natural selection but by mental and spiritual forces as well.

In the light of evolutionary biology man can now see himself as the sole agent of further evolutionary advance on this planet, and one of the few possible instruments of progress in the universe at large. He finds himself in the unexpected position of business manager for the cosmic process of evolution. He no longer ought to feel separated from the rest of nature, for he is part of it — that part which has become conscious, capable of love and understanding and aspiration. He need no longer regard himself as insignificant in relation to the cosmos. He is intensely significant. In his person, he has acquired meaning, for he is constantly creating new meanings. Human society generates new mental and spiritual agencies, and sets them to work in the cosmic process: it controls matter by means of mind.

It is easy enough to make broad statements about the steps of advance which have transformed the quality of human life and experience. We have the technical steps — the step from food-gathering to hunting; the domestication of animals and plants; the invention of the wheel and of building in stone; the development of urban life; the invention first of writing, then of alphabetic writing; and so on to the familiar triumphs of modern applied science. We also have the steps in the organization of thought and creative expression; the passage from thinking exclusively in terms of magic to thinking also in terms of gods; the origin of philosophy from mythology and of drama from ritual; the pursuit of learning for its own sake; the rise of the scientific method of inquiry. That, I repeat, is easy. What is difficulty is to discover just how any one step is effected, still more to distinguish desirable from undesirable change, and restrictive from non-restrictive improvement.

That is the job of the science of man. Perhaps, I should say, the job of the human sciences, from psychology to history, from ethnology to economics, for there is as yet no single science of man, in the sense of an organized branch of inquiry with a common body of postulates and ideas. It is, I think, fair to say that the human sciences today are somewhat in the position occupied by the biological sciences in the early 1800s; they are rapidly exploring different sectors of their field, but still looking for a central core of general principles. One idea which came into my mind during the writing of this chapter was that, in the human phase of evolution, the struggle for existence has been largely superseded, as an operative force, by the struggle for fulfilment. It is the combination of these two terms which seems to me important. Human life *is* a struggle — against frustration, ignorance, suffering, evil, the maddening inertia of things in general; but it is also a struggle *for* something, and for something which our experience tells us can be in some measure achieved, even if we personally find ourselves debarred from any measure that seems just or reasonable. And fulfilment seems to describe better than any other single word the positive side of human development and human evolution

> — the realization of inherent capacities by the individual and of new possibilities by the race; the satisfaction of needs, spiritual as well as material; the emergence of new qualities of experience to be enjoyed; the building of personalities.

Huxley offers a stirring portrayal of human beings as the torch-bearers of evolution and emphasizes, much as Popper in the previous chapter, that the arena of human evolution is now cultural. It is, as Huxley notes, an oversimplification to regard his three stages as strictly distinct and sequential since there are large areas of overlap. However, the model of progressive stages within which evolution has a very different character is a necessary framework for the question of how consciousness may have played a role in the different stages.

To suggest that consciousness is significant in prebiotic evolution implies that elementary material processes or at least anticipates rudimentary levels of consciousness. Wald, and others who take seriously the strong anthropic principle, suggest that consciousness was indeed prefigured in some way in the universe prior to the emergence of life. This is a significant departure from Western science, though it is very much the view taken by some forms of Eastern cosmogeny.[1] Huxley appears to take seriously the notion of panpsychism which, informally put, is the idea that there is a little bit of mind in everything.[2] David Bohm advances a somewhat similar view and within the last few decades, scientists and others have been increasingly concerned with the continuity of mental and physical levels of organization.[3] However, this line of thought moves away from the intended direction of the reader and this chapter concentrates on how consciousness may have figured in the biological and the psycho-social stages of evolution.

Darwin's interest in behavioural and psychological evolution makes him the founder of comparative psychology. The theory of evolution is the fundamental paradigm for the study of mind and behaviour in contemporary biology.[4] In looking at consciousness comparatively, we move from the physical sciences of the last chapter to biology and psychology. One aim of comparative psychology is to discover how a psychological taxonomy would correspond to taxonomies produced by other methods, such as anatomy. In general, comparative research shows the distribution of psychological capacity is broadly what might be expected — psychological capacities like perception, memory and learning increase as the phylogenetic tree is ascended.[5] But there can be discontinuities. Sometimes a capacity is more developed in a particular species than might be expected or sometimes a capacity is more evenly distributed over a wide range of species than a simple interpretation of the phylogenetic scale might predict.[6] The significance of these unexpected distributions depends on what the capacity in question is. If an irregularity merely reflects particular adaptations, for example the navigation skills of homing species, then this is less significant than when more general capacities such as learning, intelligence or consciousness are in question.

When such general capacities are being considered we might expect the distribution to be more in line with a hierarchical evolutionary pattern, with the more developed forms found at higher phylogenetic levels. In higher species the ratio of brain to body weight increases to a maximum value in the human case. Since the brain is the organ of mental life we might expect human beings to have the most developed capacity for mental life found within the phylogenetic tree. Indeed most writers conclude that equivalents to human conscious awareness are unlikely to be found anywhere else in the animal kingdom and that the challenge for comparative psychology is to appreciate the diversity of modes of consciousness.[7] This is clearly part of the much larger question of whether animals can in general be said to have a mental life. This is too large an issue to be addressed in this book though the question itself can act as a test for the quality of theories of mind themselves. The test is whether a theory of mind gives a definite and clear yes or no answer to the question 'do animals have minds?'. If it does, then the theory is too simple.

When theories of mind were theories of the soul, animals were set apart from human beings by the qualities of their souls. For Descartes animals were automata because they lacked the qualities of soul which gave human beings free will. For strict materialist theories of mind, human beings and animals were both automata, but automata of different complexity and the differences between them were quantitive rather than qualitative ones. With a better understanding of what the mind is and is not, the question of animal minds has developed accordingly, although many of the central issues are still unresolved such as whether the capacity to use language is exclusive to human beings.[8] As far as consciousness is concerned the most common view is that the human capacity for awareness is so markedly influenced by the emergence of language that it represents a qualitative break with the forms of consciousness found in other animals.

Donald Griffin, an authority on animal behaviour, is cautious about such a discontinuity and suggests that awareness is more widely distributed than is generally supposed. Griffin, who is best known for studies on homing behaviour and echo location, studies animal behaviour rather than anatomy. Attributing awareness on the basis of what animals do rather than brain size seems to lead to slightly different conclusions.

> The flexibility and appropriateness of animal behaviour suggest both that complex processes occur within their brains, and that these events may have much in common with our own conscious mental experiences. To the extent that this proves to be true, many of our ideas and opinions about the relationship between animals and men will require modification. The current scientific *Zeitgeist* almost totally avoids consideration of mental *experience* in other species, while restricting attention to overt and observable behavior and physiological mechanisms. To the extent that animal thinking and feeling become accessible to scientific scrutiny and analysis, ethology will be greatly broadened and enriched. The implications of these developments are profound. For one thing, they oblige us to reconsider deep-seated assumptions about human nature, and to inquire whether our kinship with other living organisms may be closer than we have been accustomed to recognize. Some of those mental attributes which we have been accustomed to view as unique prerogatives of our species may turn out to be more widely distributed, although presumably in limited and simpler forms. If so, it becomes reasonable and promising to attempt the study of mental experience in animal surrogates. This book will therefore examine both the pertinent evidence and its general significance in the hope of stimulating renewed interest in, and investigation of, the possibility that mental experiences occur in animals and have important effects on their behavior.
>
> Terms such as mental experiences, mind, awareness, belief, intention, or consciousness are obviously difficult to define; and one reason for avoiding a cognitive approach to animal behavior has been confusion about the meaning of the terms and concepts involved. Depending on how awareness is defined, its possible existence in other species can vary all the way from being trivially obvious to the most preposterous level of implausibility. At the first extreme, one might define awareness as any capacity or reaction; but this would allow the inclusion of all living organisms, plus even such a simple mechanism as a mousetrap. At another extreme, one might demand the use of written language, or the most complex levels of understanding known to human thinkers — the creative insights of Beethoven, Einstein, or Whitehead, for instance. But these requirements would eliminate many members of our own species.
>
> It is important to recognize at the outset that almost any concept can be quibbled to death by excessive insistence on exact operational definitions.

Even such widely used and clearly useful terms as 'hunger', 'memory', 'aggression', or 'metabolism' have been subjected to erudite analyses in a search for definitions that will satisy all demands and avoid every possible ambiguity. These efforts tend to come in waves, each followed by a truce of sheer exhaustion, after which the term continues to be used, but with clearer appreciation of the breadth of its connotations. Excessive concern to avoid all terms that cannot be rigorously defined suffers from the danger of retaining only verbal corpses that display rigor mortis.

It is appropriate to begin with the most obvious fact about mental experiences: *all of us have them*. Every normal person thinks about objects and events. These may be close at hand in time and space, like a toothache, a frightening antagonist, or a beloved infant. But we can also think about things that are remote from the current local situation; our thoughts may concern some distant place and they can reach far into the past or future. Some mental experiences are as simple as recalling the color of last summer's flowers or yearning for satisfying food; others are elaborate and complex, like an astronomer's concept of stellar evolution. I suggest that, for present purposes, we consider an animal to have a *mind* if it has such experiences, whether they be simple or complex.

Schaffer (1975) defines 'mind' as follows: 'as the term is used more technically . . . and in the philosophy of mind today, [it] encompasses sense perception, feeling and emotion, traits of character and personality, and the volitional aspects of human life; as well as the more narrowly intellectual phenomena'. Elsewhere Schaffer states: 'One thing that sharply distinguishes man from the rest of nature is his highly developed capacity for thought, feeling, and deliberate action. Here and there in other animals, rudiments, approximations, and limited elements of this capacity may occasionally be found; but the full-blown development that is called a mind is unmatched elsewhere in nature.' A cognitive ethologist may wonder whether perhaps the mental capabilities of animals will turn out to be more substantial and significant than Schaffer implies. To define mental experiences as uniquely human certainly discourages inquiry into the possibility of their occurrence in other species, and begs the question I am trying to examine.

Kenny, Longuet-Higgins, Lucas, and Waddington (1972) devoted a lucid, thoughtful, and stimulating series of Gifford Lectures at Edinburgh to *The Nature of Mind*, without explicitly defining the terms 'mind' or 'mental.' Ryle in a very influential book entitled *The Concept of Mind*, also avoided any specific definition of the term. But in a second series of Gifford Lectures (Kenny et al., 1973, p. 47) Kenny stated that 'to have a mind is to have the capacity to acquire the ability to operate with symbols in such a way that it is one's own activity that makes them symbols and confers meaning on them.' The communicative dances of honeybees certainly satisfy this criterion; for it is each forager's own activity that makes the waggle dance into a symbolic statement that conveys to other bees useful information about distance, direction, and desirability of something the dancer has visited.

Some philosophers may object to calling the bee dances symbolic on the ground that only thinking creatures can recognize symbols, so that use of the term *symbol* implies that bees do think, and thus tricks the reader into accepting the conclusion at issue. For the moment, I mean to point out simply that the bee dances satisfy the particular definition advanced by Kenny. In Chapters 3 and 4, I will discuss in greater detail how this animal communication system provides at least suggestive evidence of conscious thinking. Elsewhere in these Gifford Lectures, Longuet-Higgins offered quite a different sort of definition: 'An organism which can have intentions I think is one which could be said to possess a mind [provided it has] . . . the ability to form a plan, and make a decision — to adopt the plan' (Kenny et al., 1972, p. 136). Many animals behave as though they do have plans of at least a simple sort, and adjust their behavior appropriately in attempts to carry them out.

The neuropsychologist E. R. John (in Thatcher and John, 1977, pp. 294–304) defines consciousness as 'a process in which information about multiple individual modalities of sensation and perception is combined into a unified multidimensional representation of the state of the system and its environment, and integrated with information about memories and the needs of the organism, generating emotional reactions and programs of behavior to adjust the organism to its environment. . . . Consciousness about an experience is defined as information about the information in the system, that is, consciousness itself is a representational system. . . . Perhaps our philosophical quandary [concerning mind-brain dualism] arises from the assumption that organized processes in human brains are *qualitatively* different from organized processes in other nervous systems or even in simpler forms of matter. Perhaps the difference is only quantitative; perhaps we are actually not as unique as we have assumed.'

Conscious awareness and mental experience may sometimes be limited to a single sensory modality, for example vision, so that a rigid requirement that consciousness entail integration across modalities may not be justified. It might also improve John's definition to add that conscious minds deal with dynamic mental images of future, as well as past, events. But in other respects this definition is close to the cautious and tentative views of many neuroscientists, and it is important to note that it does not limit conscious awareness to our species.

An increasing number of behavioral scientists seem willing to consider the possibility that animals are sometimes aware of objects and events, but insist that self-awareness is unique to our species. This is one of a very few areas of cognitive ethology that have already been illuminated by objective, verifiable experiments. Gallup (1977) has developed an ingenious procedure which exploits the interest displayed by chimpanzees in their own mirror images. These and other Great Apes have often learned to use mirrors to examine parts of their bodies which they cannot see directly (Hayes, 1951; Hayes and Hayes, 1951). Gallup gave four chimpanzees ample opportunity to use mirrors, while two others had no such experience. In the crucial experiments, one animal at a time was deeply anesthetized and a conspicuous spot of inert, quick-drying colored material was placed on its forehead or earlobe. Upon awakening, they paid no attention to the markings, indicating that local, tactile stimulation was absent or ineffective. But when a mirror was provided, the chimpanzees familiar with their mirror images looked at themselves and then almost immediately reached for the colored spot and rubbed it or picked at it with their fingers. Those lacking experience with mirrors continued to ignore the paint marks. Certainly this experiment suggests that they recognized the mirror image as a representation of their own bodies.

So far, Gallup's type of experiment has yielded positive results only with Great Apes. Despite intensive efforts, gibbons, monkeys, and other laboratory animals have failed to react to mirror images as replicas of their own bodies. Instead, they seem to treat the mirror image as though it were another animal. Gallup therefore concludes that no other animal has the capacity for self-awareness. But one can inquire whether the capability of responding appropriately to mirrors should be equated with self-awareness. Would other tests prove more suitable for other species and provide evidence for the concept of oneself? Only further investigation can answer such questions; meanwhile, Gallup's experiments provide a clear and successful example of a well-controlled, objective, verifiable experiment in cognitive ethology.

It is instructive to separate the possibility that animals may be capable of self-awareness from the larger question of whether they have any awareness at all. If not, the question of self-awareness obviously does not arise. Therefore, let us explore the first possibility — that animals are capable of *some* kinds of awareness, but not of self-awareness. This means that we assume the animal is aware of its companions, their actions and communication signals, as well as of its own physical surroundings, the ground beneath its feet, the wind that blows against its skin, and so forth, but that it is unable to entertain the concept of 'selfness.' Yet an abundant flow of sensory input is always arriving at the animal's central nervous system from its own body. So we must postulate that this input is somehow selectively barred from reaching the animal's awareness.

This kind of 'awareness of everything but me' is conceivable, but it becomes increasingly less plausible the wider the range of awareness the animal is postulated to have of its inanimate and animate surroundings. If we allow a particular animal to be aware of a reasonably wide range of objects, events, and relationships in the world around it, while denying the possibility of self-awareness, we run the danger of redefining self-awareness in a roundabout way as a sort of perceived hole in the universe. Self-awareness has been widely held to be absent from all species but our own (for example, see Popper and Eccles, 1977). Premack (1976) and other assert without qualification that not even Great Apes are aware that they themselves will die. But direct and unequivocal evidence is nonexistent, and this whole subject challenges cognitive ethologists to seek for relevant data.

This Chapter has reviewed many of the principal issues which make it of the utmost importance to clarify the differences between mental experience in human and nonhuman animals: Do mental experiences occur in nonhuman animals? Does complex behavior such as that required for the more impressive examples of animal orientation indicate the presence of conscious thinking? When animals learn some new task or sensory discrimination, are they ever consciously aware of the facts and relationships they have learned? Can scientists investigate animal awareness in a balanced, open-minded fashion without undue reliance on appealing, but unsupported, assertions? Can mental images, consciously perceived, be detected in nonhuman animals, and if so how? Do animals experience beliefs and intentions, and are they aware of themselves?

It is now time to examine one category of animal behavior that is perhaps most closely linked to whatever they may be thinking — their communication. Animals often go to great trouble to direct messages at conspecifics, signals that produce important changes in their behavior, sometimes including exchanges of communicative signals which lead, in turn, to coordinated group action. When such social communication is versatile and appropriately adapted to a variety of situations, the possibility arises that the communicating animals are thinking about the messages they exchange with their fellows.

Ethologists have observed that several kinds of animals employ different alarm calls when confronted with different sorts of predators. For example, ground squirrels have two types of alarm calls, one for aerial and the other for ground predators . . . lemurs and squirrel monkeys also give different calls for aerial and ground predators. But the most convincing evidence available to date that alarm calls may designate specific categories of predators or dangers comes from the carefully controlled field experiments of Seyfarth, Cheney, and Marler (1980). Struhsaker (1967) had observed that East African vervet monkeys (*Cercopithecus aethiops*) give three acoustically distinct types of alarm calls when they see (1) a large mammalian predator such as a leopard (*Panthera pardus*), (2) a martial eagle (*Polemaetus bellicosus*), or (3) a dangerous snake, such as a python. The responses of other monkeys differ in an adaptive fashion. The leopard alarm call causes monkeys on the ground to climb trees, the eagle alarm call results in their looking up and running into dense bushes, and the response to snake alarm calls is to look down at the ground, often while standing bipedally on the hind legs, in which posture they can, of course, see a wider area. Seyfarth et al. conducted playback experiments under natural, but carefully controlled, conditions. The responses were analyzed from motion pictures of responses to alarm calls played back from a concealed loudspeaker. Calls of a particular group member in response to an actual predator were played back only when the individual was absent. The results confirmed that these three types of calls elicited the appropriate responses when no predator was actually present, and when all three calls had approximately the same physical intensity.

These three alarm calls of the vervet monkey are noniconic in the sense that they bear no resemblance to the predators they designate or to any sound accompanying the monkeys' responses. They are acoustically distinct in temporal pattern and frequency. Young vervet monkeys give similar calls to a much wider range of external stimuli than do adults, indicating that the exact referents are learned. For instance, infants sometimes give eagle alarm calls to harmless small birds, or even to a butterfly or falling leaf, whereas adults ignore quite similar-looking large birds, such as vultures, which do not attack monkeys.

The distinctive and sensible responses to these three classes of alarm calls seem to show that they convey three distinct messages, namely the presence of one of three kinds of danger. The calls could also be interpreted as injunctions to behave in certain ways. That is, the leopard alarm call might mean 'Climb into a tree', or the eagle alarm call 'First look up and then dive into the bushes.' But in either case these alarm calls convey distinct and adaptively advantageous messages, and thus constitute a simple, but nonetheless significant, example of semantic communication. The calls are arbitrary, in that other sounds could equally well serve the same purpose, and a considerable amount of learning is required to achieve the specificity with which adults employ them to warn their companions of distinct categories of predators. These experiments indicate that vervet monkey alarm calls share one important property with human language, namely reference to external objects and events, a feature that had been judged to be completely lacking in animal communication.

Many behavioral scientists express feelings of discomfort, or even outrage, at the inference of conscious intention in animals because previously unsuspected complexities in their orientation and communication have been discovered. On strictly logical grounds, complexity of behavior and conscious awareness are neither commensurate with, or necessarily related to, one another in any way. Inanimate mechanisms can be enormously complex and difficult to understand, but most *descriptions* of animal behavior can be modeled by mechanisms far simpler than a television receiver. The same can be said of many physiological mechanisms. For instance, a very simple electronic circuit can produce an electrical signal that closely resembles the spike potential of a neuron. But only the most naive engineer-turned-neurophysiologist would accept the existence of such a circuit as a satisfactory explanation of the functioning of nervous systems. To Loeb, the existence of a phototactic machine constructed out of wheels, electric motors, and photocells was evidence for believing that animal, and even human, behavior could be explained in terms of tropisms or forced movements. But the crippling limitations of such intellectual myopia should now be clearly apparent; the simplicity often lies not in the behavior, but in its description.

Despite the above considerations, it remains a fact that one of the principal reasons that led linguists and cognitive psychologists to abandon the strict behavioristic approach to human language and cognition was the staggering complexity of stimulus-response mechanisms that had to be postulated in order to account for human behavior on the basis of

Skinnerian formulations. Of course, it is always possible to postulate more and more complex and intricate stimulus-response relationships, but those to which one is driven in seeking to encompass human thinking and human conversations become so unwieldy that they can scarcely be justified on the grounds of scientific parsimony.

The importance of the questions discussed in this book is demonstrated by the heavy reliance of linguists and philosophers on the consciously intentional use of language as the principal distinguishing characteristic of our species. A major reason for this philosophical assertion has been the acceptance by those linguists and philosophers of the general conclusions expressed by students of animal behavior. I suggest that behavioral scientists now have the opportunity, and perhaps an obligation, to explore and discuss the limitations of this traditional, behavioristic viewpoint in the light of recent discoveries about communication behavior in animals.

When the behavioristic position is stated at its scholarly best — for example, by Lashley (1923, 1958) — it is essentially agnostic. It does not deny the existence of mental states, but argues that they are one and the same as neurophysiological processes, and that it is unprofitable to attempt any sort of scientific analysis based on introspective reports. Half a century of behavioral science has progressed on this basis, along with many discoveries in neurobiology in the broadest sense, including ethology. But what was originally an agnostic position tended to drift implicitly into a sort of *de facto* denial that mental states or consciousness exist outside our own species.

It is very easy for scientists to slip into the passive assumption that phenomena with which their customary methods cannot deal effectively are unimportant or even nonexistent. To quote Fouts (1973): 'All one needs to do is to look around and *not* see something and then conclude that the thing that was not seen in a particular species is totally absent in that species.' Here I should also like to follow the example of Holloway (1974) in quoting Daniel Yankelovich ('Smith,' 1972): 'The first step is to measure whatever can be easily measured. This is okay as far as it goes. The second step is to disregard that which can't be measured or give it an arbitrary quantitative value. This is artificial and misleading. The third step is to presume that what can't be measured easily isn't very important. This is blindness. The fourth step is to say what can't be easily measured really doesn't exist. This is suicide.'

Biological evolution is universally accepted by behavioral scientists as historical fact. Animals are used as surrogates or 'models' for behavioral investigations on the implicit assumption that principles discovered in this way are applicable to our own species. Certainly this assumption implies qualitative continuity. If, for example, all human learning were believed to be radically different in kind from that available for analysis in other animals, no one would even suggest applying to questions of human education what has been learned by studying rats, pigeons, or monkeys. Yet, when questions of communication and language arise, even hard-nosed behaviorists take for granted a large element of discontinuity. It is indefensibly circular to argue that language is unique to man and, therefore, no matter how complex animal communication turns out to be, it cannot possibly be comparable to human language.

Must we reject evolutionary continuity in order to preserve our gut feeling of human superiority? Or can we be satisfied with a merely quantitative, if enormous, difference between communication behavior in our own and other species? If we insist on a qualitative human-animal distinction in the area of communication behavior, a radical difference in kind in Adler's terms, must we support our insistence by criteria as subjective and difficult to test as those that were rejected by the founders of behaviorism?

The rigid position of the strict behaviorists has been questioned with increasing frequency. For instance, Mowrer (1960b) introduced a chapter entitled 'Images, Memory, and Attention (Observing Reactions)' with the remark that these terms have been 'and perhaps are still, in some measure *taboo*. Many of us were taught, under pain of banishment from professional psychology, never to use these terms, at least not during "working hours" . . . such language was deemed completely unsuited to the purposes of science. . . . But it is slightly ironical that those very methods of analysis and research which radical Behaviorism introduced are now leading us, ineluctably, back to concepts which Behaviorism was determined to ignore — or even destroy.'

In facing squarely the problems of dealing with the possibility that animals have mental experiences, it may be helpful to recognize that our current climate of opinion in the behavioral sciences involves a gradient of acceptability concerning the terms and concepts listed below:

O.K. ↑ PATTERN RECOGNITION
NEURAL TEMPLATE
SOLLWERT
SEARCH IMAGE
AFFECT
SPONTANEITY
EXPECTANCY
COVERT VERBAL BEHAVIOR
INTERNAL IMAGE
CONCEPT
UNDERSTANDING
INTENTION
FEELING
AWARENESS
MENTAL EXPERIENCE
MIND (MENTAL)
THOUGHT
CHOICE
FREE WILL
TABOO ↓ CONSCIOUSNESS

Individual scientists might wish to rearrange some entries in this rank order of orthodoxy, but there is no doubt that the gradient is a significant reflection of the current *Zeitgeist*. Rearranging these terms like playing cards is an entertaining game, but few radical rearrangements would leave the list a plausible one. It is also instructive to ask where one should draw a line to represent the boundary of scientific validity. Very strict behaviorists might stop after Affect, others may venture farther down the list. There are, of course, many philosophers who disagree with positivism, and they feel comfortable with a list extending beyond this one in the direction labeled Taboo.

Perhaps Jennings and Thorpe have outlined the most reasonable view, considering the limited evidence available: that the gradient is a true continuum without sharp discontinuities. Furthermore, it seems more likely than not that certain animals have mental experiences involving, to varying degrees, the attributes represented crudely by this rank-ordered list of terms.

Many branches of science have made significant and substantial progress by employing postulated entities that could not be observed directly, at least when first developed, but which were inferred from observations of their supposed causes and effects. Gravitation, electric potentials, magnetic fields, atoms, neutrinos, x-rays, chemical bonds, hormones, genes, and nerve impulses are pertinent examples. The impossibility of neatly verifying the existence of mesons or quarks has not inhibited theoretical physicists. Nor has the well-known dilemma concerning the wave and quantal properties of electromagnetic radiation led physicists to stop all investigations of quantum mechanics or particle physics simply because they cannot yet tell us whether light is waves or particles, or explain how it can have the properties of both at the same time. Likewise, paleontologists do their best to make sense out of the fossil record and sketch in evolutionary sequences or unfossilized morphologies without realistic hope of obtaining specific verification within the foreseeable future. Astrophysics is also based on concepts about events and processes immune from direct observation by any methods we can yet imagine.

Investigators of behavior have attempted to formulate comparable explanatory concepts, such as motivation, drives, or Lorenz's specific action potential. But perhaps we have been overlooking more directly pertinent concepts lying close to hand or even closer — inside our own heads. When thinking about Washoe in the act of exchanging information about objects, actions, or desires via manual gestures, or when contemplating Lindauer's swarming bees dancing about the suitability and location of cavities where the swarm might find a new home, I submit that it may actually clarify our thinking to entertain such thoughts as 'Washoe *hopes* to go out for a romp, and *intends* to influence her human companions to that end,' or 'This bee *likes* one cavity better than the other, and *wants* her swarm to occupy the preferred one.'

Of course, the use of such terms as *want* or *like* does not explain the basic causes of the observed behavior or of any mental experiences that may accompany it. Nor should the use of these or similar terms be taken to imply identity with any human mental experiences. The degree of similarity or differences would be a stimulating possibility for future investigations. Perhaps this return to a consideration of basic subjective qualities can supply a unifying framework into which many complexities of animal behavior can be fitted, perhaps we can understand how, and to what extent, animals make sense of the flow of events of which their behavior forms a part.

Most people not indoctrinated in the behaviorist tradition take it for granted that animals do have sensations, feelings, and intentions. This intuitive impression is based on our experience with patterns of animal behavior that appear sufficiently analogous to some of our own behavior to permit us to empathize. The dilemma of contemporary behavioral scientists results from our indoctrination that *as scientists* we must put such notions behind us as childish sentimentality unworthy of a rigorous investigator. Yet the behavioristic and reductionistic parsimony typified by Watson and Loeb may have led us down a sort of blind alley, at the end of which we find ourselves defending to the last, at least by implication, a denial of mental experience to animals, a denial which we cannot justify on any explicit basis except the presumed absence of communication with conscious intent.

Even Descartes, the fountainhead of the philosophical view that animals are merely machines, admitted that they could feel pain or pleasure and express passions. Yet many behaviorists believe that it makes no difference whether one thinks in terms of possible mental experiences or simply in terms of stimuli and responses, however complex. But the same argument can be applied to people. Inasmuch as we have only indirect evidence about *their* mental experiences, we may logically question whether they really exist. But if questions are raised by others about the reality of one's own subjective feelings, who is likely to fall back on a negative, or even an agnostic, response?

The communication behavior of certain animals is complex, versatile, and, to a limited degree, symbolic. The best-analyzed examples are the dances of honeybees and the signing of captive chimpanzees. These and other animal communication systems share many of the basic properties of human language, although in very much simpler form.

Language has generally been regarded as a unique attribute of human beings, different in kind from animal communication. But on close examination of this view, as it has been expressed by linguists, psychologists, and philosophers, it becomes evident that one of the major criteria on which this distinction has been based is the assumption that animals lack any conscious intent to communicate, whereas men know what they are doing. The available evidence concerning communication behavior in animals suggests that there may be no qualitative dichotomy, but rather a large quantitative difference in complexity of signals and range of intentions that separates animal communication from human language.

Human thinking has generally been held to be closely linked to language, and some philosophers have argued that the two are inseparable or even identical. To the extent that this assertion is accepted, and insofar as animal communication shares basic properties of human language, the employment of versatile communication systems by animals becomes evidence that they have mental experiences and communicate with conscious intent. The contrary view is supported only by negative evidence, which justifies, at the most, an agnostic position.

According to the strict behaviorists, it is more parsimonious to explain animal behavior without postulating that animals have any mental experiences. But mental experiences are also held by behaviorists to be identical with neurophysiological processes. Neurophysiologists have so far discovered no fundamental differences between the structure or function or neurons and synapses in man and other animals. Hence, unless one denies the reality of human mental experiences, it is actually parsimonious to assume that mental experiences are as similar from species to species as are the neurophysiological processes with which they are held to be identical. This, in turn, implies qualitative evolutionary continuity (though not identity) of mental experiences among multicellular animals.

The possibility that animals have mental experiences is often dismissed as anthropomorphic because it is held to imply that other species have the same mental experiencs a man might have under comparable circumstances. But this widespread view itself contains the questionable assumption that human mental experiences are the only kind that can conceivably exist. This belief that mental experiences are a unique attribute of a single species is not only unparsimonious; it is conceited. It seems more likely than not that mental experiences, like many other characters, are widespread, at least among multicellular animals, but differ greatly in nature and complexity.

Awareness probably confers a significant adaptive advantage by enabling animals to react appropriately to physical, biological, and social events and signals from the surrounding world with which their behavior interacts.

Opening our eyes to the theoretical possibility that animals have significant mental experiences is only a first step toward the more difficult procedure of investigating their actual nature and importance to the animals concerned. Great caution is necessary until adequate methods have been developed to gather independently verifiable data about the properties and significance of any mental experiences animals may prove to have.

It has long been argued that human mental experiences can only be detected and analyzed through the use of language and introspective reports, and that this avenue is totally lacking in other species. Recent discoveries about the versatility of some animal communication systems suggest that this radical dichotomy may also be unsound. It seems possible, at least in principle, to detect and examine any mental experiences or conscious intentions that animals may have through the experimental use of the animal's capabilities for communication. Such communication channels might be learned, as in recent studies of captive apes, or it might be possible, through the use of models or by other methods, to take advantage of communication behavior which animals already use.

Recognizing the hazards of both positive and negative dogmatism in our present state of ignorance, how can ethologists handle the unsettled (and to some, unsettling) questions of animal awareness and consciousness? Open-minded agnosticism is clearly a necessary first step. Then, when the behavior of an animal suggests awareness, conscious intention, or simple forms of knowledge and belief, a second step might be to entertain the hypothesis that the particular animal under the given conditions may be aware of a certain fact or relationship or may be experiencing some feeling or perception. Granting that such hypotheses are difficult to test by currently available procedures, the tentative consideration of their plausibility might pave the way for thoughtful ethologists to devise improved methods to study when and where animal consciousness may occur and what its content may be. The future extension and refinement of two-way communication between ethologists and the animals they study offer the prospect of developing in due course a truly experimental science of cognitive ethology.

Griffin acknowledges that there are qualitative differences between human and non-human awareness. However, he is clearly more inclined than most in this field to attribute to animals quite significant components of human consciousness such as intentionality, self-awareness and symbolic communication. As well as surveying the evidence for the distribution of conscious awareness, Griffin considers its adaptive value. This functionalist approach, emphasizing the use of particular adaptations or capacities, is a natural attitude to take to evolutionary issues. The whole process of emergence, survival and development of species is, after all, assumed to be a matter of how any particular characteristic of a species fits its niche and enables it to survive. Such an approach to consciousness would lead to the question of what it is *for*.

Now any functionalist account of a particular mental capacity in some distant period of evolutionary history faces the problem of being difficult to test empirically. We cannot return to past ages to observe behaviour directly. Nevertheless, informed speculation on the behaviour and mental life of human ancestors is a legitimate exercise of the scientific imagination. Although the author of the next reading, the psychologist Nicholas Humphrey, labels his speculations a 'just-so story', the evolutionary function he suggests for consciousness seems plausible enough.

Biologists who have thought, but not thought enough, about consciousness will be found toying with two contradictory ideas. First — the legacy of the positivist tradition in philosophy — that consciousness is an essentially private thing, which enriches the spirit but makes no material difference to the flesh, and whose existence either in man or other animals cannot in principle be confirmed by the objective tools of science. Second — the legacy of evolutionary biology — that consciousness is an adaptive trait, which has evolved by natural selection because it confers some (as yet unspecified) advantage on the individuals who possess it.

Put in this way, the contradiction is apparent. Biological advantage means an increased ability to stay alive and reproduce; it exists, if it exists at all, in the public domain. Anything which confers this kind of advantage — still more, anything whose evolution has specifically depended on it — cannot therefore remain wholly private. If consciousness *is* wholly private it cannot have evolved. Or if it has evolved, it must in Hamlet's words be but private north-north-west; when the wind is southerly it must be having public consequences. If the blind forces of natural selection have been able in the past to get a purchase on these consequences, so now should a far-seeing science be able to.

Yet scholars will, I suspect, continue to tolerate the contradiction, paying lip-service both to the privacy and to the evolutionary adaptiveness of consciousness, until they are offered a plausible account of just wherein the biological advantage lies. At present, so far from having a testable hypothesis which we could apply to species other than our own, we lack even the bones of a good story about consciousness in human beings. I offer one here: a Just-So Story.

But first some pointers to what, in the context of this story, I take 'consciousness' to mean. I rely on there already being between us the basis for a common understanding. I assume that you yourself are another conscious human being; that you have a personal conception of what consciousness is like; that you have experienced, waking and sleeping, both its presence and its absence; and that having noticed the contrasts you have already formed some notion of what consciousness is for. I assume moreover that although you may never have had occasion to pronounce on it, you will not find it difficult to recognise someone else's pronouncements (mine, below) as true of your own case.

Provided, that is, you are in fact a conscious human being, and not as it happens an unconscious robot or a philosopher from Mars. Provided, also, that you have not been too much influenced by Wittgenstein. When Wittgenstein alluded to consciousness as a 'beetle' in a box — 'No one can look into anyone else's box, and everyone says he knows what a beetle is only by looking at *his* beetle . . . it would be quite possible for everyone to have something different in his box . . . the box might even be empty' — he chose the name of a thing which has no obvious use to us, and thereby implicitly ruled out the possibility that the things in our several boxes might bear a functional resemblance to each other. But suppose the thing in the box had been called, let's say, a 'pair of scissors'. One person's pair of scissors might indeed look rather different from another's: long scissors, short scissors, scissors made of brass or steel. But scissors to be scissors have to cut. There is really no danger that what we both agree to call a 'pair of scissors' could in my case be, say, a jelly baby while in your case it is merely empty air.

From all I know about myself, what strikes me — and seems to give some kind of cutting edge to consciousness — is this. The behaviour of human beings, myself included, is in every case under the control of an internal nervous mechanism. This mechanism is responsive to and engaged with the external environment but at the same time operates in many ways autonomously, collating information, hatching plans, and making decisions between one course of action and another. Being internal and autonomous it also, for the most part, operates away from other people's view. You cannot see directly into my mechanism, and I cannot see directly into yours. Yet, *in so far as I am conscious*, I can see as if with an inner eye into my own.

During most of my waking life I have been aware that my own behaviour is accompanied by certain conscious feelings — sensations, moods, desires, volitions and so on — which together form the structure and content of my conscious mind. So regular indeed is this accompaniment, so rarely does anything happen to me without its being either preceded or paralleled by the experience of a conscious feeling, that I have long ago come to regard my conscious mind as the very same thing as the internal mechanism which controls my bodily behaviour. If I ask myself *why* I am doing something, like as not my answer will be framed in conscious mental terms: I am doing it *because* I am aware of this or that going on inside me. 'Why am I looking in the larder? Because I'm feeling hungry . . . Why am I raising my right arm? Because I wish to . . . Why am I sniffing this rose? Because I like its smell . . .'

Thus consciousness (some would say 'self-consciousness', though what other kind of consciousness there is I do not know) provides me with an explanatory model, a way of making sense of my behaviour in terms which I could not devise by any other means. And to the extent that it is successful, this is presumably because the workings of my conscious mind do in reality correspond in some formal (if limited) way to the workings of my brain. 'Hunger' corresponds to a state of my brain; 'wishing' corresponds to a state of my brain; even the organising principle of consciousness, my concept of my 'self', corresponds to an organising principle of brain states. Not that physiologists have yet come up with an analysis of brain activity along these lines. But that, for the moment, is their problem, not mine. As a child of the evolutionary process, whose ancestors have been in this business for many million years, I am, in relation to my own behaviour, like an ancient astronomer who has found a way of looking directly at the wheels and cogs which move the stars across the heavens: the stars are my behaviour, the cog-wheels are the mechanism which controls it, and the astronomer peering in on them is I my self.

So what?

So, once upon a time there were animals ancestral to man who were not conscious. That is not to say that these animals lacked brains. They were no doubt percipient, intelligent, complexly motivated creatures, whose internal control mechanisms were in many respects the equals of our own. But it is to say that they had no way of looking in upon the mechanism. They had clever brains, but blank minds. Their brains would receive and process information from their sense-organs without their minds being conscious of any accompanying sensation, their brains would be moved by, say, hunger or fear without their minds being conscious of any accompanying emotion, their brains would undertake voluntary actions without their minds being conscious of any accompanying volition . . . And so these ancestral animals went about their lives, deeply ignorant of an inner explanation for their own behaviour.

To our way of thinking such ignorance has to be strange. We have experienced so often the connection between conscious feelings and behaviour, grown so used to the idea that our feelings are actually the causes of our actions, that it is hard to imagine that in the absence of feelings behaviour could carry on at all. It is true that in rare cases human beings may show a quite unexpected competence to do things without being conscious of their inner reasons: the case, for example, of 'blindsight' where a patient with a cerebral lesion can point to a light without being conscious of any sensation accompanying his seeing (and without, as he says, knowing how he does it). But the patient himself in such a case confesses himself baffled; and you and I will not pretend that that would not be our reaction too.

Such bafflement, however, was one among the many things our unconscious ancestors were spared. Having never in their lives known inner reasons for their actions, they would not have missed them when they were not there. And whether we can imagine it or not, we should assume that, for the life-style to which they were adapted, 'unconsciousness' was no great handicap. With these animals it was their behaviour itself, not their capacity to give an inner explanation of it, which mattered to their biological survival. As the occasion demanded they acted hungry, acted fearful, acted wishful and so on, and they were none the worse off for not having the feelings which might have told them why.

None the less, these animals were the ancestors of modern human beings. They were coming our way. Though their lives may once have been comparatively brutish and relatively short, as generations passed they began to live longer, their life histories grew more complicated, and their relationships with other members of their species became more dependent, more intimate, and at the same time more unsure. Sooner or later the capacity to explain themselves and to explain others — to take on the role of a natural 'psychologist', capable of understanding and predicting his own and others' behaviour within the social group — would become something they could no longer do without. At that stage would not their lack of consciousness have begun to tell against them?

Not necessarily. At least not at first, and not to the extent that all that's said above implies. For inner explanations are not the only kind of explanations of behaviour. Debarred as our unconscious ancestors may have been from looking in directly on the workings of their brains, they could still have observed behaviour from outside: they could have observed what went into the internal mechanism and what came out, and so have pieced together an external, objectively based explanatory model. 'Why am I [Humphrey] looking in the larder?' Not, maybe, 'Because I'm feeling hungry', but rather 'Because it's five hours since Humphrey last had anything to eat' or 'Because Humphrey has shown himself to be less fidgety after a snack.'

In short, while our ancestors lacked the capacity to explain themselves by 'introspection', there was nothing to stop them doing it by the methods of 'behaviourism'. 'The behaviorist', wrote one of its first modern champions, J. B. Watson, 'sweeps aside all medieval conceptions. He drops from his scientific vocabulary all subjective terms such as sensation, perception, image, desire, purpose, and even thinking and emotion.' And who better placed to follow this recommendation than an unconscious creature for whom such conceptions could not have been further from his mind? In fact, it is we conscious human beings who have trouble being hard-headed behaviourists: it is *we* who, as B. F. Skinner has lamented, 'seem to have a kind of inside information about our behaviour. *We* have feelings about it. And what a diversion they have proved to be! . . . Feelings have proved to be one of the most fascinating attractions along the path of dalliance.'

Why, then, when ignorance of the inner reasons for behaviour might have been bliss, did human beings ever become wise? Adam, the behavioural scientist, might with Newtonian detachment have simply sat back and watched the apple fall; but no, he ate it.

What tempted him was a leap in the complexity of social interaction, calling in its turn for a leap in the psychological understanding of oneself and others. Suddenly the old-time psychology which was good enough for our unconscious ancestors — which may still apparently be good enough for Watson and for Skinner — was no longer good enough for their descendants. Behaviourism could only take a natural psychologist so far. And human beings were destined to go further.

At what point the threshold was crossed we cannot tell. But there is evidence that by three or four million years ago, and possibly much earlier, our ancestors had already embarked on what was in effect a new experiment in social living. Leaving behind the relatively dull life of their ape-like forebears — leaving behind their thick skins, large teeth and heavy bones, leaving behind their habitation in the forest and their hand-to-mouth existence as vegetarian gypsies — they sought this new life as hunter-gatherers on the African savanna. They sought it with stone tools, they sought it with fire; they pursued it with forks and hope. But above all they sought it through the company of others of their kind.

For it was membership of a co-operative social group which made the life of hunting and gathering on the plains a viable alternative to what had gone before. Life from now on was to be founded on collaboration, centred on a home base and a place in the community. This community of familiar souls would provide the context in which individuals could reap the rewards of co-operative enterprise, where they could benefit from mutual exchange of materials and ideas, and where (against all subsequent advice) they could become borrowers and lenders and then borrowers again — borrowers of time, of care, of goods and services. But most important, the community would provide them as they grew up first with a nursery and then with a general purpose school where they could learn from others the practical techniques on which the life of the hunter-gatherer depended.

But the intense social engagement which this new life-style entailed spelt trouble. For human beings would not, overnight, abandon self-interest in favour of the common good. And while it's true that each individual stood to gain by preserving the social system as a whole, each continued also to have his own particular loyalties — to himself, to his kin and to his friends. A society based, as this was, on an unprecedented degree of interdependency, reciprocity and trust, was also a society which offered unprecedented opportunities for an individual to manoeuvre and out-manoeuvre others in the group.

Thus the scene was set for a long-running drama of personal and political intrigue. Men and women were to become actors in a human comedy, played out upon the flinty apron-stage which formed their common home. It was a comedy which would be tragedy for some. It was a play of ambitions, jealousies, loves, hates, spites and charities, where success meant success in the conduct of personal relationships. And when the curtain fell it was to those who, as natural psychologists, had shown the greatest insight into human nature that natural selection would give the biggest hand.

Imagine now two different kinds of player, with very different casts of mind. One the traditional unconscious behaviourist, who based his psychology entirely upon external observation; the other a new breed of introspectionist, who took the short cut of looking directly in upon the workings of his brain.

The behaviourist starts with a blank slate. In the manner familiar to those who have followed the progress of behaviourism as a modern science, he patiently collects evidence about what he sees happening to himself and other people, he correlates 'stimuli' and 'responses', he looks for 'contingencies of reinforcement', he tries to infer the existence of 'intervening variables' . . . and thus, without prejudice, he searches for a pattern in it all.

This programme for doing psychology is not, let it be said, a hopeless one. It must have sufficed for our unconscious ancestors for many million years. It probably still suffices for many if not all non-human social animals alive today. With a bit of luck it might have sufficed for those who began to live the life of social human beings — had they but world enough and time, had there been no one else around with the gift of doing the job much better.

But now there *was* someone else around, and world, time and luck were all at once in short supply. An introspectionist had entered on the scene: someone who starts with a slate on which the explanatory pattern is already half sketched in. From earliest childhood the introspectionist has had the opportunity to observe the causal structure of his own behaviour emerging in full inner view: he has sensed the connection between stimulus and response, he has felt the positive and negative effects of reinforcement, he has been directly apprised of the intervening variables, and he has daily experienced the unifying presence of his conscious self.

In the first instance, certainly, the introspectionist's explanatory model applies only to his own behaviour, not to others'. But once a pattern of connections has been forced on his attention in his own case, the idea of that pattern will dominate his perception in other cases where the connections are not openly on show. Once an outer effect has been seen, in his own case, to have an obvious inner cause, the idea of that cause will help him to make sense of situations where the effect alone can be observed. Notice that a fire in your own private hearth causes smoke to issue from your chimney, and try *not* to imagine that the smoke coming from the house across the road implies the presence of a fire within those walls as well.

Thus the introspectionist's privileged picture of the inner reasons for his own behaviour is one which he will immediately and naturally project on other people. He can and will use his own experience to get inside other people's skins. And since the chances are that he himself is not in reality untypical of human beings in general — since the chances are that, just as from house to house there is generally no smoke without fire, so from person to person there is generally no looking in the larder without hunger, no running away without fear, no rage without anger, etc. — this kind of imaginative projection gives him an explanatory scheme of remarkable generality and power.

Let us return then to the age-old human play. Scattered among the population of unconscious behaviourists, there arose in time these conscious prodigies. Soon enough an unconscious Watson would find himself up against a conscious Iago, an unconscious Skinner would find himself laying suit to a conscious Portia . . . Natural selection was there to supervise their exits and their entrances.

It was clear where the story for the human species had to end. But for the rest of the animal kingdom? As the bias of my story must have shown, I am not yet convinced that any other species has followed the same path to consciousness as man. But studies of the social systems of other species are not far advanced, and studies of how individual animals themselves do their psychology are only now beginning. It may yet turn out that there are, in fact, non-human species whose social systems rival the complexity of man's; it may yet turn out that individuals of those species are, in fact, making use of explanatory systems which bear the hallmarks of a mind capable of looking in upon the inner workings of the brain. Stories have been wrong before. The cat, we know, does not walk by itself. But the rhino? Nothing suggests that the rhino gets inside another rhino's skin.

Meanwhile, for the obvious candidates — the social carnivores, the great apes — there will biologists who in fairness want to leave the question undecided. Undecided, but not undecidable. In medieval England a jury could bring in one of four verdicts at a trial: Guilty, Not Guilty, *Ignoramus* (we do not know), *Ignorabimus* (we *shall* not know).

'Ignoramus' may be a proper verdict for biologists. But if consciousness has evolved we shall know it by its works. 'Ignorabimus' would be a counsel of philosophical despair.

In suggesting that a function of consciousness is to help animals manage their interactions with each other Humphrey draws our attention to the social dimension of consciousness. In human evolution, it is clear, complex social interactions are likely to have been a major factor in the success of the species. The degree to which human beings can plan and act cooperatively may have developed more than in other species, even prior to the emergence of language.[9] The need to coordinate actions between individuals and promote gregariousness may also have been factors in bringing human consciousness into being.[10]

Whether we accept Humphrey's reconstruction at face value, it nevertheless raises the important issue of reflexive thought. Reflexivity is the capacity to make psychological operations the object of other psychological operations, that is, all forms of thinking about thinking. This underlies the distinctive human capacity for self-awareness. It seems very likely that reflexivity, which is more developed in human beings than in any other species, is a significant factor in making human consciousness unique.

The next reading is by Karl Popper who, in a lecture in Darwin's honour, addresses the evolution of mind in broad terms. Popper clearly accepts consciousness as causal within the evolutionary process. In fact, he identifies consciousness with choice, much as Bergson did.

The topic of my lecture is 'Natural Selection and the Emergence of Mind'. Natural selection is, obviously, Darwin's most central theme. But I shall not confine myself to this theme alone. I shall also follow Darwin in his approach to the problem of body and mind, both the mind of man and the animal mind. And I shall try to show that the theory of natural selection supports a doctrine which I also support. I mean the unfashionable doctrine of mutual interaction between mind and brain.

My lecture will be divided into four sections.

In the first section, entitled 'Darwin's *Natural Selection* versus Paley's *Natural Theology*', I shall briefly comment upon the Darwinian revolution and on today's counter-revolution against science.

The second section is entitled 'Natural Selection and its Scientific Status'.

The third section is entitled 'Huxley's Problem'. It contains the central argument of my lecture, an argument based on natural selection. It is an argument for mutual interaction between mind and brain, and against T. H. Huxley's view that the mind is an epiphenomenon. It is also an argument against the so-called identity theory, the now fashionable theory that mind and brain are identical.

The fourth section, entitled 'Remarks on the Emergence of Mind', concludes with a few speculative suggestions on what seems to be the greatest marvel of our universe — the emergence of mind and, more especially, of consciousness.

The first edition of Darwin's *Origin of Species* was published in 1859. In a reply to a letter from John Lubbock, thanking Darwin for an advance copy of his book, Darwin made a remarkable comment about William Paley's book *Natural Theology*, which had been published half a century before. Darwin wrote: 'I do not think I hardly ever admired a book more than Paley's "Natural Theology". I could almost formerly have said it by heart.' Years later in his Autobiography Darwin wrote of Paley that 'The careful study of [his] works . . . was the only part of the academical course [in Cambridge] which . . . was of the least use to me in the education of my mind.'

I have started with these quotations because the problem posed by Paley became one of Darwin's most important problems. It was *the problem of design*.

The famous *argument from design* for the existence of God was at the centre of Paley's theism. If you find a watch, Paley argued, you will hardly doubt that it was designed by a watchmaker. So if you consider a higher organism, with its intricate and purposeful organs such as the eyes, then, Paley argued, you are bound to conclude that it must have been designed by an intelligent Creator. This is Paley's argument from design.

Although Darwin destroyed Paley's argument from design by showing that what appeared to Paley as purposeful design could well be explained as the result of chance and of natural selection, Darwin was most modest and undogmatic in his claims. He had a correspondence about divine design with Asa Gray of Harvard; and Darwin wrote to Gray, one year after the *Origin of Species*: '. . . about Design. I am conscious that I am in an utterly hopeless muddle. I cannot think that the world, as we see it, is the result of chance; and yet I cannot look at each separate thing as the result of Design.' And a year later Darwin wrote to Gray: 'With respect to Design, I feel more inclined to show a white flag than to fire . . . [a] shot . . . You say that you are in a haze; I am in thick mud; . . . yet I cannot kept out of the question.'

To me it seems that the question may not be within the reach of science. And yet I do think that science has taught us a lot about the evolving universe that bears in an interesting way on Paley's and Darwin's problem of creative design.

I think that science suggests to us (tentatively of course) a picture of a universe that is inventive or even creative; of a universe in which *new things* emerge, on *new levels*.

There is, on the first level, the theory of the emergence of heavy atomic nuclei in the centre of big stars, and, on a higher level, the evidence for the emergence somewhere in space of organic molecules.

On the next level, there is the emergence of life. Even if the origin of life should one day become reproducible in the laboratory, life creates something that is utterly new in the universe: the peculiar activity of organisms; especially the often purposeful actions of animals; and animal problem solving. All organisms are constant problem solvers; even though they are not conscious of most of the problems they are trying to solve.

On the next level, the great step is the emergence of conscious states. With the distinction between conscious states and unconscious states, again something utterly new and of the greatest importance enters the universe. It is a new world: the world of conscious experience.

On the next level, this is followed by the emergence of the products of the human mind, such as the works of art; and also the works of science; especially scientific theories.

I think that scientists, however sceptical, are bound to admit that the universe, or nature, or whatever we may call it, is creative. For it has produced creative men; it has produced Shakespeare and Michelangelo and Mozart, and thus indirectly their works. It has produced Darwin, and so created the theory of natural selection. Natural selection has destroyed the proof for the miraculous specific intervention of the Creator. But it has left us with the marvel of the creativeness of the universe, of life, and of the human mind. Although science has nothing to say about a personal Creator, the fact of the emergence of novelty, and of creativity, can hardly be denied. I think that Darwin himself, who could not 'keep out of the question', would have agreed that, though natural selection was an idea which opened up a new world for science, it did not remove, from the picture of the universe that science paints, the marvel of creativity; nor did it remove the marvel of freedom: the freedom to create; and the freedom of choosing our own ends and our own purposes.

The fact that the theory of natural selection is difficult to test has led some people, anti-Darwinists and even some great Darwinists, to claim that it is a tautology. A tautology like 'All tables are tables' is not, of course, testable; nor has it any explanatory power.

I have in the past described the theory as 'almost tautological', and I have tried to explain how the theory of natural selection could be untestable (as is a tautology) and yet of great scientific interest.

Nevertheless, I have changed my mind about the testability and the logical status of the theory of natural selection; and I am glad to have an opportunity to make a recantation. My recantation may, I hope, contribute a little to the understanding of the status of natural selection.

What is important is to realize the explanatory task of natural selection; and especially to realize *what* can be explained *without* the theory of natural selection.

We may start from the remark that, for sufficiently small and reproductively isolated populations, the Mendelian theory of genes and the theory of mutation and recombination together suffice to predict, *without natural selection*, what has been called 'genetic drift'. If you isolate a small number of individuals from the main population and prevent them from interbreeding with the main population, then, after a time, the distribution of genes in the gene pool of the new population will differ somewhat from that of the original population. This will happen even if selection pressures are completely absent.

Moritz Wagner, a contemporary of Darwin, and of course of pre-Mendelian, was aware of this situation. He therefore introduced a theory of *evolution by genetic drift*, made possible by reproductive isolation through geographical separation.

In order to understand the task of natural selection, it is good to remember Darwin's reply to Moritz Wagner. Darwin's main reply to Wagner was: if you have no natural selection, you cannot explain the evolution of the apparently designed organs, like the eye. Or in other words, without natural selection, you cannot solve Paley's problem.

Many years ago I visited Bertrand Russell in his rooms at Trinity College and he showed me a manuscript of his in which there was not a single correction for many pages. With the help of his pen, he had instructed the paper. This is very different indeed from what I do. My own manuscripts are full of corrections — so full that it is easy to see that I am working by something like trial and error; by more or less random fluctuations from which I select what appears to me fitting. We may pose the question whether Russell did not do something similar, though only in his mind, and perhaps not even consciously, and at any rate very rapidly. For indeed, what seems to be instruction is frequently based upon a roundabout mechanism of selection, as illustrated by Darwin's answer to the problem posed by Paley.

More than forty years ago I proposed the conjecture that this is also the method by which we acquire our knowledge of the external world: we produce conjectures, or hypotheses, try them out, and reject those that do not fit.

What is important here has been described by Ernst Gombrich by the excellent phrase 'making comes before matching'. This phrase can be applied with profit to every case of selection, in partiuclar to the method of producing and testing hypotheses, which includes perception, and especially *Gestalt* perception. Of course, the phrase 'making comes before matching' can be applied also to Darwinian selection. The making of many new genetic variants precedes their selection by the environment, and thus their matching with the environment. The action of the environment is roundabout because it must be preceded by a partly random process that produces, or makes, the material on which selection, or matching, can operate.

One of the important ponits about this roundabout method of selection is that it throws light on the problem of downward causation to which Donald Campbell and Roger Sperry have called attention.

We may speak of downward causation whenever a higher structure operates causally upon its substructure. The difficulty of understanding downward causation is this. We think we can understand how the substructures of a system co-operate to affect the whole system; that is to say, we think that we understand upward causation. But the opposite is very difficult to envisage. For the set of substructures, it seems, interacts causally in any case, and there is no room, no opening, for an action from above to interfere. It is this that leads to the heuristic demand that we explain everything in terms of molecular or other elementary particles (a demand that is sometimes called 'reductionism').

A choice process may be a selection process, and the *selection* may be *from* some repertoire of random events, *without being random in its turn*. This seems to me to offer a promising solution to one of our most vexing problems, and one by downward causation.

The denial of the existence of mind is a view that has become very fashionable in our time: mind is replaced by what is called 'verbal behaviour'. Darwin lived to see the revival of this view in the nineteenth century. His close friend, Thomas Henry Huxley, proposed the thesis that animals, including men, are automata. Huxley did not deny the existence of conscious or subjective experiences, as do now some of his successors; but he denied that they can have any effect whatever on the machinery of the human or animal body, including the brain.

'It may be assumed', Huxley writes, '. . . that molecular changes in the brain are the causes of all the states of consciousness . . . [But is] there any evidence that these states of consciousness may, conversely, cause . . . molecular changes [in the brain] which give rise to muscular motion?' This is Huxley's problem. He answers it as follows: 'I see no such evidence . . . [Consciousness appears] to be related to the mechanism of . . . [the] body simply as a collateral product of its working . . . [Consciousness appears] to be . . . completely without any power of modifying [the] working [of the body, just] as the steam-whistle . . . of a locomotive engine is without influence upon its machinery.'

Huxley puts his question sharply and clearly. He also answers it sharply and clearly. He says that the action of the body upon the mind is one-sided; there is no mutual interaction. He was a mechanist and a physical determinist; and this position necessitates his answer. The world of physics, of physical mechanisms, is causally closed. Thus a body cannot be influenced by states of consciousness. Animals, including men, must be automata, even if conscious ones.

Thus the theory of natural selection constitutes a strong argument against Huxley's theory of the one-sided action of body on mind and for the mutual interaction of mind and body. Not only does the body act on the mind — for example, in perception, or in sickness — but our thoughts, our expectations, and our feelings may lead to useful actions in the physical world. If Huxley had been right, mind would be useless. But then, it could not have evolved, no doubt over long periods of time, by natural selection.

My central thesis here is that the theory of natural selection provides a strong argument for the doctrine of *mutual interaction* between mind and body or, perhaps better, between mental states and physical states.

I conjecture that life, and later also mind, have evolved or emerged in a universe that was, up to a certain time, lifeless and mindless. Life, or living matter, somehow emerged from non-living matter; and it does not seem completely impossible that we shall one day know how this happened.

I regard the emergence of mind as a tremendous event in the evolution of life. Mind illuminates the universe; and I regard the work of a great scientist like Darwin as important just because it contributes so much to this illumination. Herbert Feigl reports that Einstein said to him: 'But for this internal illumination, the universe would be just a rubbish heap.'

I think we have to admit that the universe is creative, or inventive. At any rate, it is creative in the sense in which great poets, great artists, and great scientists are creative. Once there was no poetry in the universe; once there was no music. But then, later, it was there. Obviously, it would be no sort of explanation to attribute to atoms, or to molecules, or even to the lower animals, the ability to create (or perhaps to proto-create) a forerunner of poetry, called proto-poetry. I think it is no better explanation if we attribute to atoms or molecules a proto-psyche, as do the panpsychists. No, the case of great poetry shows clearly that the universe has the power of creating something new. As Ernst Mayr once said, the emergence of real novelty in the course of evolution should be regarded as a fact.

We may distinguish two kinds of behavioural programmes, *closed behavioural programmes* and *open behavioural programmes*, as Mayr calls them. A closed behavioural programme is one that lays down the behaviour of the animal in great detail. An open behavioural programme is one that does not prescribe all the steps in the behaviour but leaves open certain alternatives, certain choices; even though it may perhaps determine the probability or propensity of choosing one way or another. The open programmes evolve, we must assume, by natural selection, due to the selection pressure of complex and irregularly changing environmental situations.

I can now state my conjecture as follows:

Ecological conditions like those that favour the evolution of *open behavioural programmes* sometimes also favour the evolution of the beginnings of consciousness, by favouring conscious choices. Or in other words, consciousness originates with the choices that are left open by open behavioural programmes.

Let us look at various possible stages in the emergence of consciousness.

As a possible first stage there may evolve something that acts like a centralized warning, that is, like irritation or discomfort or pain, inducing the organism to stop an inadequate movement and to adopt some alternative behaviour in its stead before it is too late, before too much damage has been done. The absence of a warning like pain will lead in many cases to destruction. Thus natural selection will favour those individuals that shrink back when they receive a signal indicating an inadequate movement; which means, anticipating the inherent danger of the movement. I suggest that pain may evolve as such a signal; and perhaps also fear.

As a second stage, we may consider that natural selection will favour those organisms that try out, by some method or other, the possible movements that might be adopted *before they are executed*. In this way, *real* trial and error behaviour may be replaced, or preceded, by *imagined* or vicarious trial and error behaviour. The imagining may perhaps initially consist of incipient efferent nervous signals, serving as a kind of model, or symbolic representation of the actual behaviour, and of its possible results.

Richard Dawkins has brilliantly developed some such speculations about the beginnings of mind in considerable detail. The main ponts about them are two. One is that these beginnings of mind or consciousness should be favoured by natural selection, simply because they mean the substitution of imagined or symbolic or vicarious behaviour for real trials which, if erroneous, may have fatal consequences. The other is that we can here apply the ideas of *selection* and of *downward causation* to what is clearly a choice situation: the open programme allows for possibilities to be played through tentatively — on a screen, as it were — in order that a *selection* can be made from among these possibilities.

As a third stage, we may perhaps consider the evolution of more or less conscious aims, or ends: of purposeful animal action, such as hunting. Unconscious instinctive action may have been purpose-directed before, but once vicarious or imagined trial and error behaviour has started, it becomes necessary, in situations of choice, to evaluate the end state of the imagined behaviour. This may lead to feelings of avoidance or rejection — to *anticipations* of pain — or to feelings of eager acceptance of the end state; and the latter feelings may come to characterize a consciousness of aim or end or purpose. In connection with open choices, a feeling may evolve of preference for one possibility rather than another; preference for one kind of food, and thus for one kind of ecological niche, rather than another.

The evolution of language and, with it, of the world 3 of the products of the human mind allows a further step: the human step. It allows us to *dissociate ourselves* from our own hypotheses, and to look upon them critically. While an uncritical animal may be eliminated together with its dogmatically held hypotheses, we may *formulate* our hypotheses, and criticize them. Let our conjectures, our theories die in our stead! We may still learn to kill our theories instead of killing each other. If natural selection has favoured the evolution of mind for the reason indicated, then it is perhaps more than a utopian dream that one day may see the victory of the attitude (it is the rational or the scientific attitude) of eliminating our theories, our opinions, by rational criticism, instead of eliminating each other.

My conjecture concerning the origin of mind and the relation of the mind to the body, that is the relation of consciousness to the preceding level of unconscious behaviour, is that its usefulness — its survival value — is similar to that of the preceding levels. On every level, making comes before matching; that is, before selecting. The creation of an expectation, of an anticipation, of a perception (which is a hypothesis) *precedes* its being put to the test.

If there is anything in this interpretation, then the process of variation followed by selection which Darwin discovered does not merely offer an explanation of biological evolution in mechanical terms, or in what has been slightingly and mistakenly described as mechanical terms, but it actually throws light on downward causation; on the creation of works of art and of science; and on the evolution of the freedom to create them. It is thus the entire range of phenomena connected with the evolution of life and of mind, and also of the products of the human mind, that are illuminated by the great and inspiring idea that we owe to Darwin.

The emergence of symbolic function is briefly mentioned at the end of Popper's lecture. He notes that this is what has allowed evolution to take what he calls 'the human step'. Later chapters will deal further with how human consciousness depends upon the socio-cultural context that symbolic capacity has created. Note, however, that symbolic function does not *constitute* human consciousness but only makes it possible. Just how symbolic function may have lead to the emergence of human consciousness is the concern of later chapters.

What we have seen in this chapter has been a small selection from a very large literature on the comparative and evolutionary context of consciousness. We see that consciousness is not just a property or function of an individual but is generated by and is dependent on interaction between individuals. We also see that the emergence of symbolic function and reflexivity has been a vital part of the process that has made human consciousness unique. Symbolic function and sociality are both crucial in the construction and maintenance of the cultural matrix surrounding human life. It is within this matrix that human consciousness arises, is formed and is maintained.

This socio-cultural process will be the focus of the later chapters of the reader. Before moving on to them, the biological and psychological support for consciousness will be examined in more detail since consciousness, as the most significant product of evolution, is inseparable from the brain and its psychological processes, which are also evolutionary products. Accordingly, the next chapter deals with readings from brain scientists while the following chapter will survey some psychological experiments and theory.

Notes

1. Wicken, J., 'The Cosmic Breath: Reflections on the Thermodynamics of Creation', *Zygon*, 19(4):487–505.
2. Darwin was also interested in this notion. See C. Smith, 'Charles Darwin, the Origin of Consciousness and Panpsychism', *Journal of the History of Biology*, 11(2): 245–67.
3. Stapp, H., 'Mind, Matter and Quantum Mechanics', *Foundations of Physics*, 12:321–56.
4. Richards, R., *Darwin and the Emergence of Evolutionary Theories of Mind and Behaviour*, 1988, University of Chicago Press. See Chapters Two and Three.
5. Walker, S., *Animal Thought*, 1983, Routledge.
6. McPhail, E. *Brain and Intelligence in Vertebrates*, 1982, Oxford University Press.
7. Jerison, H. 'On the Evolution of Mind' in D. Oakley, *Brain and Mind*, op.cit.
8. Premack, D., *Gavagai! Or the Future History of the Animal Language Controversy*, 1981, MIT Press.
9. Leakey, R. and Lewin, R., *The People of the Lake*, 1981, Pelican. See Chapter Nine.
10. Barlow, H., 'Nature's Joke: A Conjecture on the Biological Role of Consciousness' in B. Josepheson and V. Ramachandran, eds., *Consciousness in the Physical World*, 1980, Pergamon.

4

The Neuropsychology of Consciousness

READINGS

Wilder Penfield
The Mystery of the Mind

David Oakley
Animal Awareness, Consciousness and Self-image

John Eccles
Self-consciousness and the Human Person

Roger Sperry
Mind-Brain Interaction

The mind-body problem is something of a misnomer. The 'problem' is not to explain how the mind relates to the body in general but to the nervous system and to the brain in particular. This chapter presents material by neuropsychologists writing about consciousness.

As the late nineteenth century saw the anatomy and physiology of the nervous system described in increasing detail, more integrative accounts of neural and mental functions, that is, neuropsychology, began to appear. Early accounts often incorporated clinical findings in which the experience of patients was of central concern. A good example is the late nineteenth century work of John Hughlings Jackson on neurological disorders and brain injuries.[1] His treatment of different clinical syndromes was given coherence by being related to the flow of experience. Jackson's phenomenological slant is less pronounced in contemporary neuropsychology which has tended to concentrate on more quantifiable, objective measures such as the data from experiments or performance on psychological tests.

Even so, experience is still a central concern of those who study the human brain and Wilder Penfield, from whom the first reading comes, combined his role of neurosurgeon with an enduring interest in the nature of consciousness. Penfield's work contributed greatly to finding out where in the brain different psychological functions are concentrated but as these extracts show, he remained cautious about consciousness being especially associated with any particular part of the brain.

In the past fifty years we have come to recognize an ever increasing number of semi-separable mechanisms within the human brain. They are sensory and motor. There are also mechanisms that may be called psychical, such as those of speech and of the memory of the past stream of consciousness, and mechanisms capable of automatic interpretation of present experience. There is in the brain an amazing automatic sensory and motor computer that utilizes the conditioned reflexes, and there is a highest brain-mechanism that is most closely related to that activity that men have long referred to as consciousness, or the mind, or the spirit.

Throughout my own scientific career I, like other scientists, have struggled to prove that the brain accounts for the mind. But now, perhaps, the time has come when we may profitably consider the evidence as it stands, and ask the question: *Do brain-mechanisms account for the mind?* Can the mind be explained by what is now known about the brain? If not, which is more reasonable of the two possible hypotheses: that man's being is based on one element, or on two?

In the course of surgical treatment of patients suffering from temporal lobe seizures (epileptic seizures that are caused by a discharge that originates in that lobe), we stumbled upon the fact that electrical stimulation of the interpretive areas of the cortex occasionally produces what Hughlings Jackson had called 'dreamy states,' or 'psychical seizures.' Sometimes the patient informed us that we have produced one of his 'dreamy states' and we accepted this as evidence that we were close to the cause of his seizures. It was evident at once that these were not dreams. They were electrical activations of the sequential record of consciousness, a record that had been laid down during the patient's earlier experience. The patient 're-lived' all that he had been aware of in that earlier period of time as in a moving-picture 'flashback.'

On the first occasion, when one of these 'flashbacks' was reported to me by a conscious patient, I was incredulous. On each subsequent occasion, I marvelled. For example, when a mother told me she was suddenly aware, as my electrode touched the cortex, of being in her kitchen listening to the voice of her little boy who was playing outside in the yard. She was aware of the neighbourhood noises, such as passing motor cars, that might mean danger to him.

A young man stated he was sitting at a baseball game in a small town and watching a little boy crawl under the fence to join the audience. Another was in a concert hall listening to music. 'An orchestration,' he explained. He could hear the different instruments. All these were unimportant events, but recalled with complete detail.

D.F. could hear instruments playing a melody. I restimulated the same point thirty times (!) trying to mislead her, and dictated each response to a stenographer. Each time I re-stimulated, she heard the melody again. It began at the same place and went on from chorus to verse. When she hummed an accompaniment to the music, the tempo was what would have been expected.

In other cases, different 'flashbacks' might be produced from successive stimulations of the same point.

What has been said about epileptic automatism throws much light on what must be happening in the normal routine in our lives. By taking thought, the mind considers the future and gives short-term direction to the sensory-motor automatic mechanism. But the mind, I surmise, can give direction only through the mind's brain-mechanism. It is all very much like programming a private computer. The program comes to an electrical computer from without. The same is true of each biological computer. Purpose comes to it from outside its own mechanism. This suggests that the mind must have a supply of energy available to it for independent action.

I assume that the mind directs, and the mind-mechanism executes. It carries the message. As Hippocrates expressed it so long ago, 'the brain is messenger' to consciousness. Or, as one might express it now, the brain's highest mechanism is 'messenger' between the mind and the other mechanisms of the brain.

Consider, if you will, the various functional mechanisms that operate within the brain. There is one mechanism that wakens the mind and serves it each time it comes into action after sleep. Whether one adopts a dualist or a monist hypothesis, the mechanism is essential to consciousness. It comes between the mind and the final integration that takes place automatically in the sensory-motor mechanism and it plays an essential role in that mechanism. This recalls the thinking of Hughlings Jackson in regard to the word 'highest' as applied to levels of function within the brain. That this *highest mechanism*, most closely related to the mind, is truly a functional unit is proven by the fact that epileptic discharge in gray matter that forms a part of its circuits interferes with its action selectively. During epileptic interference with the function of this gray matter, consciousness vanishes and, with it, goes the direction and planning of behavior. That is to say, the mind goes out of action and comes into action with the normal functioning of this mechanism.

The human automaton, which replaces the man when the highest brain-mechanism is inactivated, is a thing without the capacity to make completely new decisions. It is a thing without the capacity to form new memory records and a thing without that indefinable attribute, a sense of humor. The automaton is incapable of thrilling to the beauty of a sunset or of experiencing contentment, happiness, love, compassion. These, like all awarenesses, are functions of the mind. The automaton is a thing that makes use of the reflexes and the skills, inborn and acquired, that are housed in the computer. At times it may have a plan that will serve it in place of a purpose for a few minutes. This automatic co-ordinator that is ever active within each of us, seems to be the most amazing of all biological computers.

By listening to patients as they describe an experiential flashback, one can understand the complexity and efficiency of the reflex coordinating and integrative action of the brain. In it, the automatic computer and the highest brain-mechanism play interactive roles, selectively inhibitory and purposeful.

Does this explain the action of the mind? Can reflex action in the end, account for it? After years of studying the emerging mechanisms within the human brain, my own answer is 'no.' Mind comes into action and goes out of action with the highest brain-mechanism, it is true. But the mind has energy. The form of that energy is different from that of neuronal potentials that travel the axone pathways.

Consider with me the beginning of life, leaving aside, for the moment, the question of the essential nature of the mind.

A baby brings with him into the world an active nervous system. He (or she) is already endowed with inborn reflexes that cause him to gasp and to cry aloud, and presently to search for the nipple, and to suck and swallow, and so set off a complicated succession of events within the body that will serve the purpose of nourishing it.

In the very first month you can see him — if you will take time to observe this wonder of wonders — stubbornly turning his attention to what interests him, ignoring everything else, even the desire for food or the discomfort of a wet diaper. It is evident that, already, he has a mind capable of focusing attention and evidently capable of curiosity and interest. Within a few months he recognizes concepts such as those of a flower, a dog, and a butterfly; he hears his mother speaking words, and shortly he is busy programming a large area of the uncommitted temporal cortex to serve the purposes of speech.

The beginning of speech is important. The first time he hears the word and imitates it, the sound will be far from his eventual pronunciation of 'dog.' A parrot can imitate, too. But it is not long before the infant takes another step. A dog appears in the stream of consciousness, whereupon the highest brain-mechanism carries a patterned neuronal message to the non-verbal concept-mechanism. The past record is scanned and a similar appearance recalled through the hippocampal system. The mind compares the two images that have thus appeared in the stream of consciousness, and sees similarity. There is a sense of familiarity or recognition. While all this is still in the stream of consciousness, another patterned neuronal message is formed, made up of the remembered concept modified by the present experience. This message is sent to the speech mechanism and the word 'dog' flashes up into consciousness.

Then he acts. A message is sent to the gray matter in the articulation area of the motor cortex. He speaks the word 'dog' aloud — and laughs, perhaps, in conscious triumph. I imagine that the parts in this sequence that have not become automatic are carried out by the highest brain-mechanism under the direction of the mind. I want to point out only that every learned-reaction that becomes automatic was first carried out within the light of conscious attention and in accordance with the understanding of the mind.

If one is to make a judgment on the basis of behavior, it is apparent that man is not alone in the possession of a mind. The ant (whose nervous system is a highly complicated structure), as well as such mammals as the beaver or the dog or chimpanzee, shows evidence of consciousness and of individual purpose. The brain, we may assume, makes consciousness possible in him too. In all of these forms, as in man, memory is a function of the brain. Animals, particularly, show evidence of what may be called racial memory. Secondly, new memories are acquired in the form of conditioned reflexes. In the case of man, these preserve the skills, the memory of words, and the memory of non-verbal concepts. Then there is, in man at least, the third important form of memory, experiential memory, and the possibility of recalling the stream of consciousness with varying degrees of completeness. In this form of memory and in speech, the convolutions, which have appeared in the temporal lobe of man as a late evolutionary addition, are employed as 'speech-cortex and interpretive-cortex.'

Perhaps I should recapitulate. I realized as early as 1938 that to begin to understand the basis of consciousness one would have to wait for a clearer understanding of the neuronal mechanisms in the higher brain-stem. They were obviously responsible for the neuronal integrative action of the brain that is associated with consciousness. Since then, electrical stimulation and a study of epileptic patterns in general have helped us to distinguish three integrative mechanisms. Each has a major area or nucleus of gray matter (within the higher brain-stem), an aggregation of nerve cells that may be activated or paralyzed.

Highest Brain-Mechanism
The function of this gray matter is to carry out the neuronal action that corresponds with action of the mind. Proof that this is a mechanism in itself depends on the facts that injury of a circumscribed area in the higher brain-stem produces invariable loss of consciousness, and that the selective epileptic discharge, which interferes with function within this gray matter, can produce the unconsciousness (seen in epileptic automatism) without paralysis of the automatic sensory-motor control mechanism located nearby.

Automatic Sensory-Motor Mechanism
The function of this gray matter is to coordinate sensory-motor activity previously programmed by the mind. This biological computer mechanism carries on automatically when the highest brain-mechanism is selectively inactivated. It causes a major convulsive attack, *grand mal*, by activation of cortical motor convolutions, when epileptic discharge occurs in its central gray matter.

The Record of Experience
The function of the central gray matter of this mechanism, as shown by electrode activation, is to recall to a conscious individual the stream of consciousness from past time.

Conscious attention seems to give to that passage of neuronal impulses permanent facilitation for subsequent passage of potentials along the neuronal connections in the same pattern. Thus, a recall engram is established. This, one may suggest, is the real secret of learning. It is effective in the establishment of experiential memory, as well as of the memory based on conditioned reflexes.

If this is so, then it was correct to assume that the record of the stream of consciousness is not in some duplicated separate memory apparatus such as that of the hippocampus apparatuses, but in the operational mechanism of the highest brain-mechanism. The hippocampal gyrus in each hemisphere is part of the mechanism of scanning and recall. Therefore, there must be a laying down in each hippocampus of some sort of duplicated key-of-access to the record of the stream of consciousness.

One can only conclude that conscious attention adds something to brain-action that would otherwise leave no record. It gives to the passage of neuronal potentials an astonishing permanence of facilitation for the later passage of current, as though a trail had been blazed through the seemingly infinite maze of neurone connections. The same principle applies to the acquisition of speech skills and the storing of non-verbal concepts. Permanent facilitation of a patterned sequence in these brain mechanisms is established only when there is a focusing of attention on the phenomenon that corresponds to it in consciousness.

If previous decision regarding the focusing of attention is made in the mind, then it is the mind that decides when the facilitating engram is to be added. One may assume that it is the highest brain-mechanism that initiates the brain action associated with that decision. One may assume, too, that the engram is simultaneously added to conditioned reflexes and to the sequential record of conscious experience.

Is there any evidence of the existence of neuronal activity within the brain that would account for what the mind does?

Before venturing to give an answer, it may be of interest to refer again to action that the mind seems to carry out independently, and then to reconsider briefly our experience with stimulation of the cortex of conscious patients and our experience of what effects are produced by epileptic discharge in various parts of the brain. This should give some clue if there is a mechanism that explains the mind.

It is what we have learned to call the mind that seems to focus attention. The mind is aware of what is going on. The mind reasons and makes new decisions. It understands. It acts as though endowed with an energy of its own. It can make decisions and put them into effect by calling upon various brain mechanisms. It does this by activating neurone-mechanisms. This, it seems, could only be brought about by expenditure of energy.

When I have caused a conscious patient to move his hand by applying an electrode to the motor cortex of one hemisphere, I have often asked him about it. Invariably his response was: 'I didn't do that. You did.' When I caused him to vocalize, he said: 'I didn't make that sound. You pulled it out of me.' When I caused the record of the stream of consciousness to run again and so presented to him the record of his past experience, he marvelled that he should be conscious of the past as well as of the present. He was astonished that it should come back to him so completely, with more detail than he could possibly recall voluntarily. He assumed at once that, somehow, the surgeon was responsible for the phenomenon, but he recognized the details as those of his own past experience. When one analyzes such a 'flashback' it is evident, as I have said above, that only those things to which he paid attention were preserved in this permanently facilitated record.

I have been alert to the importance of studying the results of electrode stimulation of the brain of a conscious man, and have recorded the results as accurately and completely as I could. The electrode can present to the patient various crude sensations. It can cause him to turn head and eyes, or to move the limbs, or to vocalize and swallow. It may recall vivid re-experience of the past, or present to him an illusion that present experience is familiar, or that the things he sees are growing large and coming near. But he remains aloof. He passes judgment on it all. He says 'things *seem* familiar,' not 'I have been through this before.' He says, 'things are growing larger,' but he does not move for fear of being run over. If the electrode moves his right hand, he does not say, 'I wanted to move it.' He may, however, reach over with the left hand and oppose his action.

There is no place in the cerebral cortex where electrical stimulation will cause a patient to believe or to decide.

There is *no* area of gray matter, as far as my experience goes, in which local epileptic discharge brings to pass wha could be called 'mind-action.'

I am forced to conclude that there is no valid evidence that either epileptic discharge or electrical stimulation can activate the mind.

Let us consider what light our positive neurophysiological evidence can throw on the nature of man's being: If there is only one fundamental element in man's being, then neuronal action within the brain must account for all the mind does. The 'indispensable sub-stratum' of consciousness is in the higher brain-stem. The highest brain-mechanism's activity seems to correspond with that of the mind. This mechanism, as it goes out of action in sleep and resumes action on waking, may switch off the mind and switch it on. It may, one can suggest, do this by supplying and by taking away the energy that might come to the mind from the brain. But to expect the highest brain-mechanism or any set of reflexes, however complicated, to carry out what the mind does, and thus perform all the functions of the mind, is quite absurd.

If that is true, what other explanation can one propose? Only that there is, in fact, a second fundamental element and a second form of energy. But, on the basis of mind and brain as two semi-independent elements, one would still be forced to assume that the mind makes its impact upon the brain through the highest brain-mechanism. The mind must act upon it. The mind must also be acted upon by the highest brain-mechanism. The mind must remember by making use of the brain's recording mechanisms. The mind is present whenever the highest brain-mechanism is functioning normally.

For my own part, after years of striving to explain the mind on the basis of brain-action alone, I have come to the conclusion that it is simpler (and far easier to be logical) if one adopts the hypothesis that our being does consist of two fundamental elements. If that is true, it could still be true that energy required comes to the mind during waking hours through the highest brain-mechanism.

It would seem that this specialized brain-mechanism switches off the power that energizes the mind each time it falls asleep. It switches on the mind when it wakens. This is the daily automatic routine to which all mammals are committed and by which the brain recovers from fatigue.

The highest brain-mechanism switches on this *semi-independent element*, which instantly takes charge during wakefulness, and switches it off in sleep. Does this seem to be an improbable explanation? It is not so improbable, to my mind, as is the alternative expectation — that the highest brain-mechanism should itself understand and reason, and direct voluntary action, and decide where attention should be turned and what the computer must learn, and record, and reveal on demand.

But in the case of either alternative, the mind has no memory of its own as far as our evidence goes. The brain, like any computer, stores what it has learned during active intervals. All of its records are instantly available to the conscious mind throughout the person's waking life, and in a distorted fashion during the dreams of the half-asleep state.

The facts and hypotheses discussed here may well be of use in many fields of specialized thinking such as religion, philosophy and psychiatry, as well as physics, chemistry, and medicine. Whether the mind is truly a separate element or whether, in some way not yet apparent, it is an expression of neuronal action, the decision must wait for further scientific evidence. We have discussed here only one body of that evidence.

Meanwhile, there is a practical problem confronting man. He must learn to control his own social evolution, and I speak, for the moment, as a concerned physician rather than as a basic scientist. This is the pressing problem of human destiny. This problem will be solved only through more fundamental understanding. Comprehension will make clear the way of wisdom.

Biophysicists might well reflect that after life appeared in the long, long story of creation, evidence of self-awareness made its first appearance along with the appearance of the complicated brain. Thus, late in the process of biological evolution, *consciousness* — a new and very different phenomenon — had presented itself. This resulted in the appearance of a new world, created by the mind of man, in which there was understanding and reason and, eventually, onrushing social evolution!

What a challenge is here for man to face, a problem no less vast than that to be glimpsed in outer space! It was a physicist, Albert Einstein, who in a moment of understanding exclaimed: 'The mystery of the world is its comprehensibility.'

I have no doubt the day will dawn when the mystery of the mind will no longer be a mystery.

In the end I conclude that there is no good evidence, in spite of new methods, such as the employment of stimulating electrodes, the study of conscious patients and the analysis of epileptic attacks, that the brain alone can carry out the work that the mind does. I conclude that it is easier to rationalize man's being on the basis of two elements than on the basis of one. But I believe that one should not pretend to draw a final scientific conclusion, in man's study of man, until the nature of the energy responsible for mind-action is discovered as, in my own opinion, it will be.

Thus, let me state again that, working as a scientist all through my life, I have proceeded on the one-element hypothesis. That is really the same as the Jacksonian alternative that Symonds and Adrian seem to have chosen, i.e., 'that activities of the highest centers and mental states are one and the same thing, or are different sides of the same thing.'

In any case, as a scientist, I reject the concept that one must be either a monist or a dualist because that suggests a 'closed mind.' No scientist should begin thus, nor carry on his work with fixed preconceptions. And yet, since a final conclusion in the field of our discussion is not likely to come before the youngest reader of this book dies, it behooves each one of us to adopt for himself a personal assumption (belief, religion), and a way of life without waiting for a final word from science on the nature of man's mind.

In ordinary conversation, the 'mind' and 'the spirit of man' are taken to be the same. I was brought up in a Christian family and I have always believed, since I first considered the matter, that there was work for me to do in the world, and that there is a grand design in which all conscious individuals play a role. Whether there is such a thing as communication between man and God and whether energy can come to the mind of a man from an outside source after his death is for each individual to decide for himself. Science has no such answers.

Penfield dealt almost exclusively with the *human* brain. It is therefore well to remember that, although he did not find any part of the brain which, when stimulated, resulted in patients doing something they felt they meant to do, the direct stimulation of the nervous systems of simple animals can elicit organized behaviour patterns.[2] The human brain is remarkably more complex than most other brains and thus the direct relationship between mind and brain that Penfield rejects may nevertheless be true elsewhere in the phylogenetic scale.

Penfield's view that consciousness involves 'a second form of energy' and that the human phenomenon contains 'two fundamental elements' is a reflection of his religious stance. He also accepts the metaphor of the brain as a computer and mind as a programme. As we shall see in later chapters, a number of neuropsychologists think this metaphor may now have developed to the point where there is no longer any need to propose such special explanations. Another point of interest in Penfield's work is the proposition that consciousness creates a particular form of memory. This idea will appear again in the reading from George Mandler in Chapter Five.

Penfield showed that there was no special area in the brain responsible for consciousness, though he did attribute responsibility to a level of integrative functions. Rather than asking *where* consciousness is in the brain, it could be asked *when* conscious events occur relative to the brain events with which they are associated. Research to date leaves open the question of how far consciousness is in step with brain events.[3]

In any case, seeking definite answers to such 'where and when' questions is premature in view of how much there is yet to discover about how the brain works. Contemporary brain science offers a picture of the brain as a complex of sub-systems, sometimes referred to as modules, with different but overlapping functions.[4] A module is defined more in functional than in anatomical terms since, depending on the specificity of the function in question, the module responsible for it might be associated with a particular part of the brain or with a number of different parts. What integrates these different brain parts is the psychological function they carry out.

Could there be a module responsible for consciousness much as there seem to be modules responsible for aspects of vision, hearing, speech and motor control? Even though we do not fully understand them by any means, we have a good idea about how these processes are likely to inter-relate, what disorders to expect if the module is damaged, what input and output connections to look for and so on. On this basis it is possible to try to locate modules in particular places in the brain by tracing connections and by identifying disorders with the sites of brain injuries. When more generalized psychological capacities such as, say, memory are in question, it becomes a harder problem to identify them with particular sites in the brain. For example, while specific memories do not appear to be stored at specific brain locations, nevertheless memory deficits can be caused by quite localized brain damage. Since the nature and function of consciousness is even more difficult to specify so it is even less likely that particular parts of the brain support it.

Accordingly, contemporary brain research treats consciousness, when it treats it at all, as a function of the interaction pattern between modules. Such a conclusion, while necessarily general in some ways, still points to research on what contribution each module has to make.[5] For example, the frontal lobe system seems to be involved in forming intentions, planning and scheduling actions[6] while the hippocampus is involved in modality-independent spatial awareness.[7] A modular approach aims to investigate the modules themselves, their interactions and thus the ways in which they contribute to human conscious experience. Observing how disorders of brain function affect different components of consciousness is a feature of this research.[8]

Taking up a theme from the previous section, we might inquire into the neuropsychological basis of *reflexive* consciousness: the capacity to make mental process mental objects in their own right and a necessary precondition to self-awareness. Brain research, taking the modular approach, is able to address this issue quite specifically. David Oakley is a neuropsychologist known for his work on the psychological functions of the cortex, and in the next reading he looks at the phylogenetic development of the brain. He examines what new levels of organization may have appeared to make the different aspects of consciousness possible. Not surprisingly, reflexivity is at the centre of his discussion.

It is common to commence discussions of this kind by claiming that there are as many definitions of awareness or consciousness as there are writers (and readers). This, of course, allows the author to propose his own definitions and to pursue his own particular interests. This author is no exception, though he would claim a particular virtue in the fact that the model of consciousness which emerges does so on the basis of ideas which were developed for other purposes, and so may have a more general heuristic value. The problem is that we all know intuitively what consciousness is and, though formal definitions never quite seem to catch its essence, our own experience of being 'conscious' quickly fills the gaps. A view expressed throughout this book is that the development of representational systems as practical internal models of the real world was a significant step forward in the design of nervous systems, and this may provide a way of resolving the problem of definition. The classification of awareness and consciousness which follows is summarized in Figure 5.1.

First, the terms 'awareness' and 'consciousness' are frequently used synonymously, but I would like to differentiate them. To the extent that an animal is capable of responding to events in its environment or within its own body it can be said to be 'aware' (from the Old English *ge* + *waer* = quite wary or heedful). Awareness thus is to be the broadest category of responsivity. Awareness in this sense may take a simple form and be based on inbuilt mechanisms and the most elementary of acquired reactions. I would then like to suggest that the emergence of neural modelling, the ability to create central representations of external events and to use them as a basis for behaviour, corresponds to the emergence of mind, and that the activity of processing information via such cognitive systems constitutes consciousness (from the Latin *conscius* = cognizant of: *cum* = with; *scire* = to know). Similar definitions of consciousness have been proposed by Jerison (1973, 1976) and Griffin (1976) and the interested reader is recommended to read their somewhat different accounts. Consciousness thus constitutes a second category of awareness.

AWARENESS				
SIMPLE AWARENESS			*CONSCIOUSNESS*	*SELF-AWARENESS*
REFLEX SYSTEMS	HOMEOSTATIC SYSTEMS	ASSOCIATION SYSTEMS	REPRESENTATIONAL SYSTEMS	RE-REPRESENTATIONAL SYSTEMS

Subdivisions of awareness and the information processing systems which underly them.

There is a third category of awareness which must be considered. Just as we are aware of events in the world around us and respond to them, we may also become aware of the products of our own consciousness. In order to avoid the clumsiness of terms like 'awareness of consciousness' I have previously suggested 'self-awareness' to refer to this awareness of our inner world (Oakley, 1979a). I shall argue later that there were practical reasons for developing a system which re-represented items from consciousness, and that this system forms the substrate of self-awareness. It is, however, the subjective aspect of self-awareness which is for humans its most striking attribute and the most difficult to deal with when we consider animals. Indeed it is only by an act of faith that we accept self-awareness in other humans. We know from our own experience that we can only report upon mental and perceptual events which enter our current window of self-awareness, and we take the introspective reports of others as describing the same subjective domain. Perhaps, as Griffin (1976) points out, we shall have to wait until the science of 'cognitive ethology' has progressed to the point at which introspective reports can be taken from animals, before we will have a full answer to the comparative question with respect to self-awareness. Griffin also notes that until such time as this is possible, all the comparative and neuro-physiological evidence of biological continuity should lead us to conclude that the mental experiences of animals are similar to our own (see also Walker, 1983). There is certainly no reason to assume that human self-awareness is a unique property of our own version of the vertebrate brain. Within the representational systems there may be developed, a representation of the individual himself or herself. The behavioural and psychological consequences of this self-representation, and in particular of its entry into self-awareness as a self-image, are considered later in this chapter.

My own view, as I indicated at the end of the previous chapter, is that the capacity to develop representations from experience is associated with the expansion of cortical systems, especially hippocampus and neocortex in mammals, and a model of the subsystems of awareness taking this into account is presented as Figure 5.2. A mammal surgically deprived of hippocampus, neocortex (neodecorticated) or both should thus still show simple awareness and so be capable of association learning, but should demonstrate a loss of representational capacities. Neodecorticated rats and rabbits have been found in fact to show habituation (Yeo and Oakley, 1983), nictitating membrane conditioning (e.g. Oakley and Russell, 1977), autoshaping (Oakley, Eames, Jacobs, Davey and Cleland, 1981), alleyway running for food reward (Oakley, 1979b) and bar-pressing on ratios of up to sixty presses per reinforcement (e.g. Oakley, 1980). In many of these studies the association learning produced by the animals without neo-cortex was more reliable and more vigorous than that of normal animals, but it was also noticeably more bound by immediate stimulus and response factors. Damage to the hippocampus, similarly, does not impair the basic processes of either Pavlovian or instrumental learning (see O'Keefe and Nadel, 1978, for a review).

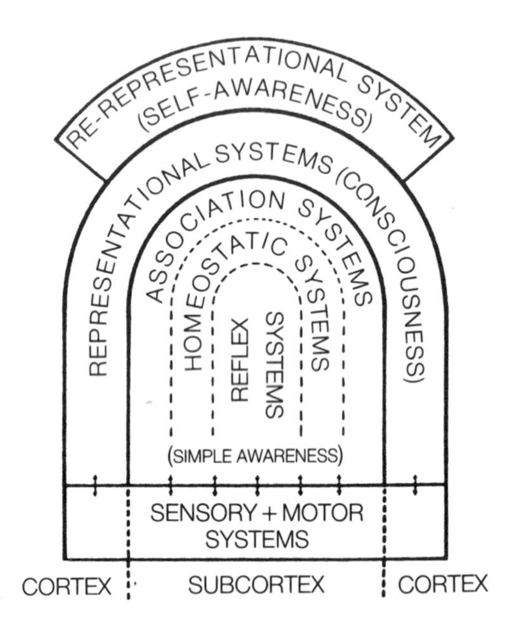

A quasi-anatomical model of awareness in vertebrate brains. Cortex refers primarily to neocortex and hippocampus in mammals and their homologues in birds and reptiles. Subcortex refers to the rest of the central nervous system, including spinal cord. Sensory and motor systems have cortical and subcortical components. Double arrows show mutual exchange of information or input/output relationships.

Abilities dependent upon representational systems, such as spatial mapping in a water maze, on the other hand, are abolished both by damage to the hippocampal system (Sutherland, Kolb and Whishaw, 1982) and by removal of neocortex (Kolb, Sutherland and Whishaw, 1983). A mammal deprived of neocortex and hippocampus can thus be said to show simple awareness but not consciousness, if the above arguments and their supporting data are accepted. It is an interesting corollary to this that it is also possible to imagine a situation in which consciousness could exist without simple awareness. Cognitive processes may perhaps continue on the basis of sensory input and stored representations provided neocortical and hippocampal systems are intact, even if reflex behavioural responsiveness and the products of association learning are absent, due possibly to diffuse subcortical damage. At the comparative level it would appear that simple awareness, in the form of the ability to form new associations in the traditional learning paradigms, is present in all animals from protozoans onwards, and is present throughout the nervous systems of those animals which have them (see Oakley, 1979a). The last point is particularly well documented in recent studies showing good evidence of instrumental learning in the isolated spinal cord in rats (Sherman, Hoeler and Buerger, 1982), and of Pavlovian conditioning of a bladder reflex in a spinal human (Ince, Brucker and Alba, 1978). In both phylogenetic and neuroanatomical comparisons, however, there is a considerable expansion in the range of behavioural reactions and associatively learned responses which are possible as central nervous systems appear and differentiate or as more rostral structures remain intact. On this basis it is possible to claim that the simple awareness shown by a paramecium is less than that of a rat, but that this is a difference of amount rather than kind.

Consciousness is qualitatively different from simple awareness in that it depends on different information processing strategies. If I am consistent I must go on to claim also that consciousness, in the form of a representational learning capacity, emerged after the evolution of the hippocampus and neocortex or their homologues, at some point in their subsequent differentiation. Taking the emergence of a differentiated neocortex as a criterion, strong candidates for consciousness would be the more differentiated reptiles, birds and mammals.

The case for attributing cognitive capacities to birds and mammals in preference to other types of vertebrates is discussed further in Walker (1983).

By argument from our own case, self-awareness reflects an ability to attend to, or become aware of, certain aspects of our own consciousness. We may become aware of at least some of the information arriving from our sensory systems, particularly of our complex visual world, of having feelings and emotions of various sorts, of entertaining and dismissing thoughts and of originating actions. It is evident also that there are limits to self-awareness and that we canot admit more than a few aspects of consciousness into our window of self-awareness at any one time.

The animal body is a very limited output system and usually only one of its range of possible responses can be executed at a given time. Equally, of the vast array of stimulus inputs and central representations available, only some will be relevant to the solution of an immediate problem. It seems reasonable to suppose that originally some small part of the brain's overall capacity, perhaps within particular cortical systems, was held in permanent reserve to handle and rework relevant information from a variety of sources and to achieve a priority access to the effector systems. This area within which selected contents of consciousness systems could be re-represented forms the basis of self-awareness.

Self-awareness, then, is being presented here as derived from a priority decision making and action system, which re-represents information from representational systems elsewhere in the brain, particularly on the basis of relevance to the task in hand. As such it will contain only a selected subset of the contents of consciousness, and it is this subset which will be reportable by the individual as forming the immediate contents of his or her mind. The subjective aspects of all this are still hard to comprehend, and it must suffice to claim that subjective self-awareness *is* the operation of the self-awareness system. I would also claim that self-awareness has high biological adaptiveness and is present to some degree whenever representational systems have been developed. It serves to ensure rapid decision making and appropriate action, on the basis of immediate data and of stored representations of the real world. On this assumption it follows that all animals which display consciousness will also display self-awareness, the nature and contents of which will depend on the types of representation which exist as the substrate of consciousness itself.

Should we continue to regard a cultivated disinterest in the possible explanatory value of subjective experience in animals as a mark of scientific respectability or is it, as Griffin (1976) suggests, an obsolete strait-jacket which limits our attempts fully to understand and explain animal behaviour? The arguments offered in this chapter, based on the biological continuity of mental capacities, would urge a greater willingness to search for appropriate evidence of subjective experience throughout the animal kingdom.

One of the tasks of representational systems is to form an accurate model of the modeller himself. A representation, that is, of the self as an actor with a range of possible relationships with the environment and with other individuals. This model may be entered into self-awareness, generating as it does so an awareness of the self as an initiator of actions, an entity in its own right with unique characteristics, with particular feelings, emotions and desires. For ourselves the sense of identity, a sense of 'me-ness', is perhaps the most powerful and influential part of the contents of self-awareness.

The evidence for the existence of a self-image in humans comes from our own introspection and from sharing that introspection with others, by means of a common language. It is not clear how we would obtain acceptable evidence of a self-image in animals. One criterion of self-image, however, is the ability to become the focus of one's own attention, and one manifestation of such an ability may be self-recognition when confronted with a mirror. In pursuit of information on self-recognition in animals, a series of studies was conducted by Gallup (1977) on a group of wild-born pre-adolescent chimpanzees. Each animal was individually housed for the experiment and given access to a full-length mirror over a period of ten days. Initially the chimpanzees, like other animals, directed social responses towards the image in the mirror. They threatened it and vocalized towards it, as they would to another chimpanzee. Within two or three days, however, these reactions were replaced by self-directed responses, and the animals were seen using their mirror image to groom themselves, remove particles of food from between their teeth and to inspect parts of their bodies which they could not otherwise see, as well as pulling faces and blowing bubbles at the mirror. On the eleventh day the chimpanzees were anaesthetized, the mirrors were removed and each animal was marked with a red, odourless, non-irritant dye above one eyebrow and on top of the opposite ear. Neither of these locations could be viewed by the chimpanzee without the use of a mirror. Once they had recovered the animals were observed closely and were seen to touch the marks very rarely. Once the mirrors were reintroduced, however, a twenty-five-fold increase in the number of touches to the marked places was seen — especially to the eyebrow. Interestingly, on a number of occasions the chimpanzees were seen touching the marked area and then looking at and licking their fingers, even though the dye was indelible. As a control, a group of chimpanzees who had no prior experience of mirrors was marked in a similar fashion. When they were introduced to a mirror for the first time they showed no particular interest in the marks on their faces. It would seem, therefore, that the first group of animals possessed an image of themselves which did not include red facial markings, and seemed at some pains thereafter to restore that image by removing their newly acquired blemishes.

The self-recognition experiment has been repeated with positive results with one other ape, the orang-utan, but, somewhat surprisingly, gorillas did not display evidence of self-recognition, though they readily groomed a marked area such as their wrist if it was visible without the mirror (Suarez and Gallup, 1981).

The self-recognition studies with mirrors appear to show that an animal may become the object of its own attention and in this sense to display a self-image. The nature of subjective experience is such that I cannot conclude with any certainty that what a chimpanzee experiences when showing self-recognition is the same as a human being in the same situation. As noted many times before, however, the same can be said of my subjective experience as compared to yours. At the moment the self-recognition data is about the closest we have come to taking an introspective report from a chimpanzee, and the answer we have received, if we choose to believe it, is consistent with the existence of a self-image and appropriate subjective accompaniments. The species differences in self-recognition may reflect differences in the underlying self-representations in the animals concerned, or they may reflect the level of cognitive skill required to detect and understand the meaning of the concordance between one's own movements and those of a reflection.

Oakley illustrates how understanding more of the brain's modular organization may help us to understand the capacities for consciousness in different species. It is less clear whether this level of analysis gives any insight into the particular quality of human consciousness, however. The reflexive capacities leading to self-awareness seem to be shared by human beings and some of the apes, yet human consciousness is clearly of a very different and more developed type. Most accounts for this difference appeal to the human capacity for language and the cultural patterns this has created.

An aspect of brain organization closely involved in the use of language is the specializations of the two cerebral hemispheres. These differences are the subject of a great deal of research on whether significantly different modes of cognition occur in the right or left hemispheres. The difference has to do, broadly speaking, with linguistic and non-linguistic processing.[9] The significance of cerebral asymmetry of function is still a controversial matter. Some psychologists choose to identify it with important distinctions between forms of consciousness such as that between the rational and the intuitive.[10] Others, while acknowledging that the two halves of the human brain have different specializations, account for it as little more than cortical division of labour.[11]

The asymmetry of the human brain is central to the position of the neuropsychologist John Eccles who is unusual in holding a dualist view of mind and brain. Eccles and Karl Popper produced *The Self and its Brain*, an extensive treatment of the mind-body issue. In a chapter entitled 'The Self-conscious Mind and the Brain', Eccles uses neuro-psychological findings as a context in which to examine the nature of mind and self.[12] What follows are excerpts from a later book by Eccles which gives a similar but more condensed treatment of his position.

It is proposed to use the term 'self-conscious mind' for the highest mental experiences. It implies knowing that one knows, which is of course initially a subjective or introspective criterion. However, by linguistic communication it can be authenticated that other human beings share in this experience of self-knowing. Dobzhansky expresses well the extraordinary emergence of human self-consciousness — of self-awareness as he calls it:

> Self-awareness is, then, one of the fundamental, possibly the most fundamental, characteristic of the human species. This characteristic is an evolutionary novelty; the biological species from which mankind has descended had only rudiments of self-awareness, or perhaps lacked it altogether. Self-awareness has, however, brought in its train somber companions — fear, anxiety, and death awareness. . . . Man is burdened by death awareness. A being who knows that he will die arose from ancestors who did not know.

This state of ultimate concern devolving from self-awareness can first be identified by the ceremonial burial customs that were inaugurated by Neanderthal man about 80,000 years ago. Karl Popper recognized the unfathomable problem of its origin: 'The emergence of all consciousness, capable of self-reflection is indeed one of the greatest of miracles.' And Konrad Lorenze refers to 'that most mysterious of barriers, utterly impenetrable to the human understanding, that runs through the middle of what is the undeniable oneness of our personality — the barrier that divides our subjective experience from the objective, verifiable physiological events that occur in our body.'

The progressive development from the consciousness of the baby to the self-consciousness in the child provides a good model for the emergent evolution of self-consciousness in the hominids. There is even evidence for a primitive knowledge of self with the chimpanzee (but not lower primates) that recognizes itself in a mirror, as shown by the use of the mirror to remove a colored mark on its face. It would seem that, in the evolutionary process, there was some primitive recognition of self long before it became traumatically experienced in death-awareness, which achieved expression in some religious beliefs manifested in the ceremonial burials. Similarly, with the child knowledge of the self usually antedates by years the first experience of death-awareness.

Each of us continually has the experience of being a person with a self-consciousness not just conscious, but knowing that we know. In defining 'person' I shall quote two admirable statements by Immanuel Kant: 'A person is a subject who is responsible for his actions' and 'A person is something that is conscious at different times of the numerical identity of its self.' These statements are minimal and basic, and they should be enormously expanded. For example Popper and Eccles have recently published a six-hundred-page book on *The Self and its Brain*. On page 144 Popper refers to 'that greatest of miracles: the human consciousness of self.'

We are not able to go much farther than Kant in defining the relations of the person to its brain. We are apt to regard the person as identical with the ensemble of face, body, limbs, and the rest that constitute each of us. It is easy to show that this is a mistake. Amputation of limbs or loss of eyes, for example, though crippling, leaves the human person with its essential identity. This is also the case with the removal of internal organs. Many can be excised in whole or in part. The human person survives unchanged after kidney transplants or even heart transplants. You may ask what happens with brain transplants. Mercifully this is not feasible surgically, but even now it would be possible successfully to accomplish a head transplant. Who can doubt that the person 'owning' the transplanted head would now 'own' the acquired body, and not vice versa! We can hope that with human persons this will remain a Gedanken experiment, but it has already been successfully done in mammals. We can recognize that all structures of the head extraneous to the brain are not involved in this transplanted ownership. For example eyes, nose, jaws, scalp, and so forth are no more concerned than are other parts of the body. So we can conclude that it is the brain and the brain alone that provides the material basis of our personhood.

But when we come to consider the brain as the seat of the conscious personhood, we can also recognize that large parts of the brain are not essential. For example removal of the cerebellum gravely incapacitates movement, but the person is not otherwise affected. It is quite different with the main part of the brain, the cerebral hemispheres. They are intimately related to the consciousness of the person, but not equally. In 95 percent of persons there is dominance of the left hemisphere, which is the 'speaking hemisphere'. Except in infants its removal results in a most severe destruction of the human person, but not annihilation. On the other hand removal of the minor hemisphere (usually the right) is attended with loss of movement on the left side (hemiplegia) and blindness on the left side (hemianopia), but the person is otherwise not gravely disturbed. Damage to other parts of the brain can also greatly disturb the human personhood, possibly by the removal of the neural inputs that normally generate the necessary background activity of the cerebral hemispheres. The most tragic example is vigil coma in which enduring deep unconsciousness is caused by damage to the midbrain.

The three-world philosophy of Popper forms the basis of our further exploration of the way in which a human baby becomes a human person. As shown in Figure 3–3, all the material world including even human brains is in the matter–energy World 1. World 2 is the world of all conscious experiences. By contrast, World 3 is the world of knowledge in the objective sense, and as such has an extremely wide range of contents. Figure 3–3 offers an abbreviated list. For example World 3 comprises the expressions of scientific, literary, and artistic ideas that have been preserved in codified form in libraries, in museums, and in all records of human culture. In their material composition of paper and ink, books are in World 1, but the knowledge encoded in the print is in World 3, and similarly for pictures, sculptures, and all other artifacts such as musical scores. Most important components of World 3 are languages for communicating thoughts and a system of values for regulating conduct and also arguments generated by discussion of these problems. In summary it can be stated that World 3 comprises the records of the intellectual efforts of all mankind through all ages up to the present — what we may call the cultural heritage.

WORLD 1	WORLD 2	WORLD 3
PHYSICAL OBJECTS AND STATES	STATES OF CONSCIOUSNESS	KNOWLEDGE IN OBJECTIVE SENSE
1. INORGANIC	Subjective knowledge	Cultural heritage
Matter and energy		coded on material
	Experience of	substrates
2. BIOLOGY	perception	philosophical
Structure and actions	thinking	theological
of all living beings	emotions	scientific
–human brains	dispositional intentions	historical
	memories	literary
3. ARTIFACTS	dreams	artistic
Material substrates	creative imagination	technological
of human creativity		
of tools		Theoretical systems
of machines		scientific problems
of books		critical arguments
of works of art		
of music		

FIGURE 3–3. *Tabular Representation of the Contents of the Three Worlds in Accordance with the Philosophy of Karl Popper. These three worlds are nonoverlapping but are intimately related, as indicated by the large open arrows at the top. They contain everything in existence and in experience. World 1 is material, Worlds 2 and 3 are immaterial.*

At birth the human baby has a human brain, but its World 2 experiences are quite rudimentary, and World 3 is unknown to it. It, and even a human embryo, must be regarded as human beings, but not as human persons. The emergence and development of self-consciousness (World 2) by continued interaction with World 3, the world of culture, is an utterly mysterious process. It can be likened to a double structure (Figure 3–4) that ascends and grows by effective cross-linkage. The vertical arrow shows the passage of time from the earliest experiences of the child up to full human development. From each World 2 position an arrow leads through to World 3 at that level up to a higher, larger level, which illustrates symbolically a growth in the culture of that individual. Reciprocally the World 3 resources of the self act back to give a higher, expanded level of consciousness of that self (World 2). Figure 3–4 can be regarded symbolically as the ladder of personhood. And so each of us has developed progressively in self-creation, and this can go on throughout our whole lifetime. The more the World 3 resources of the human person, the more does it gain in the self-consciousness of World 2 by reciprocal enrichment. What we are is dependent on the World 3 that we have been immersed in and how effectively we have utilized our opportunities to make the most of our brain potentialities.

It may seem that a complete explanation of the development of the human

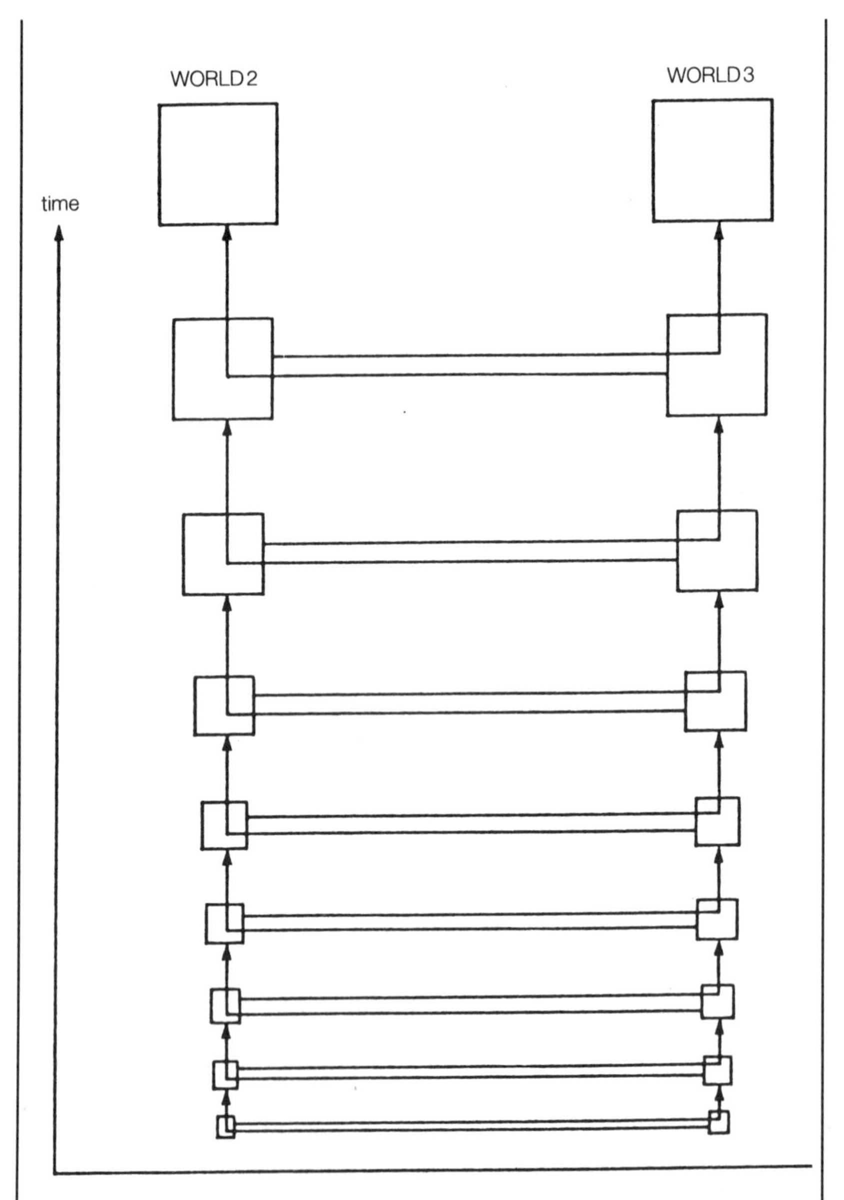

FIGURE 3–4. *Diagrammatic Representation of the Postulated Inter-relationships in the Developments of Self-consciousness (World 2) and of Culture (World 3) of a Person in Time. Development is shown by the arrows; full description in text. We may call it the ladder of personhood that we can climb up throughout life.*

person can be given in terms of the human brain. It is built anatomically by genetic instructions and subsequently developed functionally by learning from the environmental influences. A purely materialist explanation would seem to suffice with the conscious experiences as derivative from brain functioning. However, it is a mistake to think that the brain does everything and that our conscious experiences are simply a reflection of brain activities, which is a common philosophical view. If that were so, our conscious selves would be no more than passive spectators of the performances carried out by the neuronal mechanisms of the brain. Our beliefs that we can really make decisions and that we have some control over our actions would be nothing but illusions. There are, of course, all sorts of subtle cover-ups by philosophers from such a stark exposition, but they do not face up to the issue. In fact all people, even materialist philosophers, behave as if they had at least some responsibility for their own actions. It seems that their philosophy is for 'the other people, not for themselves,' as Schopenhauer wittily stated.

These considerations lead me to the alternative hypothesis of dualist-interactionism, which has been expanded at length in *The Self and its Brain*. It is really the commonsense view, namely, that we are a combination of two things or entities: our brains on the one hand; and our conscious selves on the other. The self is central to the totality of our conscious experiences as persons through our whole waking life. We link it in memory from our earliest conscious experiences. The self has a subconscious existence during sleep, except for dreams, and on waking the conscious self is resumed and linked with the past by the continuity of memory. But for memory we as experiencing persons would not exist. We have the extraordinary problem that was first recognized by Descartes: How can the conscious mind and the brain interact?

The theories of the brain-mind relationship that are today held by most philosophers and neuroscientists are purely materialistic in the sense that the brain is given complete mastery! The existence of mind or consciousness is not denied except by radical materialists but it is relegated to the passive role of mental experiences accompanying some types of brain action, as in epiphenomenalism and in psychoneural identity, but with absolutely no *effective* action on the brain. The complex neural mechanism of the brain functions in its determined materialistic fashion regardless of any consciousness that may accompany it. The 'commonsense' experiences that we can control our actions to some extent or that indicate we can express our thoughts in language are alleged to be illusory. An *effective causality* is denied to the self-conscious mind *per se* despite all the protests of the materialists to the contrary!

In contrast to these materialist or parallelist theories are the *dualist-interaction* theories. The essential feature of these theories is that mind and brain are independent entities, the brain being in World 1 and the mind in World 2, and that they somehow interact. Thus there is a frontier and across this frontier there is interaction in both directions, which can be conceived as a flow of information, not of energy. Thus we have the extraordinary doctrine that the world of matter–energy (World 1) is not completely sealed, which is a fundamental tenet of physics, but that there are small 'apertures' in what is otherwise the completely closed World 1. On the contrary, the closedness of World 1 has been safeguarded with great ingenuity in all materialistic theories of the mind. We shall later argue that this is not their strength but instead their fatal weakness.

A great point is made by all varieties of materialists that their brain–mind theories are in accord with natural law as it now stands. However, this claim is invalidated by two most weighty considerations.

First, nowhere in the laws of physics or in the laws of the derivative sciences, chemistry and biology, is there any reference to consciousness or mind. Regardless of the complexity of electrical, chemical, or biological machinery, there is no statement in the 'natural laws' that there is an emergence of this strange nonmaterial entity, consciousness or mind. This is not to affirm that consciousness does not emerge in the evolutionary process, but merely to state that its emergence is not reconcilable with the natural laws as at present understood. For example such laws do not allow any statement that consciousness emerges at a specified level of complexity of systems, which is gratuitously assumed by all materialists except radical materialists and panpsychists. The panpsychist belief that some primordial consciousness attaches to all matter, presumably even to atoms and subatomic particles, finds no support whatsoever in physics. One can also recall the poignant questions by computer lovers: At what stage of complexity and performance can we agree to endow them with consciousness? Mercifully this emotionally charged question need not be answered. You can do what you like to computers without qualms of being cruel!

Second, all materialist theories of the mind are in conflict with biological evolution. Since they all (panpsychists, epiphenomenalists, and identity theorists) assert the causal ineffectiveness of consciousness *per se* they fail completely to account for the evolutionary expansion of consciousness, which is an undeniable fact. There is first its emergence and then its progressive development with the growing complexity of the brain. Evolutionary theory holds that only those structures and processes that significantly aid in survival are developed in natural selection. If consciousness is causally impotent, its development cannot be accounted for by evolutionary theory. According to biological evolution, mental states and consciousness could have evolved and developed *only if they were causally effective* in bringing about changes in neural happenings in the brain with the consequent changes in behavior. That can occur only if the neural machinery of the brain is open to influences from the mental events of the world of conscious experiences, which is the basic postulate of dualist–interactionist theory.

Finally, the most telling criticism of all materialist theories of the mind is against its key postulate that the happenings in the neural machinery of the brain provide *a necessary and sufficient explanation of the totality both of the performance and of the conscious experience of a human being.*

For example the willing of a voluntary movement is regarded as being *completely determined* by events in the neural machinery of the brain, as also are all other cognitive experiences. But as Popper states in his Compton Lecture:

> According to determinism, any such theory such as say determinism is held because of a certain physical structure of the holder — perhaps of his brain. Accordingly, we are deceiving ourselves and are physically so determined as to deceive ourselves whenever we believe that there are such things as *arguments or reasons* which make us accept determinism. Purely physical conditions, including our physical environment make us say or accept whatever we say or accept.

This is an effective *reductio ad absurdum*. This stricture applies to all of the materialist theories. So perforce we turn to dualist–interactionist explanations of the brain–mind problem, despite the extraordinary requirement that there be effective communication in both directions.

Necessarily the dualist–interactionist theory is in conflict with present natural laws and so is in the same 'unlawful' position as the materialist theories of the mind. The differences are that this conflict has always been admitted and that the neural machinery of the brain is assumed to operate in strict accordance to natural laws except for its openness to World 2 influences.

Moreover, as Popper stated, the interaction across the frontier need not be in conflict with the first law of thermodynamics. The flow of information into the modules could be effected by a balanced increase and decrease of energy at different but adjacent micro-sites, so that there is no net energy change in the brain. The first law at this level may be valid only statistically.

A key component of the hypothesis of brain–mind interaction is that the unity of conscious experience is provided by the self-conscious mind and not by the neuronal mechanism of the neocortex. Hitherto it has been impossible to develop any theory of brain function that would explain how the immense diversity of brain events comes to be synthesized so that there is a unity of conscious experience.

It is a universal human experience that subjectively there is a mental unity recognized as a continuity from one's earliest memories. It is the basis of the concept of the self. Experimental investigations on the unity of the self have been discussed in the book *The Human Psyche* (Eccles, 1980).

By far the most important experimental evidence relating to the unity of consciousness comes from the study by Roger Sperry and his associates on commissurotomized patients. In the operation for the relief of intractable epilepsy there was a section of the corpus callosum, the great tract of nerve fibers, about 200 million, that links the two cerebral hemispheres. With the most sophisticated investigations, allowing up to two hours of continual testing, it became clear that the right hemisphere, the so-called minor hemisphere, was correlated with conscious responses at a level superior to those exhibited by any nonhuman primates. The patient's consciousness was indubitable. The perplexing question is whether the right hemisphere mediates self-consciousness, meaning that it permits the knowledge of selfhood. In the most searching investigations of Sperry and associates there was testing of the ability of the patient to identify photographs projected to the right hemisphere alone. A considerable ability was displayed, but it was handicapped by the lack of verbal expression.

The tests for the existence of self-consciousness were at a relatively simple pictorial and emotional level. We can doubt if the right hemisphere with associated consciousness has a full self-conscious existence. For example, do planning and worrying about the future take place there? Are there decisions and judgments based on some value system? These are essential qualifications for personhood as ordinarily understood and for the existence of a psyche or soul. It can be concluded that a limited self-consciousness is associated with the right hemisphere, but the person remains apparently unscathed by the commissurotomy with mental unity intact in its now exclusive left hemisphere association. After commissurotomy the right hemisphere appears to mediate a self-awareness resembling that of a very young child. There would be a small central core at a primitive level of self or ego, but with no representation of soul or psyche or personhood. It is generally agreed that the human person is not split by the commissurotomy but remains in liaison with the left (speaking) hemisphere.

It is not in doubt that each human person recognizes its own uniqueness, and this is accepted as the basis of social life and of law. When we inquire into the grounds for this belief, modern neuroscience eliminates an explanation in terms of the body. There remain two possible alternatives, the brain and the psyche. Materialists must subscribe to the former, but dualist–interactionists have to regard the self of World 2 as the entity with the experienced uniqueness. It is important to disclaim a solipsistic solution of the uniqueness of the self. Our direct experiences are of course subjective, being derived entirely from our brain and self. The existence of other selves is established by intersubjective communication.

Since materialist solutions fail to account for our experienced uniqueness, we are constrained to attribute the uniqueness of the psyche or soul to a supernatural spiritual creation. To give the explanation in theological terms: Each soul is a Divine creation, which is 'attached' to the growing fetus at some time between conception and birth. It is the certainty of the inner core of unique individuality that necessitates the 'Divine creation.' We submit that no other explanation is tenable; neither the genetic uniqueness with its fantastically impossible lottery nor the environmental differentiations, which do not *determine* one's uniqueness but merely modify it.

The crucial issue is whether Eccles gives any fundamentally new evidence for how a non-material mind is nonetheless able to intervene in the material affairs of the brain. If so, dualism has been put on a new footing; if not, Eccles' treatment of dualism is more an affirmation than a demonstration. Although Eccles cites Roger Sperry, a neuropsychologist whose work earned him a Nobel Prize in 1981, the following excerpt shows Sperry supporting Eccles' position only partially.

Are the things we hold most sacred, including the human psyche and forces of creation, best conceived in dualistic, otherworldly terms? Historically tied to interpretations of the conscious mind, the case for dualist forms of existence was effectively countered and held down by materialist science to a negligible status for many decades. Dualist concepts made a notable comeback during the 1970s, however, and again today receive vigorous support from some authorities as a viable answer to the mind–brain problem. The present critique, addressed largely to specialists, questions the logical foundations of the new dualist position.

When two eminent authorities of science and philosophy, of the stature and influence of Sir John Eccles and Sir Karl Popper, join forces to affirm dualistic beliefs in the reality of the supernatural and the existence of extraphysical, unembodied agents to challenge some of the most fundamental precepts of science, one is impelled to take more than passing notice. Regardless of one's personal convictions and reactions, the kind of public message that is conveyed, directly and indirectly, by their book *The Self and Its Brain: An Argument for Interactionism* (1977) along with Eccles' more recent volume *The Human Mystery* (1979), and the potential impact of these on the intellectual perspectives of our times become a matter of some concern. Such considerations, and the fact that my own views and writings are cited in support of some of the key concepts and as being in alighment with dualist interactionism, prompt this effort to clarify certain points that otherwise leave erroneous impressions.

Before I attempt to focus on specifics, it will help to mention broadly that whereas Sir John Eccles and I have similar outlooks with many highly congenial perspectives, aims, and values, we do, however, share certain friendly differences in regard to the nature and locus of consciousness and the support of dualism. I have always favored monism, and still do. Sir John tells me that I am a dualist and I respond: Only if the term is redefined to take on a new meaning quite different from what it traditionally has stood for in philosophy. Dualism and monism have long represented a dichotomy that offers opposing answers to one of man's most critical and enduring concerns, namely, can conscious experience exist apart from the brain? Dualism, affirming the existence of independent mental and physical worlds, says yes and opens the door to a conscious afterlife and to many kinds of supernatural, paranormal, and otherworldly beliefs. Monism, on the other hand, restricts its answers to one-world dimensions and says no to an independent existence of conscious mind apart from the functioning brain.

In recent years there has arisen some real need to change and sharpen definitions of certain philosophic terms to fit our new views in neuroscience. However, in the case of monism and dualism, I see no advantage in changing the classic definitions. We greatly need terms by which to distinguish the critical dichotomy regarding the potential separability of brain and conscious experience during life as well as after. Dualism and monism have long served this need in the past and seem best qualified to continue.

At the same time I am in strong agreement with Eccles in rejecting both materialism (or physicalism) and reductionism — or at least what these terms predominantly stood for prior to the mid-sixties. Since 1965 I have referred to myself as a mentalist and since the mid-thirties have firmly renounced reductionism in the philosophic, 'nothing but' sense to be explained below. However, in the case of the terms mentalism and the opposing materialism, and the form of dichotomy these two imply, some change and sharpening of definitions is now called for by our modified mind–brain concepts. On our new terms, which I will outline below, mentalism is no longer synonymous with dualism nor is physicalism the equivalent of monism. By our current mind–brain theory, monism has to include subjective mental properties as causal realities. This is not the case with physicalism or materialism which are the understood antitheses of mentalism, and have traditionally excluded mental phenomena as causal realities. In calling myself a mentalist, I hold subjective mental phenomena to be primary, causally potent realities as they are experienced subjectively, different from, more than, and not reducible to their physicochemical elements. At the same time, I define this position and the mind–brain theory on which it is based as monistic and see it as a major deterrent to dualism.

A first question that needs to be considered is whether the set of concepts which Eccles currently uses to support dualism (Karl Popper's arguments will be discussed separately) is significantly different from that which I proposed as an anti-dualist, monist solution. Have we independently come on two different answers for mind–brain interaction, or is it a matter of different interpretations of basically the same solution? So far as I am able to determine, the underlying concepts by which psychophysical interaction is inferred by Eccles do not differ in any relevant respect from those which I have presented as mentalist monism. In searching the arguments and evidence advanced by Eccles one finds much the same reasoning that I have used to support my own concept of consciousness. The phraseology and emphasis are somewhat different, and some different neural examples of the principles are introduced, but the conceptual model for mind–brain interaction that is inferred seems entirely consistent and certainly no distinct alternative is offered.

Eccles emphasizes with italics that 'a key component of the hypothesis is that unity of conscious experience is provided by the mind and not by the neural machinery,' and this point is again stressed in Dialogue VIII, p. 512, and again in his Gifford Lectures. Here we are in full accord. I too had made precisely the same point in 1952, stating: 'In the scheme proposed here, it is contended that unity in subjective experience does not derive from any kind of parallel unity in the brain processes. Conscious unity is conceived rather as a functional or operational derivative,' and 'there need be little or nothing of a unitary nature about the physiological processes themselves.' In his earlier thinking Eccles had given priority to quite a different concept, expressed in terms of extra-physical 'ghostly influences' affecting the course of synaptic events. I have since referred to and consistently reiterated the above explanation of mental unity in reference to the role of the cerebral commissures and to the 'graininess' problem, emphasizing that the subjective unity does not correlate with the array of excitatory details comprising the infrastructure of the brain process but rather with the holistic mental properties.

In a reflective appraisal near the end of the their volume, Eccles observes, 'As we have developed our hypothesis, we have returned to the views of past philosophies that the mental phenomena are now ascendant again over the material phenomena.' Similarly, I too from the start have described the hypothesis as one that 'puts mind back over matter' and 'would restore mind to its old prestigious position over matter.' That our key concepts for this and for mind–brain interaction in general are essentially one and the same is further indicated where Eccles ends the condensed summary of his hypothesis with the statement: 'Sperry has made a similar proposal (Sperry 1969)' and in another 'very brief summary or outline of the theory' concludes: 'Thus, in agreement with Sperry, it is postulated that the self-conscious mind exercises a superior interpretative and controlling role upon the neural events.'

When we turn to the solution of the mind–brain problem upheld by Sir Karl Popper, we find it is also basically the same, but the history of its acquisition is quite different. Prior to 1965, Popper's support of dualism rested mainly on the argument that no causal physical theory of the descriptive, argumentative functions, of language is possible. Products of the mind, like myths, abstractions, and mathematical formulas cannot be accounted for by the laws of physiology or physics. During the years in which this argument was propounded, it failed by itself to have much influence in countering physicalist objections that products of the mind have neural correlates and that the products of the mind, like other mental entities, were better interpreted in parallelistic terms as being epiphenomena, inner aspects of, or identical to their neurological correlates. As expressed by Oppenheimer and Putnam:

> It is not sufficient, for example, simply to advance the claim that certain phenomena considered to be specifically human, such as the use of verbal language in an abstract and generalized way, can never be explained on the basis of neurophysiological theory, or to make the claim that this conceptual capacity distinguishes man in principle and not only in degree from nonhuman animals.

In 1965, Popper proposed a new solution to the mind–brain relation that was exactly what he had been looking for in his earlier arguments and which has since become a major theme of his philosophy. In a lecture devoted first to a discussion of physical indeterminism, and in a departure from his prior long-time concerns with the logic of knowing, Popper added a second theme concerning some revised perspectives on evolution which he then extended to include the body–mind problem. He emerged with what seems to be basically the same view of evolution and the mind–brain relation that I too had proposed a year earlier in a James Arthur Lecture. In essence, the idea of emerging hierarchic controls is applied to the mind–brain relation. This 1965 switch in Popper's philosophy from a position in which evolutionary theory was held to be tautological, explaining almost nothing, to one in which it explains almost everything was offered with 'many apologies,' as a development for which he was obliged 'to eat humble pie.' In line with the main theme of his lecture, a plastic indeterminacy of the emergent controls was emphasized, but the degree of looseness or tightness in the controls is not a critical part of the argument.

Because these concepts concerning hierarchic organization and downward control are crucial both to the Popper and Eccles volume and to the present chapter, I restate them with exact quotes:

> Evolution keeps complicating the universe by adding new phenomena that have new properties and new forces and that are regulated by new scientific principles and new scientific laws — all for future scientists in their respective disciplines to discover and formulate. Note also that the old simple laws and primeval forces of the hydrogen age never get lost or canceled in the process of compounding the compounds. They do, however, get superseded, overwhelmed, and outclassed by the higher-level forces as these successively appear at the atomic, the molecular and the cellular and higher levels . . .
>
> . . . recall that a molecule in many respects is the master of its inner atoms and electrons. The latter are hauled and forced about in chemical interactions by the overall configurational properties of the whole molecule. At the same time, if our given molecule is itself part of a single-celled organism such as paramecium, it in turn is obliged, with all its parts and its partners, to follow along a trail of events in time and space determined largely by the extrinsic overall dynamics of *Paramecium caudatum*. When it comes to brains, remember that the simpler electric, atomic, molecular, and cellular forces and laws, though still present and operating, have been superseded by the configurational forces of higher-level mechanisms. At the top, in the human brain, these include the powers of perception, cognition, reason, judgment, and the like, the operational, causal effects and forces of which are equally or more potent in brain dynamics than are the outclassed inner chemical forces.

Note that this statement includes the basic key concepts on which the Popper and Eccles case for mind–brain interaction mainly rests, i.e., the downward causal control influence of higher emergent (mental) over lower (neural) entities, and the fact that the mental and neural events are different kinds of phenomena regulated by different kinds of laws and forces.

New developments in the mind–brain identity position, the recent 'consciousness' movement in clinical and humanistic psychology, and the counterculture developments of the 1960s have all be chronologically and otherwise associated, but also similarly fail to furnish any critical reasoning that would distinguish between the causal efficacy of consciousness and that of its neural correlates, or to otherwise refute, so far as science is concerned, the long dominant materialist–behaviorist paradigm. The one development that does this and presents a logical and plausible alternative, is the modified concept of mind as a causal, functional emergent.

It is the idea, in brief, that conscious phenomena as emergent functional properties of brain processing exert an active control role as causal determinants in shaping the flow patterns of cerebral excitation. Once generated from neural events, the higher order mental patterns and programs have their own subjective qualities and progress, operate and interact by their own causal laws and principles which are different from, and cannot be reduced to those of neurophysiology, as explained further below. Compared to the physiological processes, the conscious events are more molar, being determined by configurational or organizational interrelations in neuronal functions. The mental entities transcend the physiological just as the physiological transcend the molecular, the molecular, the atomic and subatomic, etc. The mental forces do not violate, disturb, or intervene in neuronal activity but they do supervene. Interaction is mutually reciprocal between the neural and mental levels in the nested brain hierarchies. Multilevel and interlevel causation is emphasized in addition to the one-level sequential causation more traditionally dealt with. This idea is very different from those of extra-physical ghostly intervention at synapses and of indeterministic influences on which Eccles and Popper had earlier relied. The question at issue is whether this form of psychophysical interaction is fundamentally monistic as I interpret it or whether it is dualistic as presented by Popper and Eccles.

As I interpret it, this concept of the mind–brain relation not only refutes the doctrines of behaviorism and materialism, mechanistic determinism and reductionism, as Popper and Eccles correctly infer, but also and with equal force, strongly discounts dualism. By explaining conscious experience in monistic terms we undermine dualism at its source and point of strongest support, leaving for dualism only abstract arguments like those of Plato and Popper, and observations like those from parapsychology.

It will be helpful as we proceed to have in mind some further concrete examples of the principles of emergent (holist) control as illustrated at different levels in some simpler and more familiar physical systems. I have used the example of how a wheel rolling downhill carries its atoms and molecules through a course of time and space and to a fate determined by the overall system properties of the wheel as a whole and regardless of the inclination of the individual atoms and molecules. The atoms and molecules are caught up and overpowered by the higher properties of the whole. One can compare the rolling wheel to an ongoing brain process or a progressing train of thought in which the overall properties of the brain process, as a coherent organizational entity, determine the timing and spacing of the firing patterns within its neuronal infrastructure. The

> control works both ways; hence, mind–brain 'interaction'. The subsystem conponents determine collectively the properties of the whole at each level and these in turn determine the time–space course and other relational properties of the components. The organism and its component cells and organs is another familiar example. The principles are universal. Along with the failure to qualify as dualism by definition, our proposed mind–brain model also is nondualistic in that it makes mind and brain inseparable parts of the same continuous hierarchy, the great bulk of which, by common agreement, is not dualistic. It becomes illogical to make a special exception of the principle at the one level of mind and not at those above and below. In the proposed scheme, one can proceed continuously in the same universe of discourse, following the path of evolution, from subatomic elements in the brain up through molecules, cells and nerve circuits to brain processes lacking or having conscious properties and on upward through higher compounds all within the one this-world mode of existence.
>
> The concept of consciousness as a causal emergent has been presented from the outset as a view that restores to science the common sense impression (overruled during the behaviorist–materialist era) that we do indeed have a mind and mental faculties over and above, and different from our brain physiology — just as we have cellular properties that are over and above and different from their molecular constituents.
>
> Of all the questions one can ask about conscious experience, there is none for which the answer has more profound and far-ranging implications than the question of whether or not consciousness is causal. The alternative answers lead to basically different paradigms for science, philosophy, and culture in general.
>
> When Popper and Eccles, representing modern philosophy and neuroscience, jointly proclaim arguments and beliefs in dualism, the supernatural, and unembodied worlds of existence, the repercussions quickly extend beyond professional borders to influence attitudes and faith–belief systems in society at large. The result has been a major setback for those of us who see hope for the future, and for the very aims and ideals that I think Popper and Eccles strive for, to lie in replacing old dualist perspectives, values and beliefs, dualist theologies and related mythological, supernatural guidelines of the past with a new unifying holistic–monistic interpretation of reality as an ultimate reference frame for transcendant value and higher meaning.

Sperry's view on the mind–brain relationship is that consciousness is causal in the workings of the brain. As he crisply puts it, 'mind moves matter in the brain', but he resists allowing mind any existence apart from the brain. For Sperry, mind is the pattern of activity in the brain which has its own causal dynamics. These dynamics are distinct from the electro-chemical brain events which support them since they are able to act back or 'supervene' in these events. For Eccles, the contemporary state of brain science is compatible with the idea of mind as soul-like; for Sperry it is not. This difference looks difficult to resolve.

Neuropsychology provides a powerful vocabulary for discussing the relationship of mind and brain though not yet a complete one. Taken together, the readings have suggested how the brain and the mental operations it makes possible support the human capacity for reflexive consciousness. However, this capacity cannot be accounted for by the neuropsychological organization of the brain alone. Reflexive consciousness, created as it is in the interaction between the biological structure of the human brain and the cultural context provided by human society, is also a psychological and social phenomenon. Accordingly, we will move on to consider consciousness as a psychological issue in the next chapter.

Notes

1. Jackson, H. J., *The Selected Writings of John Hughlings Jackson*, 1932, J. Taylor, ed., Hodder & Stoughton.
2. Kandel, E., 'Small Systems of Neurones', *Scientific American*, September, 1979.
3. The work of Bejamin Libet is most often cited here. For a review see pp. 193–202 in E. Harth, *Windows on the Mind*, 1985, Pelican.
4. Nauta, W. and Feirtag, M., 'The Organization of the Brain', *Scientific American*, September, 1979; A. Ellis and A. Young, *Human Cognitive Neuropsychology*, 1988, Lawrence Erlbaum Associates.
5. Oakley, D. and Eames, L., 'The Plurality of Consciousness' in D. Oakley, ed., *Brain and Mind*, 1985, Methuen.
6. Luria, A. R., 'The Functional Organization of the Brain', *Scientific American*, March, 1970.
7. O'Keefe, J. and Nadel, L., *The Hippocampus as a Cognitive Map*, 1979, Clarendon Press.
8. Newcombe, F., 'Neuropsycholog of Consciousness', in D. Oakley, ed., *Brain and Mind*, op.cit.
9. Springer, S. and Deutsch, G., *Left Brain, Right Brain*, 1985.
10. Ornstein, R., *The Psychology of Consciousness*, 1977, Harcourt Brace Jovanovitch.
11. Levy, J., 'Psychobiological Implications of Bilateral Asymmetry' in S. Dimond and J. Beaumont, eds., *Hemispheric Function and the Human Brain*, 1974, Elek.

12. The chapter in question from Popper and Eccles' *The Self and its Brain* is E8. In the dialogues which form the last part of the book, Popper and Eccles discuss their many differences; dialogues 2, 5 and 7 are particularly relevant.

5

The Psychology of Consciousness

READINGS

Psychology's treatment of consciousness has acted as an historical index of its willingness to take mental life seriously. In times when the task of psychology was to explain the richness of mental life, then consciousness was prominent. But when there was a shift to explaining *away* mentalistic concepts, consciousness was the first to go. In this chapter we look very briefly at some of this history and then concentrate on the contemporary treatment of consciousness in which it has returned to the centre of the psychological stage after a considerable absence.[1] This return may reflect the confidence felt in the theoretical and experimental resources of what is now the most widely accepted approach to the mind, cognitive science.

During the latter part of the nineteenth century, which saw the emergence of psychology as a distinct discipline, the aim was to place the study of the mind on as objective a footing as the other sciences of the time had placed the study of the material world. Nonetheless consciousness was taken to be the fundamental datum. Early psychologists as different as the experimentally-inclined Fechner and Wundt were from the phenomenologically-inclined Brentano could probably have agreed that the study of subjective experience posed no problem of principle even though their methods for going about it were very different. Likewise, although later functionalists like Angell and James argued with Titchener and others about the role of mental life and the proper means to study it, the assumption of a subjective mental life was a shared and obvious focus which made consciousness psychology's natural primary topic.[2]

The treatment of consciousness around the turn of the century involved, among other thngs, experimental studies of sensation, perception and attention, introspective techniques, clinical observations, phenomenological and philosophical analyses. Among the best known treatments of consciousness of the time is William James' *The Principles of Psychology* which remains a rich source of observation and interpretation. At various times James identified consciousness with thoughts, emotions, spiritual feelings and choice but never identified any one of these as being of the essence. In keeping with his earlier use of the stream metaphors, and recalling Whitehead's views from Chapter One, his later writings warned against identifying consciousness as any particular *thing* when in fact it was a *process*. For James, with his important place in American functionalism, his consciousness-as-process view drew attention away from particular contents of experience and towards its role in organizing adaptive behaviour. Now, however, it is possible to be more precise in what we mean by the function or functions of consciousness.

We now begin our study of the mind from within. Most books start with sensations, as the simplest mental facts, and proceed synthetically, constructing each higher stage from those below it. But this is abandoning the empirical method of investigation. No one ever had a simple sensation by itself. Consciousness, from our natal day, is of a teeming multiplicity of objects and relations, and what we call simple sensations are results of discriminative attention, pushed often to a very high degree. It is astonishing what havoc is wrought in psychology by admitting at the outset apparently innocent suppositions, that nevertheless contain a flaw. The bad consequences develop themselves later on, and are irremediable, being woven through the whole texture of the work. The notion that sensations, being the simplest things, are the first things to take up in psychology is one of these suppositions. The only thing which psychology has a right to postulate at the outset is the fact of thinking itself, and that must first be taken up and analyzed. If sensations then prove to be amongst the elements of the thinking, we shall be no worse off as respects them than if we had taken them for granted at the start.

The first fact for us, then, as psychologists, is that thinking of some sort goes on. I use the word thinking for every form of consciousness indiscriminately. If we could say in English 'it thinks,' as we say 'it rains' or 'it blows,' we should be stating the fact most simply and with the minimum of assumption. As we cannot, we must simply say that *thought goes on*.

How does it go on? We notice immediately five important characters in the process, of which it shall be the duty of the present chapter to treat in a general way:

1) Every thought tends to be part of a personal consciousness.
2) Within each personal consciousness thought is always changing.
3) Within each personal consciousness thought is sensibly continuous.
4) It always appears to deal with objects independent of itself.
5) It is interested in some parts of these objects to the exclusion of others, and welcomes or rejects — *chooses* from among them, in a word — all the while.

In considering these five points successively, we shall have to plunge *in medias res* as regards our vocabulary, and use psychological terms which can only be adequately defined in later chapters of the book. But every one knows what the terms mean in a rough way; and it is only in a rough way that we are now to take them.

I have already said that the breach from one mind to another is perhaps the greatest breach in nature. The only breaches that can well be conceived to occur within the limits of a single mind would either be *interruptions, time*-gaps during which the consciousness went out altogether to come into existence again at a later moment; or they would be breaks in the *quality* or content, of the thought, so abrupt that the segment that followed had no connection whatever with the one that went before. The proposition that within each personal consciousness thought feels continuous, means two things:

1. That even where there is a time-gap the consciousness after it feels as if it belonged together with the consciousness before it, as another part of the same self;
2. That the changes from one moment to another in the quality of the consciousness are never absolutely abrupt.

Consciousness, then, does not appear to itself chopped up in bits. Such words as 'chain' or 'train' do not describe it fitly as it presents itself in the first instance. It is nothing jointed; it flows. A 'river' or a 'stream' are the metaphors by which it is most naturally described. *In talking of it hereafter, let us call it the stream of thought, of consciousness, or of subjective life.*

Looking back we see that the mind is at every stage a theatre of simultaneous possibilities. Consciousness consists in the comparison of these with each other, the selection of some, and the suppression of the rest by the reinforcing and inhibiting agency of attention. The highest and most elaborated mental products are filtered from the data chosen by the faculty next beneath, out of the mass offered by the faculty below that, which mass in turn was sifted from a still larger amount of yet simpler material, and so on. The mind, in short, works on the data it receives very much as a sculptor works on his block of stone. In a sense the statue stood there from eternity. But there were a thousand different ones beside it, and the sculptor alone is to thank for having extricated this one from the rest. Just so the world of each of us, howsoever different our several views of it may be, all lay embedded in the primordial chaos of sensations, which gave the mere *matter* to the thought of all of us indifferently. We may, if we like, by our reasonings unwind things back to that black and jointless continuity of space and moving clouds of swarming atoms which science calls the only real world. But all the while the world *we* feel and live in will be that which our ancestors and we, by slowly cumulative strokes of choice, have extricated out of this, like sculptors, by simply rejecting certain portions of the given stuff. Other sculptors, other statues from the same stone! Other minds, other worlds from the same monotonous and inexpressive chaos! My world is but one million alike embedded, alike real to those who may abstract them. How different must be the worlds in the consciousness of ant, cuttle-fish, or crab!

Some of the techniques used by the psychologists of James' period, particularly introspection, gave rise to serious difficulties. Using reported experiences as data lead to irresolvable differences, no matter how well trained were the people from whom these reports came. Perhaps as a reaction, psychology between 1910 and 1920 began to take a more objective and anti-mentalistic line on both methods and subject matter. Not so very long after this we find Watson dismissing consciousness as unnecessary mentalistic baggage and, much later, Skinner sets psychology the agenda of relating behaviour to external rather than internal events. Consciousness became, along with most of mental life, almost a non-topic.[3]

While Watson and Skinner both recommended that consciousness be taken off the psychological agenda, they did so for different reasons. Watson believed that all mentalistic concepts would be abandoned when research showed that what they 'explained' was perfectly explicable without them. Skinner, on the other hand, while equally committed to avoiding mentalism, did so on more pragmatic grounds.[4] His view was that if psychology attempted to deal with mental concepts that were both hard to define and impossible to observe, it would stagnate. So, apart from any principled stand on the reality of mental life, neo-behaviourism preferred to deal with observable contingencies between behaviour and reinforcement since in this way it was possible to make progress. Consequently subjective mental life was taken to be outside the scope of a properly scientific psychology.

The exclusion of experience persisted in mainstream experimental psychology for some time, though consciousness remained the necessary focus of psychoanalysis and humanistic psychology. In this respect, the psychology of the mid-twentieth century resembled the physical sciences of the nineteenth. While psychoanalysis has tended to be isolated by a restrictive set of assumptions and a relatively closed theoretical vocabulary, it is of some significance that in contemporary psychology we can now find a number of attempts to develop the links between psychoanalysis, neuropsychology and cognitive science.[5] However, in mainstream psychology, with more potential for contact with the larger body of physical and biological sciences, consciousness was marginalized for some three or four decades.

But, the position of consciousness as a mainstream topic is now very different from what it was at the height of the behaviourist era. Post-war developments in science and technology, combined with a dissatisfaction with the limitations of behaviourism, led to what Gardner and others have termed the 'cognitive revolution' of the late 1950s.[6] Mentalism not only became respectable but tractable.[7]

What made this possible was the language developed within cognitive science for the description of mental life. This derived from linguistics, computer science, information theory and neuropsychology. With the refinement of this language, psychology has become willing to consider consciousness again. Mental life is seen as reflecting a level of symbolic activity in the brain. The mind deals in mental structures which represent our knowledge of the world. Processes, similar in some ways to the logical operations inside computers, are said to be responsible for the storage, manipulation and use of these structures and, taken together, these structures and processes are the essence of mental life.

Cognitive science aims to treat mental events as objectively as behaviourism treated the external and hence more accessible products of mental life. Even though the classic period of cognitive science is now coming to an end, its theories and techniques have come to have a central place in contemporary psychology and we may thus ask how and if they can be applied to consciousness. Can a science which claims to be both genuinely mentalistic and objective at the same time get a purchase on this most essential psychological function?

The next reading, which reviews the cognitive approach to consciousness, is by George Mandler who was among the first to suggest a return to the study of consciousness, declaring it to be '. . . respectable, useful and even necessary'.[8] In the present reading, Mandler notes the chequered career of consciousness in psychology and offers a functional basis for investigating it. He asks what it is consciousness makes possible in human mental life and comes up with some clear and plausible suggestions.

The study of consciousness has started to regain its rightful place in the psychological enterprise. There may not be general agreement with the proposition that consciousness is the central problem of a human cognitive psychology; however, enough people have started to worry about the problem, and many others feel obliged at least to consider it.

The individual experiences feelings, attitudes, thoughts, images, ideas, beliefs, and other contents of consciousness, but these contents are not accessible to anyone else. Briefly stated, it is not possible to build a phenomenal psychology that is shared. A *theory* of phenomena may be shared, but once private consciousness is expressed in words, gestures, or in any way externalized, it becomes necessarily a transformation of the private experience. No theory external to the individual (i.e., one that treats the organism as the object of observation, description, and explanation) can, at the same time, be a theory that uses private experiences, feelings, and attitudes as data. Events and objects in consciousness can never be available to the observer without having been restructured, reinterpreted, and appropriately modified. The content of consciousness, as philosophers and psychologists have told us for centuries, is not directly available as a datum in psychology.

A new approach to the occurrence of consciousness, developed by A. J. Marcel in the context of perceptual phenomena, has great promise for the development of a new model of consciousness. I have tried to indicate some of the steps, starting with Marcel's initial insights. Marcel is concerned with structures and the conditions under which they reach the conscious state. However, in contrast to the view that structures *become* conscious so that consciousness is simply a different state of a structure, Marcel sees consciousness as a constructive process in which the phenomenal experience is a specific construction to which previously activated schemas have contributed. Marcel relates his position to the rejection of the identity assumption, which postulates that conscious states are to be seen as merely another state of a pre-conscious structure. The identity position characterizes practically all current views of consciousness, which postulate that some preconscious state "breaks through", "reaches," "is admitted," "crosses a threshold," "enters," into consciousness. A constructivist position states, in contrast, that most conscious states are constructed out of those pre-conscious structures in response to the requirements of the moment.

We can be conscious only of experiences that are constructed out of activated schemas. We are not conscious of the process of activation or the constituents of the activated schemas. A constructed conscious experience depends on the activated schemas of one or more of the constituent processes and features. The schemas that are available to constructive consciousness must be adequately activated and must not be inhibited. The resulting phenomenal experience is "an attempt to make sense of as much data as possible at the highest or most functionally useful level possible. . . ."

If I become aware of a dog, I may not be conscious of his teeth unless a relevant growl shifts the current conscious construction. The teeth may well have activated the relevant schemas, but these do not enter automatically into consciousness. We are customarily conscious of the important aspects of the environs, but never conscious of all the evidence that enters the sensory gateways or of all our potential knowledge of the event. A number of experiments have shown that people may be aware of what are usually considered higher order aspects of an event without being aware of its constituents. Thus, subjects are sometimes able to specify the category membership of a word without being aware of the specific meaning or even the occurrence of the word itself. A similar disjunction between the awareness of categorical and event-specific information has been reported for some clinical observations.

This approach to consciousness suggests highly selective constructions that may be either abstract/general or concrete/specific, depending on what is appropriate to current needs and demands. It is also consistent with arguments that claim immediate access to complex meanings of events. These higher order "meanings" will be readily available whenever the set is to find a relatively abstract construction, a situation frequent in our daily interactions with the world. In general, it seems to be the case that "we are aware of [the] significance [of a set of cues] instead of and before we are aware of the cues."

What are the most obvious occasions for conscious constructions? First, we are often conscious in the process of acquiring new knowledge and behavior. While not all new learning is conscious, the construction of complex action sequences and the acquisition or restructuring of knowledge require conscious participation. In the adult, thoughts and actions typically are conscious before they become well integrated and subsequently automatic. Learning to drive a car is a conscious process, wheras the skilled driver acts automatically and unconsciously. It follows that conscious evaluations of one's actions should more often reflect those mental and behavioral events that are in the process of being acquired or learned and less often the execution of automatic sequences. In fact, there is some evidence that subjective estimates of mental workload are correlated with indices of performance on unpracticed novel tasks but unrelated to the performance of well-practiced, familiar ones.

The sequence from conscious to unconscious is not ubiquitous. It is reversed in the infant and apparently is reversed in simple adult functions, such as in perceptual learning and in the acquisition of some simple motor skills, where skills learned unconsciously may only subsequently be represented in consciousness. The products of such acquisitions also may be divided into conscious and unconscious ones, a distinction that is found in the division between declarative and procedural knowledge. In addition, shifts from unconscious to conscious processing occur frequently. For example, the pianist will acquire skills in playing chords and trills and in reading music that are at first consciously represented but then become unconscious. However, the analytic (conscious) mode is used when the accomplished artist practices a particular piece for a concert, when conscious access becomes necessary to achieve the proper emphases, phrasings, and tempi. One wonders to what extent this process is similar to that seen in the psychoanalytic encounter, where automatic (unadaptive?) ways of dealing with the world are the object of a conscious theory of their function and then become accessible for conscious repair and change.

Second, conscious processes are frequently active during the exercise of choices and judgments, particularly with respect to action requirements. These choices, often novel ones, require the consideration of possible outcomes and consequences and frequently involve what the behaviorist literature calls "covert trial and error." However, it seems unreasonable to postulate that a conscious state exists only when selections are required. How do we account for our continuous consciousness of the surround? I suggest, as a first approximation, that a state of consciousness exists that is constructed out of the most general structures currently being activated by current concerns and environmental requirements. It provides, in consciousness, a specification in abstract terms of where we are and what we are doing there. Choice and selectivity will then produce changes in that current "reflection" of the state of the world.

Third, conscious processes exercise an important function during "trouble-shooting." Thus, relevant aspects of the world are brought into consciousness when automatic structures somehow fail in their functions, when a particular habitual way of acting fails, or when a thought process cannot be brought to an appropriate conclusion. Experienced drivers become "aware" of where they are and what they are doing when something new and different happens; when a near miss, a police car, or an unexpected traffic light is suddenly registered. The troubleshooting function of consciousness permits repair of current troublesome or injudicious processing and subsequent choice from among other alternatives. These arguments stress the role of consciousness in action, in contrast to a contemplative, reflective view of conscious states. A similar apporach has been suggested for the role of consciousness in the execution and voluntary initiation of actions and in its association with "predictive inadequacy or failure."

These various suggestions, already distilled from a longer list I presented a decade ago, can be brought under a general rule for changes in conscious states. *When current conscious constructions do not account for the state of the world, then a new conscious state will be initiated.* The adequacy of current constructions is usually indexed by discrepancies and competition among alternatives arising either out of external activations (and demands) or out of intrapsychic interchanges. A current conscious state will be changed if it does not account for (make sense of) a situation in which the available alternatives fail to meet some criterion for action or problem solving. Such a state will, of course, also change whenever an external event indexes the inadequacy of current thought or action. Our expectations may be violated when the environment keeps changing or when some external piece of evidence cannot be assimilated. Change is defined in terms of our current conscious state and the particular events that are acceptable (expected) within that state. A jogger who may not notice (be conscious of) others along the path because they have always been encountered before will become aware of an elephant. On the other hand the jogger who revels in the loneliness of the long distance runner may well become aware of intrusive others, whether elephants or not. When the environment is constant, we respond to internal demands and use those for conscious constructions. Daydreamers are unaware of their surrounds, until a shout or a raindrop demand to be accounted for in the stream of consciousness.

In other words, the contents of consciousness change whenever there is a change in the state of the world, defined as any change in the sensory evidence or in intentions, instructions, context, or situation. Spatiotemporal attentional adjustments take place whenever there is movement in the environment, when there is a change in the current state of the world.

These changes can be indexed by changes in the conscious state or by (unconscious) adjustments to the locus of changes (e.g., in eye movements). I discuss later the latter mechanisms under the rubric of attention. However, I should say here that the first kind of mechanism, the general state-of-the-world consciousness, may also be constructed in response to the demands and variables that govern strictly attentional events. The world is always in a state of flux, and change in the environment may demand new conscious constructions. The results of such activity will be the most abstract, general kind of representation consonant with the intentions and requirements of the moment, unless some specific change is so intense and relevant to current functioning that it requires more concrete, special focusing and new conscious constructions. From that point of view, the continuous state of consciousness is simply a reflection of a dynamic, changing environment. As I sit in a restaurant, enjoying good food and good company, I am conscious of food or conversation at a level that reflects my current interest and intentions, until somebody spills soup over my back and consciousness radically focuses to that event.

It appears that one of the functions of the conscious construction is to bring two or more (previously unconscious) mental contents into direct juxtaposition. The phenomenal experience of choice seems to demand exactly such an occurrence. We usually do not refer to a choice unless there is a "conscious" choice between two or more alternatives. The attribute of "choosing" is applied to a decision process regarding two items on a menu, several television programs, or two or more careers, but not to the process that decides whether to start walking across a street with the right or the left foot, whether to scratch one's ear with a finger or with the ball of the hand, or whether to take one or two sips from a cup of hot coffee. The former cases involve the necessity of deciding between two or more alternatives, whereas the latter involve only the situationally predominant action. However, given a cup of hot coffee, I may "choose" to take one very small sip, or I may "choose" to start with my right foot in a 100-meter race, given certain information that it will improve my time for the distance. In other words, for some alternatives that are usually selected unconsciously, special conditions, such as possible consequences and social factors, may make it possible for conscious constructions to be involved and to make those choices "conscious" too.

Choosing is carried out by complex unconscious mechanisms that have direct connections with and relations to action systems and other executive systems. Consciousness permits the redistribution of activations, so that the choice mechanism operates on the basis of new values of schemas and structures that have been produced (activated) in the conscious state. The mechanisms that select certain actions among alternatives are not themselves conscious, but the conditions for new choices are created consciously, thus giving the appearance of conscious free choices and operations of the will. What consciousness does permit is the running through of potential actions and choices, the coexistence of alternative outcomes, the changing of weightings of currently active schemas in the direction of one that promises greater likelihood of success, and so forth. The simultaneous presence in a conscious state of several different mental contents makes possible the establishment of new associations, and the emergence of previously quiescent cognitive structures, now activated by the conscious structures. In problem-solving activities, our consciousness of various alternatives, of trying out solutions, is often taken for the process that determines the final outcome. Although these conscious activities are no doubt related to the unconscious activations and processing that they influence, they are at best not the only forces that directly determine actions. Their similarity, in many situations, to those unconscious ones leads one to conclude that thought determines actions directly, but thought, defined as conscious mental contents, is in one sense truly epiphenomenal and in another determinative of action. It determines further unconscious processes but is several steps removed from the actual processes that pervade our mental life. Conscious thoughts are approximations of those unconscious events, in fact the best available, but they are no substitute for the representations and processes that need to be postulated for an eventual understanding of human action and thought.

The account I give here is intended primarily to account for changes in the contents of consciousness; it does not address itself to the other sense of consciousness, i.e., the distinction between being conscious or not as a continuing state. The state of "being conscious" (as distinct from being unconscious and unreceptive to any internal or external evidence) implies some continuously activated mental structures that define the current state of our world and the expectations that such structures always generate.

There are many senses of the notion of adaptation, the most widely used one being the adaptive consequence of evolutionary processes. However, the consequences of evolutionary processes are not necessarily adaptive, nor are adaptive functions of an organism necessarily the consequences of selective evolutionary processes focused on the structure or function in question. Utility and universality alone do not argue for evolutionary selection. Just consider that human beings who write universally use their hands, but nobody would argue that evolutionary processes have selected the human hand for its writing utility. Furthermore, the evolution of complex functions is typically the result of the evolution of a wide variety of, sometimes entirely unrelated, functions and structures. Consciousness is probably one such complex function. No single event in our evolutionary history is likely to have resulted in all the aspects of consciousness discussed here. The brief discussion to follow implies no claims about behavioral evolution. It extends the discussion of the uses and functions of consciousness by focusing on those aspects that are, on the average, likely to have made us better suited for the world in which we live and for the events with which we have to cope.

1. As discussed earlier, the most widely addressed function of consciousness is its role in choice and the selection of action systems. This function

permits the organism more complex considerations of action-outcome contingencies that alter the probability of one or another set of actions. It also permits the consideration of possible actions that the organism has never before performed, thus eliminating the overt testing of possible harmful alternatives.

Consciousness makes possible the modification and interrogation of long-range plans as well as of immediate-action alternatives. In the hierarchy of actions and plans, this makes it possible to organize disparate action systems in the service of a higher plan. For example, in planning to drive to some new destination, we might consider subsets of the route; or, in devising a new recipe, the creative chef considers the interactions of several known culinary achievements. Within the same realm, consciousness makes it possible to retrieve and consider modifications in long-range planning activities. These, in turn, might be modified in light of other evidence, either from the immediate environment or from long-term storage.

2. In considering actions and plans, consciousness participates in retrieval from long-term memory, even though the retrieval mechanisms themselves are not conscious. Thus, frequently, although not always, the retrieval of information from long-term storage is initiated by relatively simple commands. These may be simple instructions like, "What is his name?" or "Where did I read about that?" or more complex instructions like, "What is the relation between this situation and previous ones I have encountered?" Answers seem to be just as simple as these questions appear to be. The actual, and complex, retrieval process is "hidden." Such rapid access to stored information illustrates the adaptive use of consciousness in making complex processes more easily accessible.

3. Current states of the world, as well as thoughts and actions, are represented in consciousness and can make use of available structures to construct storable representations for later reference and use. Many investigators have suggested that these encodings of experience always take place in the conscious state. Processes such as mnemonic devices and other strategies for preserving experiences for later reference apparently require the intervention of conscious structures.

In the social process, prior problem solutions and other memories may be brought into the conscious state and, together with an adequate system of communication, such as human language, generate the benefits of cooperative social efforts. Other members of the species may receive solutions to problems, thus saving time, if nothing else; they may be apprised of unsuccessful alternatives or may, more generally, participate in the cultural inheritance of the group. Such social problem solving and remembering require selection and comparison among alternatives retrieved from long-term storage, all of which apparently takes place in consciousness. Given the often abstract and general nature of conscious contents, the cultural transmission from conscious construction to verbal communication may in fact be extremely efficient. Cultural knowledge can be transmitted and new insights may socially available by the use of general instructions and conclusions, instead of piecemeal communication of detailed skills and minute serial instructions. Just imagine how the mundane matter of teaching somebody how to drive a car or make an omelet would be complicated if we could not communicate first our general sense of "how to do it." Without further elaboration, it is obvious that the transaction described here illustrate the intricate relationship between consciousness and language.

We note that both general information and specific sensory inputs may be stored. The re-presentation at some future time makes possible decision processes that depend on comparisons between current and past events and on the retrieval of relevant or irrelevant information for current problem solving.

4. We have already seen that consciousness provides a "troubleshooting" function for structures normally not represented in consciousness. There are many systems that cannot be brought into consciousness; most systems that analyze the environment in the first place probably have that characteristic. In most of these cases, only the product of cognitive and mental activities is available to consciousness. In contrast, many systems are generated and built with the cooperation of conscious processes but later become nonconscious or automatic. The latter systems apparently may be brought into consciousness, particularly when they are defective in their particular function. We all have had experiences of automatically driving a car, typing a letter, or even participating in a cocktail party conversation, and of being suddenly brought up short by some failure such as a defective brake, a stuck key, or a "You aren't listening to me." At that time the particular representations of action and memories involved are brought into play in consciousness, and repair work gets under way. The adaptive advantage of acting automatically when things go smoothly, but being able to act deliberately when they do not, may therefore also be ascribed to the conscious process.

Many of these functions permit us to react reflectively instead of automatically, a distinction that has frequently been made between human and lower animals. All of them permit more adaptive transactions between the organism and its environment.

I now return to the question with which I began this exploration: What advantages does this proposed mechanism confer on the organism it inhabits? What are we able to do with it that we could not do without it? I do not think the additional abilities are either mysterious or overwhelming, but, like many changes to our evolutionary makeup, they increment our more primitive abilities. The proposed mechanism requires no novel mechanism; it just extends the activation process — the result follows automatically. It does create conditions for making internal search processes as efficacious as external ones; we need not scan the world in order to benefit from the various possibilities that surround us. Third, it significantly speeds up decision processes and problem solving by adding information that might otherwise have to be obtained rather laboriously from the external world.

The alternative to self-activation via consciousness requires additional searches, either of the environment or of stored information. Organisms not endowed with this conscious feedback mechanism or not yet able to take advantage of it are clearly more dependent on environmental information and very often on trial and error behavior. Infants, who cannot recall in the sense used here, show such dependence; and, without entering into speculations about infant consciousness, it can be argued within the present context that neither the relevant action and event schemas nor the higher order task and intentional structures are available to the newborn. Lower animals develop strategies of increasing the activation of current alternatives in their attempt to solve problems. For example, some animals display so called vicarious trial and error (VTE) behavior at choice points. VTEs consist of visual sampling of possible choices, sometimes including short forays to one side or the other before a choice occurs. VTEs can be seen as precursors of imary (and consciousness), maximizing data input in the absence of representation in consciousness. Humans too are known to use similar strategies when unable to generate reasonable hypotheses about the current state of the world.

The mechanisms that select the relevant and important from the many potential sources of information in the environment are central to an organism's ability to deal with its world. Momentary states of consciousness are the final step in the reduction of information that floods the organism. Having arrived at a view of mental representation that invokes an efficient and economical system of parallel distributed processes, we can now invoke consciousness to provide us with a slow, restricted, and serial processing state. For example, when decisions need to be restricted to few candidates or when problem solving reaches a stalemate, consciousness provides a deliberate stage for further appropriate action. States of consciousness not only reduce the flow of information to manageable proportions, but also make sense out of what is available and make subsequent actions more adaptive.

Most if not all our knowledge and previous experience is stored (available, recorded) in structures that are both vast and differentially accessible. Some distinction needs to be made between that information and what is currently being used, worked on, and accessed. It is to a large part that function that is being exercised by consciousness. If one were aware (conscious) of *everything one knows*, or even of everything that is relevant (closely related to) some current experience, one would be swamped withinformation and unable to act. Thus the distinction needs to be made; and, in the course of the evolution of the mind, the distinction emerged in the human species as that between conscious and unconscious mental contents, as it possibly developed also in other complex animals faced with the same dilemma.

It is possible that other useful functions of consciousness emerged as a consequence of the emergence of consciousness, and not as a direct selective adaptation. In particular, the relation of thought and language may well have been such a later development, in part because of the usefulness of language for summarizing and capsulating event and thoughts.

> It is certainly the case that the interplay between human conscious thought and language has overwhelmed our view of both of them, often asserting that one could not exist without the other.
>
> One of the adaptive functions of consciousness, though not necessarily a result of specific natural selection, is the function of consciousness that chunks and abstracts knowledge into serviceable units for the social communication of knowledge.
>
> It permits the communicator to restrict the message to its salient aspects, and it also makes easier the reception of a limited rather than overinclusive "message." In addition, it permits the participants in social communication (and the transmission of culture) to construct reasonably concise and common mental models of social knowledge.

The encouragement from Mandler and others to take consciousness seriously has been gradually taken on board over the past decade or so. Cognitive science also takes the computer metaphor for the mind very seriously. Indeed, so powerful a metaphor has this become that some psychologists feel that it is no longer a metaphor but something to be treated literally; the essence of mental life has, at last, been discovered.[9] Though it is now clear that this claim is over-optimistic, it is equally clear that a symbolic capacity is fundamental to human intelligence. The related proposal that computers might have a mental life, even one that includes consciousness, is proposed quite seriously[10] though it is also opposed equally as seriously.[11]

This debate aside, it is clear that cognitive science has a special place in the debate over the psychological status of consciousness. This is because of the degree of specificity it is possible to achieve once a psychological theory is set out in objective information-processing terms.[12] The last reading in this chapter demonstrates this with respect to reflexivity. Philip Johnson-Laird, who has carried out influential work in language and mental imagery, uses a computational approach to address reflexivity and self-awareness.

> This chapter argues that the problems of consciousness are only likely to be solved by adopting a computational approach and . . . outlines a theory of consciousness based on three main assumptions: the computational architecture of the mind consists in a hierarchy of parallel processors; the processor at the top of the hierarchy is the source of conscious experience; this processor — the operating system — has access to a model of itself, and the ability to embed models within models recursively. If the thesis is correct, the problems of consciousness can be solved once we understand

what it means for a computer program to have a high-level model of its own operations.

What should a theory of consciousness explain? This is perhaps the first puzzle about consciousness, because unlike, say, the mechanism of inheritance, it is not clear what needs to be accounted for. One might suppose that a theory should explain what consciousness is and how it can have particular objects; that is, the theory should account for the qualitative aspects of the phenomenal experience of awareness. The trouble is that there are no obvious criteria by which to assess such a theory. Indeed, this lack of criteria lent respectability to the behaviouristic doctrine that consciousness is not amenable to scientific investigation; i.e. that it is a myth that the proper study of nerve, muscle, and behaviour will ultimately dispel. A prudent strategy is therefore both to take a different approach to consciousness and to suggest a more tractable set of problems for the theory to solve. My approach here will be to assume that consciousness is a computational matter that depends on how the brain carries out certain computations, not on its physical constitution. I will outline [a] principal [problem] that a theory of consciousness should solve. I will then propose a theory of mental architecture — a conjecture about how the mind is organized — and I will show that it could provide an answer to.

The problem of self-awareness. You can be aware of the task that you are carrying out and to be absorbed in it as to forget yourself. Alternatively, you can be self-aware and be conscious of what you yourself are currently doing. Sometimes, perhaps, you can even be aware that you are aware that you are aware . . . and so on. Self-awareness is, of course, essential for a sense of one's integrity, continuity, and individuality.

From the simple nerve networks of coelenterates to the intricacies of the human brain, there appears to be a uniform computational principle: asynchronous parallel processing.

If many processors compute in parallel, they can divide up the task between them whenever there are no dependencies between the computations. Such a division of labour not only speeds up performance, but it also allows several processors to perform the same sub-task so that should one of them fail the effects will not be disastrous, and it enables separate groups of processors to specialize in different sub-tasks. The resulting speed, reliability, and specialization have obvious evolutionary advantages.

But parallel computation has its dangers too. One processor may be waiting for data from a second processor before it can begin to compute, but the second processor may itself be waiting for data from the first. The two processors will thus be locked in a 'deadly embrace' from which neither can escape. Any simple nervous system with a 'wired in' program that behaved in this way would soon be eliminated by natural selection.

Higher organisms, however, can develop new programs — they can learn — and therefore there must be mechanisms, other than those of direct selective pressure, to deal with pathological configurations that may arise between the processors. A sensible design is to promote one processor to monitor the operations of others and to override them in the event of deadlocks and other pathological states of affairs. If this design feature is replicated on a large scale, the resulting architecture is an hierarchical system of parallel processors: a high-level processor that monitors lower level processors, which in turn monitor the processors at a still lower level, and so on down to the lowest level of processors governing sensory and motor interactions with the world. A hierarchical organization of the nervous system has indeed been urged by neuroscientists from Hughlings Jackson to H. J. Jerison, and Simon [1969] has argued independently that it is an essential feature of intelligent organisms.

In a simple hierarchical computational device, the highest level of processing could consist of an operating system.

The notion that the mind has an operating system verges, as we shall see, on the paradoxical, but it has some relatively straightforward consequences. The operating system must have considerable autonomy, though it must also be responsive to demands from other processors. It must be switched on and off by the mechanisms controlling sleep, though other processors continue to function. It must depend on a second level of processors for passing down more detailed control instructions to still lower levels, and for passing up interpreted sensory information. Doubtless, there are interactions between processors at the same or different levels, and facilities that allow priority messages from a lower level to interrupt computations at a higher level.

The brain is a parallel computer that is organized hierarchically. Its operating system corresponds to consciousness and it receives only the results of the computations of the rest of the system. Such a system can begin to account for the division between conscious and unconscious processes (the problem of awareness) and it can also allow the lower level processors a degree of autonomy (the problem of control).

Why is it that the contents of the operating system — more precisely, its working memory — are conscious, but all else in the hierarchy of processors is unconscious? In other words, what is it about the operating system that gives rise to the subjective experience of awareness? Of course, it is logically possible that each processor is fully aware but cannot communicate its awareness to others — an analogous view was defended by William James in order to explain the phenomena of hysteria without having to postulate an unconscious mind. The view that I shall defend, however, is that the operating system's potential capacity for self-awareness is what gives rise to consciousness. Whenever any computational device is able to assess how it itself stands in relation to some state of affairs, it is — according to my hypothesis — conscious of that state of affairs. What needs to be elucidated is therefore how a computational device could assess its own relation to some state of affairs. It is to this problem that I now turn.

Reflection on the human capacity for self-reflection leads inevitably to the following observation: you can be aware of yourself. You also understand yourself to some extent, and you understand that you understand yourself, and so on. . . . The idea is central to the subjective experience of consciousness, yet it seems as paradoxical as the conundrum of an inclusive map. [If a large map of England were traced out in accurate detail in the middle of Salisbury Plain, than it should contain a representation of itself within the portion of the map depicting Salisbury Plain (which in turn should contain a representation of itself [which in turn should contain a representation of itself (and so on *ad infinitum*)])]. Such a map is impossible because an infinite regress cannot occur in a physical object. Leibniz dismissed Locke's theory of the mind because there was just such a regress within it. However, a computational procedure for representing a map can easily be contrived to call itself recursively and thus to go on drawing the map within itself on an ever diminishing scale. The procedure could in principle run for ever: the values of the variables, though too small to be physically represented in a drawing, would go on diminishing perpetually.

There is a similar computational solution to the paradox of self-awareness. Ordinarily, when you perceive the world, vision delivers to your operating system a model that makes explicit the locations and identities of the objects in the scene. The operating system, however, can call on procedures that construct a model that makes explicit that it itself is perceiving the world: the contents of its working memory now contain a model representing it perceiving the particular state of affairs represented by the model constructed by perception. In other words, the visual model is embedded within a *model* of the operating system's current operation. Should you be aware that you are aware that you are perceiving the world, then there is a further embedding: the operating system's working memory contains a model of it perceiving the state of affairs represented by the model of it perceiving the world. Since the hierarchy of embedded models exists simultaneously, the operating system can be aware that it is aware of the world. Granted the limited processing capacity of the operating system and its working memory, there is no danger of an infinite regress.

In self-awareness, there is a need for an element in the model of the current state of affairs to refer to the system itself and to be known so to refer. In the formation of intentions and metacognitions, there is a more complex requirement. To have a conscious intention, for instance, the operating system must elicit a representation of a possible state of affairs. An essential part of the process is precisely an awareness that the system itself is able to make such decisions. The system has to be ale to represent the fact that the system can generate a representation of a state of affairs, and decide to work towards bringing it about.

Hence conscious intentions depend on having access to an element, not merely that refers to the system itself, but that represents the specific abilities of the system, and in particular its ability to plan and to act to achieve plans.

Metacognitive abilities similarly depend on access to such *models* of the system's capabilities, predilections, and preferences. You can reason about how you reason because you have access to a model of your reasoning performance.

The recursive aspect of this ability is hardly problematic. The crux of the problem of consciousness resides in the other requirement: the operating system must have a partial model of itself in order to begin the process of recursion underlying intentionality.

No one knows what it means to say that an automaton or computer program has a model of itself. The question has seldom been raised and certainly has yet to be answered.

What is . . . needed is a program that has a *model* of its own high-level capabilities. This model would be necessarily incomplete, according to the present theory, and it might also be slightly inaccurate, but it would none the less be extremely useful. People do indeed know much about their own high-level capabilities: their capacity to perceive, remember, and act; their mastery of this or that intellectual or physical skill; their imaginative and ratiocinative abilities. They obviously have access only to an incomplete model, which contains no information about the inner workings of the web of parallel processors. It is a model of the major options available to the operating system.

The present approach to consciousness depends on putting together the three main components that I have outlined; hierarchical parallel processing, the recursive embedding of models, and the high-level model of the system itself. Self-awareness depends on a recursive embedding of models containing tokens denoting the self so that the different embeddings are accessible in parallel to the operating system. Metacognitions and conscious intentions depend on a recursive embedding of a model of elements of the self within itself, and of course the ability to use the resulting representation in thought.

This approach assumes that human behaviour depends on the computations of the nervous system. The class of procedures that I have invoked are, with the exception of a program that has a high-level model of itself, reasonably well understood. The immediate priority is therefore to attempt to construct such a program. It is often said that the computer is merely the latest in a long line of inventions — wax tablets, clockwork, steam engines, telephone switchboards — that have been taken as metaphors for the brain. What is often overlooked is that no one has yet succeeded in refuting the thesis that any explicit description of an algorithm is computable. If that thesis is true, then all that needs to be discovered is what functions the brain computes and how it computes them. The computer is the last metaphor for the mind.

Johnson-Laird shows how cognitive science helps to make theories of consciousness explicit and to re-establish it as a legitimate object of study for experimental psychology.

This reader is approaching a breakpoint. The readings so far have dealt with philosophy and science and consequently have tended to look at the material support for consciousness and at the problems in defining and studying it. But consciousness is also a human experience supported by human action within a social framework. Accordingly, the remaining chapters of the book, to which the next short chapter is a bridge, deal with the social and cultural context of consciousness.

Notes

1. See the introduction to A. Marcel and E. Bisiach, eds., *Consciousness in Contemporary Science*, 1989, Oxford University Press.
2. For a review, see Chapter Five of W. O'Neil, *The Beginnings of Modern Psychology*, 1968, Penguin.
3. Watson, J. B., *Behaviourism*, 1930, Harpers.
4. Skinner, B. F., *About Behaviourism*, 1974, Random House.
5. Haldane, J., 'Psychoanalysis, Cognitive Psychology and Self-Consciousness' in P. Clark and C. Wright, eds., *Mind, Psychoanalysis and Science*, 1988, Basil Blackwell. See also S. Turkel, 'Artificial Intelligence and Psychoanalysis: A New Alliance' in M. Graubard, ed., *The Artificial Intelligence Debate*, 1988, MIT Press.
6. Gardner, H., *The Mind's New Science: A History of the Cognitive Revolution*, 1985, Basic Books.
7. Baars, B. *The Cognitive Revolution in Psychology*, 1986, Guilford Press.
8. G. Mandler, 'Consciousness, Respectable, Useful and Probably Necessary' in R. Solso, ed., *Information Processing and Cognition*, 1975, Lawrence Erlbaum Associates.
9. A. Newell, 'Physical Symbol Systems', *Cognitive Science*, 1980, vol. 4.
10. Feigenbaum, E. and McCorduck, P., *The Fifth Generation: Artificial Intelligence and Japan's Computer Challenge to the World*, 1983, Addison-Wesley.
11. Weitzenbaum, J., 'The Computer in Your Future', *New York Review of Books*, 30 October 1983.
12. Shallice, T., 'Information-processing Models of Consciousness: Possibilities and Problems' in A. Marcel and E. Bisiach, op. cit.

6

From Sentience to Symbol: Emergence and Transition

READINGS

Leslie White
Four Stages in the Evolution of Minding

Chris Sinha
A Socio-naturalistic Approach to Human Development

A. R. Luria
Cognitive Development: Its Social and Cultural Foundations

With the possible exception to the first two chapters, the readings up to this point have been a generally inward-looking inquiry into consciousness. That is, they have emphasized psychological and physiological processes occurring within the brain. We hope that what has been presented shows something of the biological and psychological systems which support human consciousness.

The readings in Chapter Four showed that the neuropsychological underpinnings of consciousness are not so much the business of a particular part of the brain but rather depend upon integrative patterns of activity involving the brain as a whole. Where do these patterns come from? What is the source of their coherence and persistence? Finding answers to questions like these will help to understand a lot more about what makes human consciousness the unique phenomenon that it is.

The beginnings of an answer are found in recognizing that while the human brain has a unique capacity to acquire consciousness it does not do so in isolation. Nor is consciousness the appearance of some intrinsic property of the brain or the provenance of some special external agency in which is prefigured the particular character of an individual's consciousness. Rather consciousness emerges, in a graded progression over both ontogenetic and phyologenetic development.

Consciousness is both the product of a process and a process itself. Indeed, Chapter Three demonstrated that not only had an evolutionary process brought the brain itself into being, but also that the support for consciousness lay in patterns of interactions between individuals as well as in patterns of events within any particular brain. Clearly, an account of human awareness cannot be a mere inward-looking inquiry into brain processes. It will require an examination of this necessary interactive complement to the biological and psychological support for consciousness that the brain provides.

Some psychologists, Watson for example, appeared to think that to understand consciousness would, eventually, require no more than a thorough scientific analysis of what makes the brain itself tick since this would be the basis for understanding social interaction as well. Others are less reductive but still see brain function as the arena in which fundamental knowledge of consciousness must be sought. Sperry, for example, suggests that there are 'emergent functional properties of brain processing' which are able to 'exert an active control role as causal determinants in shaping the flow of cerebral excitation'.

For others, the unique phenomenon of human consciousness requires a correspondingly unique explanation. These usually involve suggesting that within the brain there are causes operating that are qualitatively discontinuous from those found elsewhere in the natural order. Penfield, for example, describes the brain as the 'messenger' of consciousness. He and Eccles suggest that a special, non-material 'second form' of energy or organization inhabits the brain in much the same way as Descartes suggested the brain housed a soul.

It is hoped the readings thus far show that understanding consciousness requires neither neuropsychological reduction, nor special explanations. Rather, we begin to see, as Bohm has suggested, that consciousness is supported by continuous patterns of activity integrating physical, biological and psychological levels of order. These patterns of activity cannot be considered to reside in the brain alone although every human brain recreates them in development. They are carried in the matrix of social and cultural conditions which, in necessary conjunction with the biological potential of the brain, bring human consciousness into being. This socio-cultural matrix and how the human being relates to it is the focus of the coming chapters.

The present chapter concerns the evolutionary step which perhaps more than any other led to the appearance of human consciousness. This step is the emergence of symbols. What the transition from sentience to symbols brought into being is, broadly speaking, the concern of most of the following chapters. The emergence of the human capacity for symbolic thought and communication, as a number of readings have already stressed, was one of the crucial evolutionary preconditions required before the human socio-cultural matrix could begin to develop.

The reading from White suggests a sequence of evolutionary stages that might have preceeded the emergence of symbolic capacity. He traces a progressive development in the relationship an organism may have with the environment. This progression starts from simple sentience which, as the author notes, is not so far removed from the sensitivity of the inorganic world. It ends with the preconditions of human culture, namely, a world of human meaning based on symbols.

By "minding" we mean the reaction of a living organism to some thing or event in the external world. It is therefore a process of interaction between an organism and a thing or event lying outside it. Minding is a function of the thing or event to which the organism reacts as well as of the organism itself: $O \times E = M$, in which O is the organism, E is the thing or event to which the organism reacts, and *M* is minding. Minding varies as either *O* or *E* varies.

Interaction between two bodies means a relationship between them. Minding may therefore be understood in terms of relationships between organisms, on the one hand, and events in the external world, on the other. We may deal with these relationships in terms of the *meanings* that things and events in the external world have for organisms: an organism approaches, withdraws from, or remains neutral toward some object in its vicinity, depending upon its meaning or significance to the organism. If it is beneficial (food), it may approach; if it is injurious, it may withdraw; and if it is neutral, the organism will remain indifferent. The concepts with which minding may be analyzed and interpreted are, therefore, *action*, *reaction*, *interaction*, *relationship*, and *meaning*. How are these relationships established? How are meanings determined?

Let us begin with inanimate bodies. According to the theory of gravitation, every particle of matter in the universe attracts every other particle, i.e., a relationship obtains between them; each has meaning for all the others. These meanings are determined by their respective masses and by the distances which intervene between them: "directly as the mass, inversely as the square of the distance." Material particles attract or repel each other in such phenomena as capillary attraction, surface tension, and electromagnetic events. But all relationships among inanimate bodies can probably be reduced to three kinds: attraction, repulsion, and indifference. And in all instances, no doubt, these relationships are determined by the inherent properties of the bodies concerned, their topological relations, and their settings (presence or absence of catalysts).

When we cross the line that divides inanimate and animate bodies and come to living organisms, we find that the simplest reactions are precisely like those of inanimate bodies. The organism's reaction is positive (+), negative (−), or neutral (0). That is, it approaches, withdraws, or does nothing, depending upon the meaning that the object has for it.

In this simplest type of reaction, which we may for convenience call Type I (Fig. 1), the meaning that the thing or event has to the organism is determined by the intrinsic properties of both organism and thing or event. Or, to put the matter otherwise, the relationship between organism and thing-or-event is determined by their respective intrinsic properties.

FIG. 1.—The simple reflex: Type I behavior

The organism approaches if the stimulus is positive (e.g., food), withdraws if it is injurious, and remains indifferent if it is neutral. But, whether a thing is *food* or not depends upon the intrinsic properties of the organism as well as of the thing; edibility is a function of the eater as well as of the thing eaten; what is food to one organism may be not-food to another. And so it is with injurious things or things neutral. In every instance in this simplest type of interaction the relationship between organism and thing-or-event is determined by the intrinsic properties of both.

The next stage in the evolution of minding is characterized by the conditioned reflex, and we may use the classic experiment of Pavlov with the dog and the electric bell to illustrate it. A hungry dog salivated when he smelled food; he was indifferent to the sound of an electric bell. But when stimulated by odor and bell simultaneously for a number of times, the sound of the bell alone was sufficient to excite his salivary glands and evoke the response. We may call this kind of behavior Type II, and represent it diagrammatically in Figure 2.

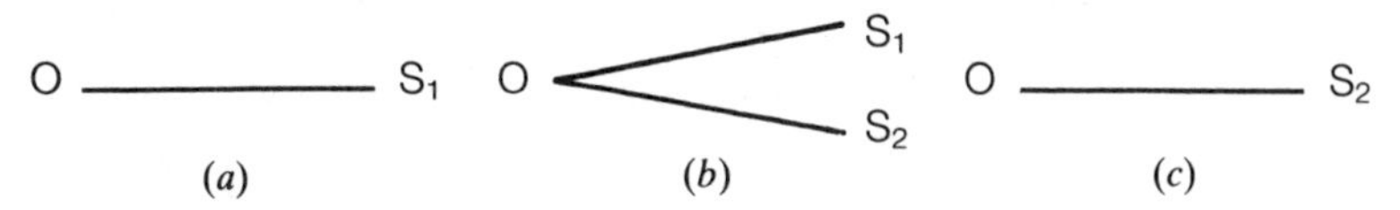

FIG. 2.—The conditioned response or reflex: Type II behavior

The process of condition takes place in three stages. In the first stage, *a*, in Figure 2, we have the same kind of situation that we have in Type I: the organism, *O*, and a significant stimulus, S_1, the odor of food, with a simple relationship between them in terms of their respective intrinsic properties. In stage *b*, we introduce S_2, the sound of the bell, which becomes related to S_1, the odor of food, on the one hand, and to *O*, the dog, on the other. Initially, S_2 is related to S_1 in time and in space, and, as a consequence of association, S_2 and S_1 become related to each other through the neurosensory-glandular system of the dog. A relationship between S_2 and the dog (*O*) is established at the same time and in the same way. When the relationship between S_2 and the dog has been established, S_1 may drop out. In stage *c* we again have a simple, direct relationship between the organism and a single stimulus.

Type II resembles Type I and grows out of it. It begins with a simple Type I reaction, and it ends with the *form* of Type I reaction. But Type II differs from Type I in a fundamental respect: Type II is characterized by a relationship between organism and stimulus which is *not* dependent upon their intrinsic properties. To be sure, the substitution of one stimulus for another could not have been effected, had not the dog been an organism capable of this kind of behavior. But the salivary-gland-meaning of the electric bell is in no sense intrinsic in the sound waves that it emits. Type II behavior ends with the *form* of Type I: the reaction takes place *as if* the relationship between dog and bell were intrinsic in them. But the response is fundamentally different in kind.

The next stage in the evolution of minding, Type III, may be illustrated by the example of a chimpanzee using a stick to knock down a banana which is suspended from the roof of his cage beyond the reach of his hand. We illustrate it in Figure 3, in which O = chimpanzee, E_1 is the banana, and E_2 is the stick.

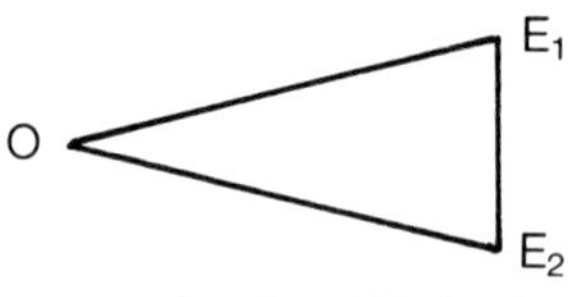

FIG. 3.—Type III behavior

Type III minding is like Type II in one respect: the organism is related simultaneously to two things-or-events in the external world (as in Fig. 2, *b*). But Type III differs from Type II in a number of important respects. In the first place, the two things, E_1 and E_2, are significant from start to finish in Type III, whereas they are significant in only one of the three stages of the process in Type II (Fig. 2, *b*). Second, the relationships established in Type III are dependent in their entirety upon their respective intrinsic properties: the chimpanzee is a banana-eating, stick-wielding animal; the banana is a knock-downable-with-a-stick thing and eatable by a chimpanzee; the stick is wieldable by a chimpanzee and can be used to knock down a banana. Third, the relationship between E_1 and E_2 in Type III is established directly and extraorganismically, whereas they are related indirectly and intraorganismically (within the neurosensory-glandular system of the dog) in Type II. And, finally, the relationships established in Type III are determined intra-organismically, i.e., by the chimpanzee himself, "of his own free will and choice," so to speak, whereas the relationships established in Type II are not determined by the organism but by its relationships to other factors — the experimenter, or circumstances, such as chance association.

We may distinguish two kinds of roles of organisms in the process of minding. Either the organism determines the configuration of behavior which it executes, or it does not; it plays either a dominant or a subordinate role. Thus, in Type I, it is not the organism alone that determines its behavior. It behaves as it does because (1) of its own intrinsic properties and (2) because of the intrinsic properties of the stimulus, *E*. It has neither alternatives nor choice. The flower turns its face to the sun because it must; it can do nothing else. It is something that it undergoes as well as something that it does. Its behavior is subordinate to the intrinsic properties of itself and its stimulus.

The organism plays a subordinate role in Type II, also. The dog "has nothing to say" about how he shall respond to the sound of the bell; this is determined by the experimenter (chance associations may be the determining factor in other processes of conditioning). Here, also, the organism has neither initiative nor choice.

It is different with Type III minding. Here the organism plays a dominant role. It is the chimpanzee who decides what to do and how to do it. He has initiative, alternatives, and choice. He may use the stick to reach and knock down the food, or he may, as they sometimes do, use it to pole-vault ceilingward and snatch the food when it comes within his reach. Or he may decide to build a tower of boxes from whose summit he can reach his prize. This is what we mean when we say that the pattern of action, the configuration of behavior, is determined intra-organismically: the chimpanzee solves his problem by insight and understanding, formulates a plan, then puts it into execution. He is a sublingual architect and builder.

In Type III, then, we are again dealing with relationships determined by the intrinsic properties of organisms and things, but here the organism plays a dominant, instead of a subordinate, role in the formulation and execution of patterns of behavior.

Type IV minding is well illustrated by articulate speech and may be diagrammed as in Figure 4. *O* is again the organism, this time a human being; E_1 is a hat; and E_2 is the word "hat." Again we have a triangular configuration as in Type III: there is a mutual and simultaneous relationship between the organism and two things-or-events in the external world. And, as in Type III, the configuration is determined intra-organismically, by the organism itself, of its own will and choice; Type IV, like Type III, is characterized by alternatives and choice. But Type IV differs from Type III in that in the former the relationships are not dependent upon the intrinsic properties of the elements involved, as they are in Type III. That is to say, there is no necessary or inherent relationship between the object hat and the combination of sounds *hat*. In this respect, Type IV resembles Type II: both are independent of the intrinsic properties of the factors involved. But Type IV differs from Type II in a fundamental way: the organism plays a dominant role in Type IV, a subordinate role in Type II.

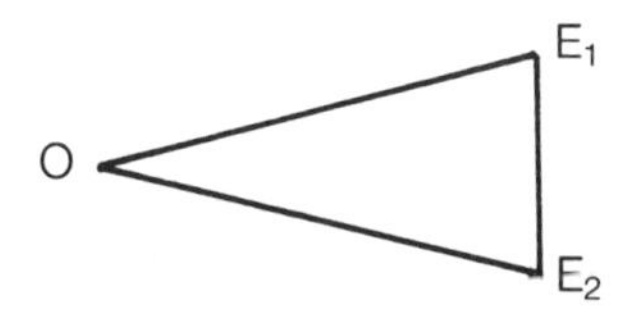

FIG. 4.—Type IV: symboling

Looking back over our four types of minding, we notice similarities and differences among them. Types I and III are dependent upon the intrinsic properties of the elements involved in the configurations of behavior; Types II and IV are not so dependent. In Types I and II, the organism plays a subordinate role in the formulation and execution of patterns of behavior; in Types III and IV, it plays a dominant role. We may summarize these facts diagrammatically as follows:

	Organism plays a subordinate role	Organism plays a dominant role
Dependent upon intrinsic properties	Type I	Type III
Independent of intrinsic properties	Type II	Type IV

FIG. 5.—Comparison of four stages of minding

Another feature of our series of stages is that our types are *kinds*, not *degrees*, of minding. An organism is either capable of conditioned reflex behavior, Type II, or it is not; there are no gradations between Types I and II. Similarly, an organism is capable of Type III or Type IV behavior, or it is not. We are confronted by a series of leaps, not by an ascending continuum.

The question might be raised at this point, Does our series of stages constitute a biological evolutionary sequence or merely a logical one? It has been derived deductively, in a sense, from a consideration of a basic concept: relationship. It has been postulated that the relationship established in the process of minding is either dependent upon the intrinsic properties of organism and things-and-events in the external world, or it is not. Second, the relationship or, more specifically, the pattern of behavior in which the relationship is expressed, is determined by the organism or it is not. This gives us four categories of minding, and it might appear at first glance that our series of stages is more artificial than real.

But this is not the whole story. We did not begin with factors selected at random. Our premises and postulates were in fact derived from a careful scrutiny and analysis of the behavior of very real organisms. The fact that the series proceeds from the simple to the complex would suggest that it constitutes an evolutionary, as well as a logical, sequence. But there are other facts that make it quite clear that we do indeed have here an evolutionary sequence in a biological sense.

Let us begin, first of all, by classifying all living species with reference to our four types of minding as far as our information will permit (Fig. 6). All organisms are capable of Type I: simple reflexes or tropisms. We know from observation that some species are capable of Type I only. Type II has grown out of Type I, as we have seen; therefore, it may be assumed that organisms capable of Type II are capable of Type I also; this assumption is validated by observation. But, obviously, there are fewer species capable of Type II than are capable of Type I. We know, also from observation, that there are organisms capable of Type II that are incapable of Type III, but all organisms capable of Type III are capable of Types II and I also. Only one species, man, is capable of Type IV, and this species is capable of Types III, II, and I, also. Hence the following generalizations: (1) our series of stages is incremental and cumulative, a new stage being added to the one, or ones, that preceded it; (2) the number of species capable of a given type of behavior diminishes as we proceed from Type I to Type IV; and (3) organisms classified according to our series of types of minding are thereby arranged in a biological evolutionary series, the lowest and simplest organisms being at the bottom, the higher and more complex at the top.

Specifically, we cannot assign each and every species to one or another stage in our series of types of minding for the simple reason that we do not have the requisite information. But we have every reason to believe that we could do this if we possessed full information as to their behavior and that no species would fail to be accommodated by our series.

We do not know at what point in the evolutionary scale organisms become capable of Type II behavior. Snails, it appears, are capable of conditioned reflex behavior, which puts this ability fairly low in the scale of biological evolution, but what other species belong here is a question that we cannot answer.

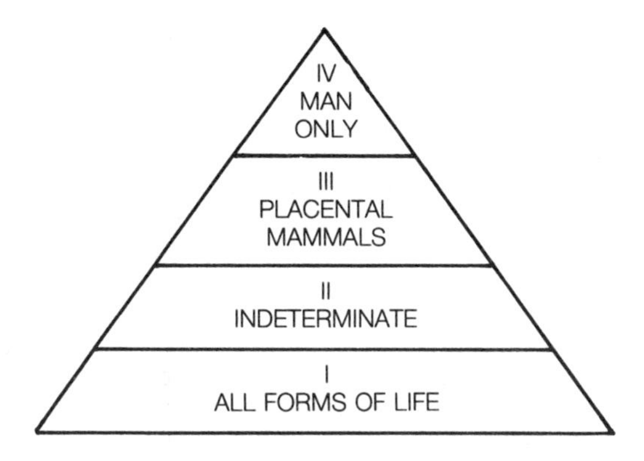

FIG. 6.—Classification of organisms with respect to types of minding

We know that apes are capable of Type III minding, and there appears to be much evidence that dogs and elephants also possess this ability. It seems reasonable to suppose that all placental mammals fall within this class, but whether marsupials and monotremes belong here or not is a question to which comparative psychology provides us with no answers.

Our series of types of minding constitute a progressive series of advantages for the life-process, for living organisms: they emancipate organisms from limitations imposed upon them by their environments, on the one hand, and confer positive control over the environment, on the other. This is another significant indication of the biological evolutionary character of our sequence of stages.

Every living organism exists in a setting which imposes limitations upon its behavior in many ways: gravitation, temperature, atmospheric pressure, and humidity are universal factors that have to be reckoned with and which circumscribe the behavior of the organism. Food, enemies, and incidental obstacles further condition behavior. As we have seen, organisms capable of only Type I minding are to a great extent subordinate in their behavior to factors of their environment: the petals of a flower close at night to reopen after sunrise; the paramecium approaches or withdraws from a stimulus; and so on. The organism has neither alternatives nor choice; it must do what its own inherent properties and those of its environment dictate. All living organisms are dynamic (thermodynamic) systems, to be sure; but the behavior of organisms capable of only Type I is something that they *undergo* as well as something that they do.

The advent of the conditioned reflex brought about a revolution in minding. At a single stroke it emancipated organisms from many limitations imposed upon them by their environment. To be sure, the organism can live only with boundaries circumscribed by such factors as gravity and temperature, but it need no longer limit its behavior to responses to stimuli as determined by their intrinsic properties. A stimulus (a thing or event) can have only one meaning in Type I minding; it may mean any one of many things in Type II. Thus the sound of the electric bell has only auditory meaning in Type I; it may mean food, danger, sex, or something else in Type II. The conditioned reflex was, so to speak, an emancipation proclamation for evolving life-forms: it emancipated organisms from limitations imposed upon them by the natural properties of things in their environment; it multiplied the number of kinds of responses that could be made to a given thing as a stimulus.

But, under Type II, the organism was still subordinate to outside things and circumstances. With a conditioned reflex, the organism can acquire new meanings for things or events (Pavlov's dog acquired a new meaning for the sound of the bell), *but it cannot determine what this new meaning shall be*; this is done by an experimenter or by other circumstances. The organism still plays a passive role with reference to his environment. The advent of Type III gives him positive control over it. The extent of this control is not, of course, complete or absolute, *but it is control*, and it was this that was lacking in Types I and II. In Type III the organism rises above his environment. It now has alternatives, and it can make choices. It is the ape himself who decides how he is to reach the banana and with what means. In Type III the organism plays a dominant role in his interaction with his environment.

In Type IV we have the emancipation from the limitations of intrinsic properties of external things or events which was won in Type II, combined with the ascendance to control over environment that was achieved in Type III. In Type IV, it is the organism, man, who determines what meaning the sound of the electric bell shall have, and he can give it any meaning he chooses. Emancipation *and* control are thus united in Type IV.

The career of life is a struggle between thermodynamic processes: a building-up process coping with a running-down, breaking-down process. Life is simply the name that we give to a thermodynamic process that moves in a direction opposite to that specified by the Second Law of Thermodynamics for the universe as a whole. Life sustains itself by the capture and utilization of free energy, obtained in the form of food. Life is enabled to extend itself by the multiplication of numbers and to evolve new and higher forms because of its ability to capture and utilize free energy in increasing amounts. To do this, it must overcome the obstacles of its natural environment. The emancipation from limitations imposed by the intrinsic properties of things, on the one hand, and the assertion of positive control over environment, on the other, that are won by Types II and III, respectively, and which are combined to produce Type IV have been the ways and means by which biological evolution has been achieved.

Type IV behavior, or minding, is characterized by freely and arbitrarily bestowing meaning upon a thing or an event and in grasping meanings thus bestowed. These meanings cannot be comprehended by the senses alone. Holy water is not the same kind of thing, from the standpoint of human experience, as mere H_2O; it has a distinctive quality, or attribute, in addition to hydrogen and oxygen in molecular organization. This new quality, or meaning, that holy water has, which distinguishes it from ordinary water, was bestowed upon it by human beings, and this meaning can be grasped and appreciated by other human beings. It is this ability to originate and bestow, on the one hand, and to grasp, on the other, meanings that cannot be comprehended with the senses that we have termed "symboling."* Symboling is a kind of behavior that is characterized by traffic in non-sensory meanings.

Things that we call "symbols" have often been confused with things that we call "signs." Some psychologists who have worked with rats or apes have described their behavior in terms of symbols. The red triangles that mean food, the green circles that mean an electric shock, the blue or yellow poker chips that are used in the "chimp-o-mat," etc., have been called symbols because their meaning or significance is not inherent in them; their meanings have been assigned to them. Therefore, they conclude, they are just like our symbols — holy water, words, crucifixes, fetishes, etc. This reasoning is unsound because it has allowed a similarity to obscure a fundamental difference between these two kinds of situations.

The red circles and the blue poker chips have indeed acquired a meaning which is not inherent in their physical structure and composition, just as holy water or the combination of sounds *see* has acquired a meaning. This is the similarity. But there is a fundamental difference also. It is not the rats and apes that have determined the meanings which the red circles or the blue poker chips have acquired, and, what is more, *they are incapable of doing this*. Only man is capable of freely originating such kinds of meanings and of bestowing them upon things or events. The apes and rats can *acquire* meanings for things, but they cannot originate or determine them. They are acquired by the mechanism of the conditioned reflex, on the level of Type II; the ability to originate and to bestow meanings is found only on the level of Type IV. The behavior of the rats responding to red circles or the poker-chip-using apes is not, therefore, symbol behavior but *sign* behavior.

Each stage in the evolution of minding has given life a new dimension:

---

* We have discussed this subject at some length in "The Symbol: The Origin and Basis of Human Behavior," *Philosophy of Science*, VII (1940), 451–63. This essay was reprinted, in slightly revised form, in Leslie A. White, *The Science of Culture* (New York: Farrar, Straus & Co., 1949; New York: Grove Press, 1958). It has also been reprinted in *ETC., a Review of General Semantics*, I (1954), 229–37; *Language, Meaning, and Maturity*, ed. S. I. Hayakawa (New York, 1954); *Readings in Anthropology*, ed. E. Adamson Hoebel *et al.* (New York, 1955); *Readings in Introductory Anthropology*, ed. Elman R. Service (Ann Arbor, Mich., 1956); *Sociological Theory*, ed. Lewis A. Coser and Bernard Rosenberg (New York, 1957); and *Readings in the Ways of Mankind*, ed. Walter Goldschmidt (1957).

Type II, the conditioned reflex, emancipated living organisms from limitations imposed upon them by the intrinsic properties of the environment; Type III gave organisms positive control over things and events in the external world. But Type IV, symboling, has been especially significant, at least as far as human beings are concerned; indeed, it was symboling that made human beings of certain primates.

Both human behavior and culture are expressions and products of symboling. Human behavior consists of acts and things, dependent upon symboling, considered in terms of their relationship to the human organism.

"Culture" is the name of a flow of things and events dependent upon symboling considered in an extra-somatic context.

Symboling has brought a certain kind of things and events into existence. They constitute a continuum, a flow of tools, customs, and beliefs, down through the ages. Into this flow, this extra-somatic continuum called "culture," every human individual and group is born. And the behavior of these human beings is a function of this extra-somatic continuum: an organism born into Tibetan culture behaves in one way (as a Tibetan); an individual born into Scandinavian culture behaves in another way. Thus the determinants of human behavior, insofar as the individuals may be considered as typical or average, are no longer the properties of the biological organism; the determinants are to be found in the extra-somatic tradition (culture). It is not the nature of the lips, palate, teeth, tongue, etc., that determine whether the human organism will speak Tibetan or Swedish; it is the linguistic tradition that determines this. Therefore, in contrast with all other kinds of living organisms, if we wish to learn why a typical individual — a typical Crow Indian or a typical Englishman — behaves as he does, we must concern ourselves not with their bodies, their neuromuscular-sensory-glandular systems, but with the cultures into which they have been born and to which they respond. Similarly, if we wish to learn why the Japanese behave differently in 1960 than in 1860, we must concern ourselves with the changes that have taken place in their culture.

In the scientific study of behavior, or minding, therefore, we must concern ourselves with organisms when we are dealing with Types I, II, and III. But when we come to Type IV and man, a new world has been created, an extra-somatic, cultural environment, and it is this which determines the behavior of peoples living within it and not their bodily structures. This means that we must have a new science: a science of culture rather than a science of psychology if we are to understand the determinants of *human* behavior.

White identifies different levels of or stages in the capacity to recognize and respond to the environment. What he is tracing is in part the evolution of learning, selectivity and displacement. At what he terms type I level behaviour, we find organisms having a level of sentience not so far removed from that of the inorganic world, in so far as the inorganic world could be said to have any sentience at all. By the time we reach type IV behaviour we are looking at symbolic capacities which are essentially human.

White's elegantly simple account remains a powerful summary even though it has been somewhat overtaken by more recent research. For example, the work on the symbolic capacity of apes brings into question White's claim that type IV behaviour is strictly limited to human beings. This, however, does not invalidate his related point about signs and symbols, which is that only human beings are capable of freely originating arbitrary symbolic meanings and attaching them to things or events.

The creation of symbolic meanings is not just a linguistic matter. The evolution of human culture is not merely the evolution of language but also of pre-linguistic social and technological practices. This is an important factor in the process by which human infants become conscious. This is illustrated in the following brief excerpt from an essay on what its author, Chris Sinha, calls a socio-naturalistic approach to human development.

> The human environment, however has always and already been *intentionally shaped* by previous generations of human agents into a material culture embedded within social practices and institutions. Thus, the physical environment of the human infant is *meaningful in its material structure* and *represents* human consciousness and intentionality. Representation, in the epigenetic viewpoint, is no longer to be seen as merely mental: consciousness and knowledge are inscribed, not just in brains and nervous systems, but also in artefacts, institutions, practices, symbols, utterances and languages. Representation, like behaviour, extends beyond the boundary of the individual organism. The human infant, in development, is engaged in an *accommodatory effort after meaning*, whereby culture and representation is assimilated anew by every generation.

Symbolic function has brought into being the realm of cultural meanings within which human consciousness develops. Sinha reminds us that this realm is not only linguistic but also that it includes technological and social activity. These extracts do not, of course, present any substantial picture of Sinha's theory.[1] They do, however, give some idea of its breadth and its relationship to other theories of cognitive development and cognition as a whole.

In emphasizing the technological and physical environment as a significant part of the system of cultural meanings which create human consciousness, Sinha addresses an issue which has been a characteristic of theories of cognitive development in Soviet psychology. These have taken the cultural and social matrix of human consciousness as a fundamental starting point, as the following extract by A. R. Luria illustrates.

> It seems surprising that the science of psychology has avoided the idea that many mental processes are social and historical in origin, or that important manifestations of human consciousness have been directly shaped by the basic practices of human activity and the actual forms of culture.
>
> Beginning in the middle of the nineteenth century, psychology tried to view itself as an independent science aspiring to an objective analysis of the physiological mechanisms involved in behavior.
>
> During the course, however, of psychology's attempt to make itself an exact science, it has looked for laws of mental activity "within the organism." It has regarded association, or apperception, the structural nature of perception, or conditioned reflexes underlying behavior as either natural and unchanging properties of the organism (physiological psychology) or as manifestations or intrinsic properties of the mind (idealistic psychology). The notion that the intrinsic properties and laws of mental activity remain unchanging has also led to attempts to set up a positivist social psychology and sociology based on the premise that social activities display mental properties operating within individuals.
>
> We cannot doubt that scientific psychology made considerable progress during the past century and contributed greatly to our knowledge of mental activity. Nonetheless, it has generally ignored the social origin of higher mental processes. The patterns it describes turn out to be the same for animals and for human beings, for humans of different cultures and different historical eras, and for elementary mental processes and complex forms of mental activity.
>
> Moreover, the laws of logical thought, active remembering, selective attention, and acts of the will in general, which form the basis for the most complex and characteristic higher forms of human mental activity, successfully resisted causal interpretation, and thus remained beyond the forefront of the progression of scientific thought.
>
> It was not by accident that Bergson spoke of the laws of "memory of the spirit" in addition to the natural laws of "memory of the body," while neo-Kantian philosophers distinguished (in addition to the laws of association that could be analyzed by natural science) laws of "symbolic forms" which functioned as manifestations of the "spiritual world" and had neither an origin nor a theory: they could be described but not accounted for. Despite objective progress, therefore, a major field of knowledge

remained divorced from causal explanations, and could not be studied in any meaningful way. This situation called for decisive steps to reexamine the basic approaches to mental activity in order to make psychology a truly scientific discipline decisively rejecting any kind of dualism and thus opening the way for a causal analysis of even the most complex mental phenomena. This reexamination implied the abandonment of subjectivism in psychology and the treatment of human consciousness as a product of social history.

The first attempts to approach human mental processes as the products of evolution were taken in the second half of the nineteenth century by Charles Darwin and his successor Herbert Spencer. These scientists attempted to trace the ways in which complex forms of mental activity develop and how elementary forms of biological adaptation to environmental conditions become more complex through the evolutionary process. The evolutionary approach, which was quite valid for a comparative study of mental development in the animal world, found itself in something of a blind alley when it tried to study the evolution of human mental activity.

This becomes intelligible to us, however, only if we understand people's actual living conditions and the language they use. This was the approach to human mental processes at the time that our work began.

The research reported here, undertaken forty years ago under Vygotsky's initiative and in the context of unprecendented social and cultural change, took the view that higher cognitive activities remain sociohistorical in nature, and that the structure of mental activity — not just the specific content but also the general forms basic to all cognitive processes — change in the course of historical development. For this reason our research remains valuable even today.

Soviet psychology, using the notion of consciousness as "conscious existence" (*das bewusste Sein*) as a starting point, has rejected the view that consciousness represents an "intrinsic property of mental life," invariably present in every mental state and independent of historical development. In line with Marx and Lenin, Soviet psychology maintains that consciousness is the highest form of reflection of reality; it is, moreover, not given in advance, unchanging and passive, but shaped by activity and used by human beings to orient themselves to their environment, not only in adapting to conditions but in restructuring them.

It has become a basic principle of materialistic psychology that mental processes depend on active life forms in an appropriate environment. Such a psychology also assumes that human action changes the environment so that human mental life is a product of continually *new* activities manifest in social practice.

The way in which the historically established forms of human mental life correlate with reality has come to depend more and more on complex social practices. The tools that human beings in society use to manipulate that environment, as well as the products of previous generations which help shape the mind of the growing child, also affect these mental forms. In his development, the child's first social relations and his first exposure to a linguistic system (of special significance) determine the forms of his mental activity. All these environmental factors are decisive for the sociohistorical development of consciousness. New motives for action appear under extremely complex patterns of social practice. Thus are created new problems, new modes of behavior, new methods of taking in information, and new systems of reflecting reality.

From the outset, the social forms of human life begin to determine human mental development. Consider the development of conscious activity in children. From birth on, children live in a world of things social labor has created: products of history. They learn to communicate with others around them and develop relationships with things through the help of adults. Children assimilate language — a ready-made product of sociohistorical development — and use it to analyze, generalize, and encode experience. They name things, denoting them with expressions established earlier in human history, and thus assign things to certain categories and acquire knowledge. Once a child calls something a "watch" (*chasy*), he immediately incorporates it into a system of things related to time (*chas*); once he calls a moving object a "steamship" (*parovoz*), he automatically isolates its defining properties — motion (*vozit'*) by means of "steam" (*par*). Language, which mediates human perception, results in extremely complex operations: the analysis and synthesis of incoming information, the perceptual ordering of the world, and the encoding of impressions into systems. Thus words — the basic linguistic units — carry not only meaning but also the fundamental units of consciousness reflecting the external world.

But the world of particular objects and verbal meanings that humans receive from earlier generations organizes not just perception and memory (thus ensuring the assimilation of experiences common to all humankind); it also establishes some important conditions for later, more complex developments in consciousness. Men can deal even with "absent" objects, and so "duplicate the world," through words, which maintain the system of meanings whether or not the person is directly experiencing the objects the words refer to. Hence a new source of productive imagination arises: it can reproduce objects as well as reorder their relationships and thus serve as the basis for highly complex creative processes. Men use a complex system of syntactical relations among the individual words in sentences, and are then able to formulate complex relationships among entities, and to generate and transmit thoughts and opinions. Because of the hierarchical system of individual sentences, of which verbal and logical constructions are a typical example, humans have at their disposal a powerful objective tool that permits them not only to reflect individual objects or situations but to create objective logical codes. Such codes enable a person to go beyond direct experience and to draw conclusions that have the same objectivity as the data of direct sensory experience. In other words, social history has established the system of language and

> logical codes that permit men to make the leap from the sensory to the rational; for the founders of materialistic philosophy, this transition was as important as that from nonliving to living matter.
>
> Human consciousness thus ceases to be an "intrinsic quality of the human spirit" with no history or intractibility to causal analysis. We begin to understand it as the highest form of reflection of reality that sociohistorical development creates: a system of objectively existing agents gives birth to it and causal historical analysis makes it accessible to us.
>
> The views expressed here are important not merely because they deal with human consciousness as a product of social history and point the way to a scientific historical analysis; they are also important because they deal with the process of broadening the limits of consciousness and of creating codes as a result of human social life. Moreover, some mental processes cannot develop apart from the appropriate forms of social life. This last observation is decisive for psychology and has opened up new and unforeseen prospects.
>
> Now the psychologist can not only describe the different and changing forms of conscious life of both the child and the adult; he can also analyze changes in the structure of those mental processes underlying mental activity at different stages of development and discover the hitherto unsuspected changes in their "interfunctional relationships." He can thus trace the historical development of mental systems.

Luria points out that the transition from sentience to symbolic function allowed evolution to proceed from the sensory to the rational. This is an evolutionary transition as important as that from the pre-biological to the biological phase and one which, as Huxley noted in Chapter Three, places human beings in the unique position of being aware of evolution and thus of being the first organisms that are, in some sense, responsible for their evolutionary progress.

The transition from sentience to symbols is the pivotal event in the evolution of human consciousness. It may also be just as crucial in the development of the individual. A variety of psychological theories model the transition from sensory motor interaction with the environment to interactions on a symbolic basis. This, it is clear, is the necessary foundation on which the human psychological subject, the person, is built.

There could hardly be a more significant psychological transition. The social creation of the psychological subject leads out from the biological and psychological support for consciousness into all other human concerns. It may be noted here that this transition and the associated emegence of the psychological subject are also the focuses of psychological traditions from other cultures. During the 1960s some of these were sought by psychologists, and strange alliances were formed to research the integration of Western and other psychological traditions.[2] This research has persisted and while some of it has proved superficial, some of it has produced substantial findings.[3] There is every reason to explore complementary approaches to consciousness and a later chapter presents perspectives from othe cultural traditions. For the present, however, the next chapter deals with the emergence of human consciousness in the social context.

Notes

1. This is most fully developed in his recent book, *Language and Representation*, 1988, Harvester Wheatsheaf.
2. Good surveys of the period may be found in R. Ornstein, Ed., *The Nature of Consciousness*, 1973, Freeman; and in C. Tart, *Altered States of Consciousness*, 1969, Wiley.
3. Johansson, R., *The Dynamic Psychology of Early Buddhism*, 1979, Curzon Press.

7

The Social Context of Consciousness

READINGS

George Herbert Mead
The Mechanism of Social Consciousness

B. N. Melzer
Mead's Social Psychology

Peter L. Berger and Thomas Luckmann
The Social Construction of Reality

S. Duval and R. Wicklund
A Theory of Objective Self Awareness

How can an organism represent experience, and more importantly how can an organism represent itself in its own experience? Furthermore, what are the consequences of doing so? The readings in this chapter begin to address these questions and those in subsequent sections can generally be described as exploring their corollaries.

We have seen in the extract by White (Chapter Six) that a case can be made for a phylogenetic elaboration of sentience and thus of experience. White calls this 'the evolution of minding' where minding refers to the organism's mode of sensory adaptation to the environment. He makes the important point that the mode determines the environment for the organism. This is to propose a truly dialectical relationship between organism and environment. White is suggesting it not only at the obvious physical level where evolutionary theorists have shown that the environment has consequences for the survival of natural forms and that the activities of these forms have consequences for the environment. He is also proposing that the physical world provides the sensory stimulation for sensation and that the sense organs determine which energy sources of the environment will become sensation and experience.

The question of how sentience can devleop to the point where it contains a representation of the organism which is at the same time the vehicle for that sentience is addressed by readings in this chapter. How can an organism grasp its own existence reflexively and be both subject and object of its thoughts—which is what we mean by self-consciousness?

We look at the human use of symbols in this chapter. We shall follow the view of the philosopher and social psychologist, George Herbert Mead, that the symbols we use to represent sensory experience as objects communicable with others and to ourselves emerge in the course of social interaction. This is why we emphasize in this and subsequent chapters the essential social nature of human consciousness.

For our use of the word 'symbol' we can return to the White article. Symbols represent or stand for some other object or event and in this respect they do a similar job to signs. There is a crucial difference, however, between a sign and a symbol which has to do with their respective relationships to their referents. For signs the relationship is natural. In other words the sign stands in some logical, temporal or spatial relationship to its referent. Thus clouds may be a sign of rain, the first swallow a sign of summer or for Pavlov's dogs the bell a sign of food on the way. In the case of symbols the choice of symbol and consequently the relationship to its referent is conventional and therefore more or less arbitrary. The symbol stands for something only because people have agreed and learned that it should do so. Thus the ink marks 'rain' stand as a symbol for drops of condensed water falling from clouds or the sound we utter when speaking the word 'summer' stands for that part of the year. This latter example reminds us that the most important and distinctively human symbols are the vocal ones of spoken language. Symbols are truly permissible only between consenting adults!

We shall see in the first of the extracts in this chapter that symbols permit the representation of subjective experience as objects exchangeable with others. As such they offer unique facilities whereby the mental state of one individual can become if not the same as, then definitely correspondent upon, that of another. This will turn out to be crucial to the understanding of the self as an object in one's own thoughts, for something special happens when the mental state of another becomes the basis of one's own mental state through the exchange of symbols.

Mead's approach to the study of human consciousness reflects the era in which he was writing — the first three decades of this century. Writing in the wake of Darwin, he brought an evolutionary and phylogenetic perspective to consciousness. This entailed that whatever human consciousness was, it had evolved from whatever animals did. With this evolutionary perspective went a functional view of animal behaviour with consciousness seen not to lie outside the field of adaptive, selected behaviour. With this point of view, and in common with a number of important philosophers such as William James and John Dewey who were writing at the end of the nineteenth century, went two more aspects of Mead's approach to psychology. These were a behaviouristic approach to consciousness and, as an offshoot of this, a pragmatist philosophical orientation which viewed the validity and meaning of behaviour and ideas in terms of their consequences for experience.

We should add that the term 'behaviouristic' when applied to Mead's work refers to something broader than its use in describing the psychology of, for example, J. B. Watson or B. F. Skinner. For Watson and Skinner behaviour refers to externally visible movement and thus precludes consciousness. Mead includes the internal behaviour involved in the passage of symbol to its consequences. His contribution to the study of consciousness amounts to no less than a behavioural theory of human consciousness. He starts with the behaviour of animals which in our use of the terms are definitely sentient, arguably conscious but definitely not *self*-conscious. From this point an account of the emergence of self-consciousness is advanced. In chronological order, the principal elements of the account are the direct instinctive stimulation of one animal by another in the course of social behaviour; the devleopment of gestural communication where animals recognize and adjust their behaviour to initial stages of acts; and symbolic communication where signs of impending acts are replaced by symbols, the most important and distinctively human of which are the vocal significant symbols of spoken language.

These elements are discussed in the first reading of the chapter by Mead. In it the emergence of the apprehension of self is charted from the so-called field of gestures to the self-conscious self where as the final sentence proposes, 'Inner consciousness is socially organized by the importation of the social organization of the outer world.' Thus the account is a social one placing social interaction as the precursor of human consciousness and definitely not *vice-versa*. The emergence of self from the symbolic designation of the organism in terms of the symbols at large in the verbal community, and the consequent emergence of mind as an internal dialogue between unreflected experience and its symbolic representation states are both processes dependent upon and conditioned by social interaction.

THE MECHANISM OF SOCIAL CONSCIOUSNESS

The organization of consciousness may be regarded from the standpoint of its objects and the relation of these objects to conduct. I have in mind to present somewhat schematically the relation of social objects or selves to the form of social conduct, and to introduce this by a statement of the relation of the physical object to the conduct within which it appears.

A physical object or percept is a construct in which the sensuous stimulation is merged with imagery which comes from past experience. This imagery on the cognitive side is that which the immediate sensuous quality stands for, and insofar satisfies the mind. The reason for this satisfaction is found in the fact that this imagery arises from past experience of the result of an act which this stimulus has set going. Thus the wall as a visual stimulus tends to set free the impulse to move toward it and push against it. The perception of the wall as distant and hard and rough is related to the visual experience as response to stimulation. A peculiar stimulus value stands for a certain response value. A percept is a collapsed act in which the result of the act to which the stimulus incites is represented by imagery of the experience of past acts of a like nature.

Insofar as our physical conduct involves movements toward or away from distant objects and their being handled when we come into contact with them, we perceive all things in terms of distance sensation — color, sound, odor — which stand for hard or soft, big or little, objects of varying forms, which actual contact will reveal.

Our conduct in movement and manipulation, with its stimulations and responses, gives the framework within which objects of perception arise — and this conduct is in so far responsible for the organization of our physical world. Percepts — physical objects — are compounds of the experience of immediate stimulation and the imagery of the response to which this stimulation will lead. The object can be properly stated in terms of conduct.

I have referred to percepts as objects which arise in physical experience because it is a certain phase of conduct which, with its appropriate stimuli and responses, gives rise to such products, i.e., movement under the influence of distant stimuli leading to contact experiences of manipulation.

Given a different type of conduct with distinguishable stimulations and responses, and different objects would arise — such a different field is that of social conduct. By social conduct I refer simply to that which is mediated by the stimulations of other animals belonging to the same group of living forms, which lead to responses which again affect these other forms — thus fighting, reproduction, parental care, much of animal play, hunting, etc., are the results of primitive instincts or impulses which are set going by the stimulation of one form by another, and these stimulations again lead to responses which affect other forms.

It is of course true that a man is a physical object to the perception of another man, and as real as is a tree or a stone. But a man is more than a physical object, and it is this more which constitutes him a social object or self, and it is this self which is related to that peculiar conduct which may be termed social conduct.

Most social stimulation is found in the beginnings or early stages of social acts which serve as stimuli to other forms whom these acts would affect. This is the field of gestures, which reveal the motor attitude of a form in its relation to others; an attitude which psychologists have conceived of as predominantly emotional, though it is emotional only insofar as an ongoing act is inhibited. That certain of these early indications of an incipient act have persisted, while the rest of the act has been largely suppressed or has lost its original value, e.g., the baring of the teeth or the lifting of the nostrils, is true, and the explanation can most readily be found in the social value which such indications have acquired. It is an error, however, to overlook the relation which these truncated acts have assumed toward other forms of reactions which complete them as really as the original acts, or to forget that they occupy but a small part of the whole field of gesture by means of which we are apprised of the reactions of others toward ourselves. The expressions of the face and attitudes of body have the same functional value for us that the beginnings of hostility have for two dogs, who are maneuvering for an opening to attack.

This field of gesture does not simply relate the individual to other individuals as physical objects, but puts him *en rapport* with their actions, which are as yet only indicated, and arouses instinctive reactions appropriate to these social activities. The social response of one individual, furthermore, introduces a further complication. The attitude assumed in response to the attitude of another becomes a stimulus to him to change his attitude, thus leading to that conversation of attitudes which is so vividly illustrated in the early stages of a dog fight. We see the same process in courting and mating, and in the fondling of young forms by the mother, and finally in much of the play of young animals.

It has been recognized for some time that speech belongs in its beginnings, at least, to this same field of gesture, so-called vocal gesture. Originally indicating the preparation for violent action, which arises from a sudden change of breathing and circulation rhythms, the articulate sounds have come to elaborate and immensely complicate this conversation of attitudes by which social forms so adjust themselves to each other's anticipated action that they may act appropriately with reference to each other.

Articulate sounds have still another most important result. While one feels but imperfectly the value of his own facial expression or bodily attitude for another, his ear reveals to him his own vocal gesture in the same form that it assumes to his neighbor. One shakes his fist primarily only at another, while he talks to himself as really as he talks to his vis-à-vis. The genetic import of this has long been recognized. The young child talks to himself, i.e., uses the elements of articulate speech in response to the sounds he hears himself make, more continuously and persistently than he does in response to the sounds he hears from those about him, and displays greater interest in the sounds he himself makes than in those of others. We know also that this fascination of one's own vocal gestures continues even after the child has learned to talk with others, and that the child will converse for hours with himself, even constructing imaginary companions, who function in the child's growing self-consciousness as the processes of inner speech — of thought and imagination—function in the consciousness of the adult.

To return to the formula given above for the formation of an object in consciousness, we may define the social object in terms of social conduct as we defined the physical object in terms of our reactions to physical objects. The object was found to consist of the sensuous experience of the stimulation to an act plus the imagery from past experience of the final result of the act. The social object will then be the gestures, i.e., the early indications of an ongoing social act in another plus the imagery of our own response to that stimulation. To the young child the frowns and smiles of those about him, the attitude of body, the outstretched arms, are at first simply stimulations that call out instinctive responses of his own appropriate to these gestures. He cries or laughs, he moves toward his mother, or stretches out his arms. When these gestures in others bring back the images of his own responses and their results, the child has the material out of which he builds up the social objects that form the most important part of his environment. We are familiar with this phase of a baby's development, being confident that he recognizes the different members of the group about him. He acts then with confidence toward them since their gestures have come to have meaning for him. His own response to their stimulations and its consequences are there to interpret the facial expressions and attitudes of body and tones of voice. The awakening social intelligence of the child is evidenced not so much through his ready responses to the gestures of others, for these have been in evidence much earlier. It is the inner assurance of his own readiness to adjust himself to the attitudes of others that looks out of his eyes and appears in his own bodily attitudes.

If we assume that an object arises in consciousness through the merging of the imagery of experience of the response with that of the sensuous experience of the stimulation, it is evident that the child must merge the imagery of his past responses into the sensuous stimulation of what comes to him through distance senses. His contact and kinesthetic experiences must be lodged in the sensuous experiences that call them out if they are to achieve objective character in his consciousness.

It will be some time before he can successfully unite the different parts of his own body, such as his hands and feet, which he sees and feels, into a single object. Such a step must be later than the formation of the physical objects of his environment. The form of the object is given in the experience of things, which are not his physical self. When he has synthesized his various bodily parts with the organic sensations and affective experiences, it will be upon the model of objects about him. The mere presence of experiences of pleasure and pain, together with organic sensations, will not form an object unless this material can fall into the scheme of an object — that of sensuous stimulation plus the imagery of the response.

In the organization of the baby's physical experience the appearance of his body as a unitary thing, as an object, will be relatively late, and must follow upon the structure of the objects of his environment. This is as true of the object that appears in social conduct, the self. The form of the social object must be found first of all in the experience of other selves. The earliest achievement of social consciousness will be the merging of the imagery of the baby's first responses and their results with the stimulations of the gestures of others. The child will not succeed in forming an object of himself — of putting the so-called subjective material of consciousness within such a self — until he has recognized about him social objects who have arisen in his experience through this process of filling out stimulations with past experiences of response. And this is indeed our uniform experience with children. The child's early social precepts are of others. After these arise incomplete and partial selves — or "me's" — which are quite analogous to the child's percepts of his hands and feet, which precede his perception of himself as a whole. The mere presence of affective experience, of imagery, of organic sensations, does not carry with it consciousness of a self to which these experiences belong. Nor does the unitary character of the response which tends to synthesize our objects of perception convey that same unitary character to the inner experience until the child is able to experience himself as he experiences other selves.

It is highly probable that lower animals never reach any such objective reference of what we term subjective experiences to selves, and the question presents itself — what is there in human social conduct that gives rise to a "me," a self which is an object? Why does the human animal transfer the form of a social object from his environment to an inner experience?

The answer to the question is already indicated in the statement of vocal gesture. Certainly the fact that the human animal can stimulate himself as he stimulates others and can respond to his stimulations as he responds to the stimulations of others, places in his conduct the form of a social object out of which may arise a "me" to which can be referred so-called subjective experiences.

Of course the mere capacity to talk to oneself is not the whole of self-consciousness, otherwise the talking birds would have souls or at least selves. What is lacking to the parrot are the social objects which can exist for the human baby. Part of the mechanism for transferring the social objects into an inner experience the parrot possesses, but he has nothing to import into such an inner world. Furthermore, the vocal gesture is not the only form which can serve for the building-up of a "me," as is abundantly evident from the building-up gestures of the deaf mutes. Any gesture by which the individual can himself be affected as others are affected, and which therefore tends to call out in him a response as it would call it out in another, will serve as a mechanism for the construction of a self. That, however, a consciousness of a self as an object would ever have arisen in man if he had not had the mechanism of talking to himself, I think there is every reason to doubt.

If this statement is correct the objective self of human consciousness is the merging of one's responses with the social stimulation by which he affects himself. The "me" is a man's reply to his own talk. Such a "me" is not then an early formation, which is then projected and ejected into the bodies of other people to give them the breadth of human life. It is rather an importation from the field of social objects into an amorphous, unorganized field of what we call inner experience. Through the organization of this object, the self, this material is itself organized and brought under the control of the individual in the form of so-called self-consciousness.

It is a commonplace of psychology that it is only the "me" — the empirical self — that can be brought into the focus of attention — that can be perceived. The "I" lies beyond the range of immediate experience. In terms of social conduct this is tantamount to saying that we can perceive our responses only as they appear as images from past experience, merging with the sensuous stimulation. We cannot present the response while we are responding. We cannot use our responses to others as the materials for construction of the self — this imagery goes to make up other selves. We must socially stimulate ourselves to place at our own disposal the material out of which our own selves as well as those of others must be made.

The "I" therefore never can exist as an object in consciousness, but the very conversational character of our inner experience, the very process of replying to one's own talk, implies an "I" behind the scenes who

> answers to the gestures, the symbols, that arise in consciousness. The "I" is the transcendental self of Kant, the soul that James conceived behind the scene holding on to the skirts of an idea to give it an added increment of emphasis.
>
> The self-conscious, actual self in social intercourse is the objective "me" or "me's" with the process of response continually going on and implying a fictitious "I" always out of sight of himself.
>
> Inner consciousness is socially organized by the importation of the social organization of the outer world.

The elaboration of these ideas into a social and behavioural view of human consciousness appears in *Mind, Self, and Society* which is made up from the edited lecture notes of some of the students taking Mead's course in social psychology at the University of Chicago shortly before his death in 1931. We have chosen extracts from 'Mead's Social Psychology' by Meltzer to represent Mead's social view of consciousness as 'mind' and 'self': a social behavioural theory of human consciousness. The extract emphasizes this fact by the order of its treatment of the topics society, self and mind.

> According to Mead, all group life is essentially a matter of cooperative behavior. Mead makes a distinction, however, between infrahuman society and human society. Insects — whose society most closely approximates the complexity of human social life — act together in certain ways because of their biological make-up.
>
> In the case of human association, the situation is fundamentally different.
>
> Human behavior is not a matter of responding directly to the activities of others. Rather, it involves responding to the *intentions* of others, *i.e.*, to the future, intended behavior of others — not merely to their present actions.
>
> We can better understand the character of this distinctively human mode of interaction between individuals by contrasting it with the infrahuman "conversation of gestures."
>
> Let us take another illustration by Mead: Two hostile dogs, in the pre-fight stage, may go through an elaborate conversation of gestures (snarling, growling, baring fangs, walking stiff-leggedly around one another, etc.). The dogs are adjusting themselves to one another by responding to one another's gestures. (A gesture is that portion of an act which represents the entire act; it is the initial, overt phase of the act, which epitomizes it, *e.g.*, shaking one's fist at someone.) Now, in the case of the

dogs the response to a gesture is dictated by pre-established tendencies to respond in certain ways. Each gesture leads to a direct, immediate, automatic, and unreflecting response by the recipient of the gesture (the other dog). Neither dog responds to the *intention* of the gestures. Further, each dog does not make his gestures with the intent of eliciting certain responses in the other dog. Thus, animal interaction is devoid of conscious, deliberate meaning.

To summarize: Gestures, at the non-human or non-linguistic level, do not carry the connotation of conscious meaning or intent, but serve merely as cues for the appropriate responses of others. Gestural communication takes place immediately, without any interruption of the act, without the mediation of a definition or meaning. Each organism adjusts "instinctively" to the other; it does not stop and figure out which response it will give. Its behavior is, largely, a series of direct automatic responses to stimuli.

Human beings, on the other hand, respond to one another on the basis of the intentions or meanings of gestures. This renders the gesture *symbolic, i.e.*, the gesture becomes a symbol to be interpreted; it becomes something which, in the imaginations of the participants, stands for the entire act.

Thus, individual A begins to act, *i.e.*, makes a gesture: for example, he draws back an arm. Individual B (who perceives the gesture) completes, or fills in, the act in his imagination; *i.e.*, B imaginatively projects the gesture into the future: "He will strike me." In other words, B perceives what the gesture stands for, thus getting its meaning. In contrast to the direct responses of the chicks and the dogs, the human being inserts an interpretation between the gesture of another and his response to it. Human behavior involves responses to *interpreted* stimuli.*

We see, then, that people respond to one another on the basis of imaginative activity. In order to engage in concerted behavior, however, each participating individual must be able to attach the same meaning to the same gesture. Unless interacting individuals interpret gestures similarly, unless they fill out the imagined portion in the same way, there can be no cooperative action. This is another way of saying what has by now become a truism in sociology and social psychology: Human society rests upon a basis of *consensus, i.e.*, the sharing of meanings in the form of common understandings and expectations.

---

* The foregoing distinctions can also be expressed in terms of the differences between "signs," or "signals," and symbols. A sign stands for something else because of the fact that it is present at approximately the same time and place with that "something else." A symbol, on the other hand, stands for something else because its users have agreed to let it stand for that "something else." Thus, signs are directly and intrinsically linked with present or proximate situations; while symbols, having arbitrary and conventional, rather than intrinsic, meanings, transcend the immediate situation. (We shall return to this important point in our discussion of "mind.") Only symbols, of course, involve interpretation, self-stimulation and shared meaning.

In the case of the human being, each person has the ability to respond to his own gestures; and thus, it is possible to have the same meaning for the gestures as other persons. (For example: As I say "chair," I present to myself the same image as to my hearer; moreover, the same image as when someone else says "chair.") This ability to stimulate oneself as one stimulates another, and to respond to oneself as another does, Mead ascribes largely to man's vocal-auditory mechanism. (The ability to hear oneself implies at least the potentiality for responding to oneself.) When a gesture has a shared, common meaning, when it is — in other words — a *linguistic* element, we can designate it as a "significant symbol."

The imaginative completion of an act — which Mead calls "meaning" and which represents mental activity — necessarily takes place through *role-taking*. To complete imaginatively the total act which a gesture stands for, the individual must put himself in the position of the other person, must identify with him. The earliest beginnings of role-taking occur when an already established act of another individual is stopped short of completion, thereby requiring the observing individual to fill in, or complete, the activity imaginatively. (For example, a crying infant may have an image of its mother coming to stop its crying.)

As Mead points out, then, the relation of human beings to one another arises from the developed ability of the human being to respond to his own gestures. This ability enables different human beings to respond in the same way to the same gesture, thereby sharing one another's experience.

This latter point is of great importance. Behavior is viewed as "social" not simply when it is a response to others, but rather when it has incorporated in it the behavior of others. The human being responds to himself as other persons repond to him, and in so doing he imaginatively shares the conduct of others. That is, in imagining their response he shares that response.†

To state that the human being can respond to his own gestures necessarily implies that he possesses a *self*. In referring to the human being as having a self, Mead simply means that such an individual may act socially toward himself, just as toward others. He may praise, blame, or encourage himself; he may become disgusted with himself, may seek to punish himself, and so forth. Thus, the human being may become the object of his own actions. The self is formed in the same way as other objects — through the "definitions" made by others.

The mechanism whereby the individual becomes able to view himself as an object is that of role-taking, involving the process of communication,

---

† To anyone who has taken even one course in sociology it is probably superfluous to stress the importance of symbols, particularly language, in the acquisition of all other elements of culture. The process of socialization is essentially a process of symbolic interaction.

especially by vocal gestures or speech. (Such communication necessarily involves role-taking.) It is only by taking the role of others that the individual can come to see himself as an object. The standpoint of others provides a platform for getting outside oneself and thus viewing oneself. The development of the self is concurrent with the development of the ability to take roles.

The crucial importance of language in this process must be underscored. It is through language (significant symbols) that the child acquires the meanings and definitions of those around him. By learning the symbols of his groups, he comes to internalize their defnitions of events or things, including their definitions of his own conduct.

It is quite evident that, rather than assuming the existence of selves and explaining society thereby, Mead starts out from the prior existence of society as the context within which selves arise.

Genesis of the Self. The relationship between role-playing and various stages in the development of the self is described below:

1. *Preparatory Stage* (not explicitly named by Mead, but inferable from various fragmentary essays). This stage is one of meaningless imitation by the infant (for example, "reading" the newspaper). The child does certain things that others near it do without any understanding of what he is doing. Such imitation, however, implies that the child is incipiently taking the roles of those around it, *i.e.*, is on the verge of putting itself in the position of others and acting like them.
2. *Play Stage*. In this stage the actual playing of roles occurs. The child plays mother, teacher, storekeeper, postman, streetcar conductor, Mr. Jones, etc. What is of central importance in such play-acting is that it places the child in the position where it is able to act back toward itself in such roles as "mother" or "teacher." In this stage, the, the child first begins to form a self, that is, to direct activity toward itself — and it does so by taking the roles of others. This is clearly indicated by use of the third person in referring to oneself instead of the first person: "John wants. . .," "John is a bad boy."

   However, in this stage the young child's configuration of roles is unstable; the child passes from one role to another in unorganized, inconsistent fashion. He has, as yet, no unitary standpoint from which to view himself, and hence, he has no unified conception of himself. In other words, the child forms a number of separate and discrete objects of itself, depending on the roles in which it acts toward itself.
3. *Game Stage*. This is the "completing" stage of the self. In time, the child finds himself in situations wherein he must take a number of roles simultaneously. That is, he must respond to the expectations of several people at the same time. This sort of situation is exemplified by the game of baseball — to use Mead's own illustration. Each player must visualize the intentions and expectations of several other players. In such situations the child must take the roles of groups of individuals as over against particular roles. The child becomes enabled to do this by abstracting a "composite" role out of the concrete roles of particular

persons. In the course of his association with others, then, he builds up a *generalized other*, a generalized role or standpoint from which he views himself and his behavior. This generalized other represents, then, the set of standpoints which are common to the group.

Having achieved this generalized standpoint, the individual can conduct himself in an organized, consistent manner. He can view himself from a consistent standpoint. This means, then, that the individual can transcend the local and present expectations and definitions with which he comes in contact. An illustration of this point would be the Englishman who "dresses for dinner" in the wilds of Africa. Thus, through having a generalized other, the individual becomes emancipated from the pressures of the peculiarities of the immediate situation. He can act with a certain amount of consistency in a variety of situations because he acts in accordance with a generalized set of expectations and definitions that he has internalized.

The "I" and the "Me." The self is essentially a social process within the individual involving two analytically distinguishable phases: The "I" and the "Me."

The "I" is the impulsive tendency of the individual. It is the initial, spontaneous, unorganized aspect of human experience. Thus, it represents the undirected tendencies of the individual.

The "Me" represents the incorporated other within the individual. Thus, it comprises the organized set of attitudes and definitions, understandings and expectations — or simply meanings — common to the group. In any given situation, the "Me" comprises the generalized other and, often, some particular other.

Every act begins in the form of an "I" and usually ends in the form of the "Me." For the "I" represents the initiation of the act prior to its coming under control of the definitions or expectations of others (the "Me"). The "I" thus gives *propulsion* while the "Me" gives *direction* to the act. Human behavior, then, can be viewed as a perpetual series of initiations of acts by the "I" and of acting-back-upon the act (that is, guidance of the act) by the "Me." The act is a resultant of this interplay.

The "I," being spontaneous and propulsive, offers the potentiality for new, creative activity. The "Me," being regulatory, disposes the individual to both goal-directed activity and conformity. In the operation of these aspects of the self, we have the basis for, on the one hand, social control and, on the other, novelty and innovation. We are thus provided with a basis for understanding the mutuality of the relationship between the individual and society.‡

---

‡ At first glance, Mead's "I" and "Me" may appear to bear a close affinity with Freud's concepts of Id, Ego, and Superego. The resemblance is, for the most part, more apparent than real. While the Superego is held to be harshly frustrating and repressive of the instinctual, libidinous, and aggressive Id, the "Me" is held to provide necessary direction — often of a *gratifying* nature — to the otherwise undirected impulses constituting the "I." Putting the matter in figurative terms:

Implications of Selfhood. Some of the major implications of selfhood in human behavior are as follows:

1. The possession of a self makes of the individual a society in miniature. That is, he may engage in interaction with himself just as two or more different individuals might. In the course of this interaction, he can come to view himself in a new way, thereby bringing about changes in himself.
2. The ability to act toward oneself makes possible an inner experience which need not reach overt expression. That is, the individual, by virtue of having a self, is thereby endowed with the possibility of having a mental life: He can make indications to himself — which constitutes *mind.*
3. The individual with a self is thereby enabled to direct and control his behavior. Instead of being subject to all impulses and stimuli directly playing upon him, the individual can check, guide, and organize his behavior. He is, then, *not* a mere passive agent.

All three of these implications of selfhood may be summarized by the statement that the self and the mind (mental activity) are twin emergents in the social process.

Development of Mind. As in the instance of his consideration of the self, Mead rejects individualistic psychologies, in which the social process (society, social interaction) is viewed as presupposing, and being a product of, mind. In direct contrast is his view that mind presupposes, and is a product of, the social process. Mind is seen by Mead as developing correlatively with the self, constituting (in a very important sense) the self in action.

Mead's hypothesis regarding mind (as regarding the self) is that the mental emerges out of the organic life of man through communication. The mind is present only at certain points in human behavior, *viz.*, when significant symbols are being used by the individual. This view dispenses with the substantive notion of mind as existing as a box-like container in the head, or as some kind of fixed, ever-present entity. Mind is seen as

Freud views the Id and the Superego as locked in combat upon the battleground of the Ego; Mead sees the "I" and "Me" engaged in close collaboration. This difference in perspective may derive from different preoccupations: Freud was primarily concerned with tension, anxiety, and "abnormal" behavior; Mead was primarily concerned with behavior generically.

It is true, on the other hand, that the Id, Ego, and Superego — particularly as modified by such neo-Freudians as Karen Horney, Erich Fromm, and H. S. Sullivan — converge at a few points with the "I" and "Me." This is especially evident in the emphasis of both the Superego and "Me" concepts upon the internalization of the norms of significant others through the process of identification, or role-taking.

Incidentally, it should be noted that both sets of concepts refer to processes of behavior, *not* to concrete entities or structures. See, also, the discussion of "mind" which follows.

a *process*, which manifests itself whenever the individual is interacting with himself by using significant symbols.

Mead begins his discussion of the mind with a consideration of the relation of the organism to its environment. He points out that the central principle in all organic behavior is that of continuous adjustment, or adaptation, to an environing field. We cannot regard the environment as having a fixed character for all organisms, as being the same for all organisms. All behavior involves selective attention and perception. The organism accepts certain events in its field, or vicinity, as stimuli and rejects or overlooks certain others as irrelevant to its needs. (For example, an animal battling for life ignores food.) Bombarded constantly by stimuli, the organism selects those stimuli or aspects of its field which pertain to, are functional to, the acts in which the organism is engaged. Thus, the organism has a hand in determining the nature of its environment. What this means, then, is that Mead, along with Dewey, regards all life as ongoing activity, and views stimuli — not as initiators of activity — but as elements selected by the organism in the furtherance of that activity.

Perception is thus an activity that involves selective attention to certain aspects of a situation, rather than a mere matter of something coming into the individual's nervous system and leaving an impression. Visual perception, *e.g.*, is more than a matter of just opening one's eyes and responding to what falls on the retina.

The determination of the environment by the biologic individual (infrahumans and the unsocialized infant) is not a cognitive relationship. It is selective, but does not involve consciousness, in the sense of reflective intelligence. At the distinctively human level, on the other hand, there is a hesitancy, an inhibition of overt conduct, which is *not* involved in the selective attention of animal behavior. In this period of inhibition, mind is present.

For human behavior involves inhibiting an act and trying out the varying approaches in imagination. In contrast, as we have seen, the acts of the biologic individual are relatively immediate, direct, and made up of innate or habitual ways of reacting. In other words, the unsocialized organism lacks consciousness of meaning. This being the case, the organism has no means for the abstract analysis of its field when new situations are met, and hence no means for the reorganization of action-tendencies in the light of that analysis.§

Minded behavior (in Mead's sense) arises around problems. It represents, to repeat an important point, a temporary inhibition of action wherein the

---

§ The reader should recognize here, in a new guise, our earlier distinction between signs and symbols. Signs have "intrinsic" meanings which induce direct reactions; symbols have arbitrary meanings which require interpretations by the actor prior to his response or action. The former, it will be recalled, are "tied to" the immediate situation, while the latter "transcend" the immediate situation. Thus, symbols may refer to past or future events, to hypothetical situations, to nonexistent or imaginary objects, and so forth.

individual is attempting to prevision the future. It consists of presenting to oneself, tentatively and in advance of overt behavior, the different possibilities or alternatives of future action with reference to a given situation. The future is, thus, present in terms of images of prospective lines of action from which the individual can make a selection. The mental process is, then, one of delaying, organizing, and selecting a response to the stimuli of the environment. This implies that the individual *constructs* his act, rather than responding in predetermined ways. Mind makes it possible for the individual purposively to control and organize his responses. Needless to say, this view contradicts the stimulus-response conception of human behavior.

When the act of an animal is checked, it may engage in overt trial and error or random activity. In the case of blocked human acts, the trial and error may be carried on covertly, implicitly. Consequences can be imaginatively "tried out" in advance. This is what is primarily meant by "mind," "reflective thinking," or "abstract thinking."

What this involves is the ability to indicate elements of the field or situation, abstract them from the situation, and recombine them so that procedures can be considered in advance of their execution. Thus, to quote a well-known example, the intelligence of the detective as over against the intelligence of the bloodhound lies in the capacity of the former to isolate and indicate (to himself and to others) what the particular characters are which will call out the response of apprehending the fugitive criminal.

The mind is social in both origin and function. It arises in the social process of communication. Through association with the members of his groups, the individual comes to internalize the definitions transmitted to him through linguistic symbols, learns to assume the perspectives of others, and thereby acquires the ability to think. When the mind has risen in this process, it operates to maintain and adjust the individual in his society; and it enables the society to persist. The persistence of a human society depends, as we have seen, upon consensus; and consensus necessarily entails minded behvior.

The mind is social in function in the sense that the individual continually indicates to himself in the role of others and controls his activity with reference to the definitions provided by others. In order to carry on thought, he must have some standpoint from which to converse with himself. He gets this standpoint by importing into himself the roles of others.

By "taking the role of the other," as I earlier pointed out, we can see ourselves as others see us, and arouse in ourselves the responses that we call out in others. It is this conversation with ourselves, between the representation of the other (in the form of the "Me") and our impulses (in the form of the "I") that constitutes the mind. Thus, what the individual actually does in minded behavior is to carry on an internal conversation. By addressing himself from the standpoint of the generalized other, the individual has a universe of discourse, a system of common symbols and meanings, with which to address himself. These are presupposed as the context for minded behavior.

Mead holds, then, that mental activity is a peculiar type of activity that goes on in the experience of the person. The activity is that of the person responding to himself, of indicating things to himself.

To repeat, mind originates in the social process, in association with others. There is little doubt that human beings lived together in groups before mind ever evolved. But there emerged, because of certain biological developments, the point where human beings were able to respond to their own acts and gestures. It was at this point that mind, or minded behavior emerged. Similarly mind comes into existence for the individual at the point where the individual is capable of responding to his own behaviour, *i.e.*, where he can designate things to himself.

Summarizing this brief treatment of mind, mental activity, or reflective thinking, we may say that it is a matter of making indications of meanings to oneself as to others. This is another way of saying that mind is the process of using significant symbols. For thinking goes on when an individual uses a symbol to call out in himself the responses which others would make. Mind, then, is symbolic behavior.¶ As such, mind is an emergent from non-symbolic behavior and is fundamentally irreducible to the stimulus-response mechanisms which characterize the latter form of behavior.

It should be evident that Mead avoids both the behavioristic fallacy of reduction and the individualistic fallacy of taking for granted the phenomenon that is to be explained.

¶ A growing number of linguists, semanticists, and students of speech disorders are becoming aware of the central role of symbols in the *content*, as well as the process of thought. Edward Sapir and Benjamin Whorf have formulated "the principle of linguistic relatively," which holds that the structure of a language influences the manner in which the users of the language will perceive, comprehend, and act toward reality. Wendell Johnson, in the field of semantics, and Kurt Goldstein, in the study of aphasia, are representative investigators who have recognized the way in which symbols structure perception and thought. Mead's theory clearly foreshadows these developments.

The view of consciousness emerging from and therefore dependent upon the social process is a fairly radical one. One consequence of seeing consciousness this way is that different modes of social organization might be expected to produce qualitatively different modes of consciousness. Attention is directed to the dependence of the organization of human consciousness on broader aspects of social organization. The study of social organization is called sociology and the branch of sociology known as the sociology of knowledge might be expected to have something to say about the social organization of consciousness. Berger and Luckmann's 'The Social Construction of Reality' offers a discussion of social practices crucial to the representation of the self in self-consciousness. In fact, the treatise concerns the role of organized social practices in the creation of all objects of consciousness out of subjective experience, including the self. Consciousness is seen as the apprehension of objects, and cannot be experienced other than in this process: consciousness *is* the experience of *objects*. Where objects come from is a complex question. William James did not doubt their primacy in consciousness:

> A mind which has become conscious of its own cognitive function plays what we have called 'the psychologist' upon itself. It not only knows the things that appear before it; it knows that it knows them. This stage of reflective condition is, more or less explicitly, our habitual adult state of mind.
> It cannot however be regarded as primitive. The consciousness of objects must come first. (James, 1890, pp. 272–273)

At first sight it might seem as though Berger and Luckmann, with their provocative title, are proposing that objects arise solely as a consequence of social interaction but in fact one of their major themes is of the dialectical relationship between the tendency of the physical and biological world to present itself as objects and the tendency of human beings to recognize objects in experience. Objects are never entirely one thing or the other, even though, according to Berger and Luckmann, human beings have a tendency to be forgetful of their own roles in the creation of objects of thought. The role of symbols, chiefly in the form of language, is central in the creation and recognition of objects. Berger and Luckmann employ the term 'subjective' to describe given sensory experience (this is close to our use of the term 'sentience') and 'objective' to describe the casting of that experience into objects. They then describe language as the process which makes the subjective objective and the objective subjective. That is, it casts our own subjective experience into objects and allows the objects of others' (and our own) utterances to register reasonably predictably in our own subjective experience.

We have chosen three short extracts from this book. The first deals with consciousness as objects and discusses the various relatively discrete domains in which they exist and interrelate. One such domain has particular importance, that in which we trade objects with other people for everyday practical purposes and which we take as the reality of everyday life. The second extract deals with our tendency to forget our own creation of the objects of this everyday reality, a process the authors refer to as 'reification' — where the objects of experience seem to have a character and origin of their own, independent of the institutions in which they have arisen. By institutions Berger and Luckmann mean accepted ways of dealing with and understanding the practical problems of everyday life; they see such institutions as evolving in the first place from 'habitualized' activity the original purpose of which is lost to memory. The third extract deals with the role of language in the translation of objective experience to subjective experience and back again and points towards the later section on language and consciousness.

Consciousness is always intentional; it always intends or is directed towards objects. We can never apprehend some putative substratum of consciousness as such, only consciousness of something or other. This is so regardless of whether the object of consciousness is experienced as belonging to an external physical world or apprehended as an element of an inward subjective reality. Whether I (the first person singular, here as in the following illustrations, standing for ordinary self-consciousness in everyday life) am viewing the panorama of New York City or whether I become conscious of an inner anxiety, the processes of consciousness involved are intentional in both instances. The point need not be belaboured that the consciousness of the Empire State Building differs from the awareness of anxiety. A detailed phenomenological analysis would uncover the various layers of experience, and the different structures of meaning involved in, say, being bitten by a dog, remembering having been bitten by a dog, having a phobia about all dogs, and so forth. What interests us here is the common intentional character of all consciousness.

Different objects present themselves to consciousness as constituents of different spheres of reality. I recognize the fellowmen I must deal with in the course of everyday life as pertaining to a reality quite different from the disembodied figures that appear in my dreams. The two sets of objects introduce quite different tensions into my consciousness and I am attentive to them in quite different ways. My consciousness, then, is capable of moving through different spheres of reality. Put differently, I am conscious of the world as consisting of multiple realities. As I move from one reality to another, I experience the transition as a kind of shock. This shock is to be understood as caused by the shift in attentiveness that the transition entails. Waking up from a dream illustrates this shift most simply.

Among the multiple realities there is one that presents itself as the reality *par excellence*. This is the reality of everyday life. Its privileged position entitles it to the designation of paramount reality. The tension of consciousness is highest in everyday life, that is, the latter imposes itself upon consciousness in the most massive, urgent and intense manner. It is impossible to ignore, difficult even to weaken in its imperative presence. Consequently, it forces me to be attentive to it in the fullest way. I experience everyday life in the state of being wide-awake. This wide-awake state of existing in and apprehending the reality of everyday life is taken by me to be normal and self-evident, that is, it constitutes my natural attitude.

I apprehend the reality of everyday life as an ordered reality. Its phenomena are prearranged in patterns that seem to be independent of my apprehension of them and that impose themselves upon the latter. The reality of everyday life appears already objectified, that is, constituted by an order of objects that have been designated *as* objects before my appearance on the scene. The language used in everyday life continuously provides me with the necessary objectifications and posits the order within which these make sense and within which everyday life has meaning for me. I live in a place that is geographically designated; I employ tools, from can-openers to sports cars, which are designated in the technical vocabulary of my society; I live within a web of human relationships, from my chess club to the United States of America, which are also ordered by means of vocabulary. In this manner language marks the coordinates of my life in society and fills that life with meaningful objects.

A final question of great theoretical interest arising from the historical variability of institutionalization has to do with the manner in which the institutional order is objectified: To what extent is an institutional order, or any part of it, apprehended as a non-human facticity? This is the question of the reification of social reality.

Reification is the apprehension of human phenomena as if they were things, that is, in non-human or possibly supra-human terms. Another way of saying this is that reification is the apprehension of the products of human activity *as if* they were something other than human products — such as facts of nature, results of cosmic laws, or manifestations of divine will. Reification implies that man is capable of forgetting his own authorship of the human world, and, further, that the dialectic between man, the producer, and his products is lost to consciousness. The reified world is, by definition, a dehumanized world. It is experienced by man as a strange facticity, an *opus alienum* over which he has no control rather than as the *opus proprium* of his own productive activity.

It will be clear from our previous discussion of objectivation that, as soon as an objective social world is established, the possibility of reification is never far away. The objectivity of the social world means that it confronts man as something outside of himself. The decisive question is whether he still retains the awareness that, however objectivated, the social world was made by men — and, therefore, can be remade by them. In other words, reification can be described as an extreme step in the process of objectivation, whereby the objectivated world loses its comprehensibility as a human enterprise and becomes fixated as a non-human, non-humanizable, inert facticity. Typically, the real relationship between man and his world is reversed in consciousness. Man, the producer of a world, is apprehended as its product, and human activity as an epiphenomenon of non-human processes. Human meanings are no longer understood as world-producing but as being, in their turn, products of the 'nature of things'. It must be emphasized that reification is a modality of consciousness, more precisely, a modality of man's objectification of the human world. Even while apprehending the world in reified terms, man continues to produce it. That is, man is capable paradoxically of producing a reality that denies him.

Reification is possible on both the pre-theoretical and theoretical levels of consciousness. Complex theoretical systems can be described as reifications, though presumably they have their roots in pre-theoretical reifications established in this or that social situation. Thus it would be an error to limit the concept of reification to the mental constructions of intellectuals. Refication exists in the consciousness of the man in the street and, indeed, the latter presence is more practically significant. It would also be a mistake to look at reification as a perversion of an originally non-reified apprehension of the social world, a sort of cognitive fall from grace. On the contrary, the available ethnological and psychological evidence seems to indicate the opposite, namely, that the original apprehension of the social world is highly reified both phylogenetically and ontogenetically. This implies that an apprehension of reification *as* a modality of consciousness is dependent upon an at least relative *de*-reification of consciousness, which is a comparatively late development in history and in any individual biography.

The formation within consciousness of the generalized other marks a decisive phase in socialization. It implies the internalization of society as such and of the objective reality established therein, and, at the same time, the subjective establishment of a coherent and continuous identity. Society, identity *and* reality are subjectively crystallized in the same process of internalization. This crystallization is concurrent with the internalization of language. Indeed, for reasons evident from the foregoing observations on language, language constitutes both the most important content and the most important instrument of socialization.

When the generalized other has been crystallized in consciousness, a symmetrical relationship is established between objective and subjective reality. What is real 'outside' corresponds to what is real 'within'. Objective reality can readily be 'translated' into subjective reality, and vice versa. Language, of course, is the principal vehicle of this ongoing translating process in both directions. It should, however, be stressed that the symmetry between objective and subjective reality cannot be complete. The two realities correspond to each other, but they are not coextensive. There is always more objective reality 'available' than is actually internalized in any individual consciousness, simply because the contents of socialization are determined by the social distribution of knowledge. No individual internalizes the totality of what is objectivated as reality in his

> society, not even if the society and its world are relatively simple ones. On the other hand, there are always elements of subjective reality that have not originated in socialization, such as the awareness of one's own body prior to and apart from any socially learned apprehension of it. Subjective biography is not fully social. The individual apprehends himself as a being both inside *and* outside society. This implies that the symmetry between objective and subjective reality is never a static, once-for-all state of affairs. It must always be produced and reproduced *in actu*. In other words, the relationship between the individual and the objective social world is like an ongoing balancing act.

Mead's discussion of consciousness with its implication of symbols, when applied to the description and explanation of social behaviour, has been called 'symbolic interactionism': the study of human social interaction as the exchange of symbols. It has been almost the exclusive interest of sociologically trained and inclined social psychologists. With the publication in 1972 by Duval and Wicklund of *A Theory of Objective Self-Awareness* experimental social psychologists were given a means of operationally defining states of awareness in terms of whether attention was directed 'inwards' to the self or 'outwards' at the immediate environment.

Duval and Wickland offer 'subjective self-awareness' as a state where attention is directed outwards at the immediate environment and 'objective self-awareness' as a state where consciousness is focused directly upon the self. Their choice of 'subjective' and 'objective' is made to distinguish whether the self is acting as a subject with respect to his or her surroundings, i.e. apprehending it, or whether it is being an object of that process, i.e. being apprehended. This seems at first reading to differ from Berger and Luckmann's use of 'objective' and 'subjective' where 'subjective' refers to private incommunicable experience and the 'objective' refers to outwardly communicable experience. However, Berger and Luckmann would class these terms as objective awareness because the immediate surroundings and the self are perceived in terms of the objects they present.

Duval and Wicklund go on to employ their distinction as a means of predicting a variety of social–psychological phenomena, such as that when attention is focused inwards people may become aware of the shortcomings of their own performance compared with various external standards. Duval and Wicklund reveal their view of self-consciousness to be purely psychological in their denial that one need internalize the point of view of another before one can be an object in one's own thoughts, i.e. be objectively self aware. They assume that the object, like the nature of the self, can be discovered without any interaction with other people. The latter part of the reading deals with their view on this and reveals their asocial view of selfhood.

We have formulated a theory on the basis of a distinction between two alternative forms of conscious attention. The theory assumes that states of awareness are directed either toward an aspect of oneself or toward the external environment, and with this distinction it becomes possible to understand numerous phenomena in terms of the self evaluation that results from attention directed toward the self.

We have not created a new distinction in psychology by postulating that conscious attention can be divided into a dichotomy. In 1934 Mead argued that the uniqueness of the self lies in its possibility of being an object unto itself whereas no other event in the universe is reflexive in the same sense. According to Mead, when an indivdual's experience is absorbed or preoccupied with objects around him, the self is the subject of consciousness, but when the person gets outside himself experientially by taking the point of view of the other he becomes an object to himself. Piaget (1966) draws a distinction similar to that of Mead in discussing his concept of egocentrism. Characteristic of egocentrism are an inability to perceive self contradictory actions, an absence of a need to verify one's statements, an assumption by the individual that others can understand him, and an assurance by the person that he is correct in all matters. In contrast to egocentrism is the self conscious state which is characterized by the opposite of the traits of egocentrism. Thus Piaget has proposed a distinction between egocentrism and self consciousness which entails the ignorance of one's unique point of view or the opposite — a consciousness that one's point of view is unique, and therefore subject to error.

Given this brief background, some theoretical terms will be defined. "Subjective self awareness" is a state of consciousness in which attention is focused on events external to the individual's consciousness, personal history, or body, whereas "objective self awareness" is exactly the opposite conscious state. Consciousness is focused exclusively upon the self and consequently the individual attends to his conscious state, his personal history, his body, or any other personal aspects of himself. Our distinction requires a sharp demarcation between the "self" and the "nonself."

The terms "objective" and "subjective" were chosen because they capture the directional nature of consciousness. When attention is directed inward and the individual's consciousness is focused on himself, he is the object of his own consciousness — hence "objective" self awareness. When attention is directed away from himself he is the "subject" of the consciousness that is directed toward external objects, thus the term "subjective" self awareness. Because attention is directed away from the self in subjective self awareness it may be argued that the person is not self aware in any usual sense of the term, and that is correct. But he is self aware in the sense that he experiences the peripheral feedback from his actions and various other feelings that arise from within the body. His self awareness is the feeling of being the source of forces directed outward, but he cannot focus attention on himself as an object in the world. In short, the subjectively self aware person is aware of himself only insofar as he experiences himself as the source of perception and action. These feelings, which are the substance of subjective self awareness, are experienced simultaneously with conscious attention that is directed outward.

Our distinction between two states of awareness assumes that conscious attention cannot be focused simultaneously on an aspect of the self and on a feature of the environment. When a person's attention is directed toward a consideration of his personal virtues, it is impossible at that same instant to focus conscious attention toward driving nails into a board.

Polanyi (1958) is . . . in agreement: ". . . our attention can hold only one focus at a time . . . ." Attention may oscillate between the internal and external, and the oscillation may be rapid enough so that attention approximates the appearance of taking two directions at once, but it is a theoretical assumption that the possibility of directed attention toward an aspect of the self and toward external events simultaneously is impossible. When applying the theory it will be convenient to speak of "increased" objective or subjective self awareness, and when such language is used, it should not be taken to mean that the objective–subjective self awareness dimension is continuous. "Increased objective self awareness" simply denotes a greater proportion of time spent in the objective state, and "increased subjective self awareness" denotes just the opposite.

It is a theoretical assumption that when attention is focused on the self, there will be an automatic comparison of the self with standards of correctness, and a perceived discrepancy and resultant self evaluation will be the result to the extent that the self is not identical with the mental representation.

The person's evaluations can be along any of the multitude of possible self related dimensions. For example, once he focuses on himself as a singer (perhaps through hearing his tape recorded voice) he will evaluate himself according to the disparity between his voice as he perceives it and a voice to which he aspires, and the greater the disparity, the more negative his self evaluation. Similarly, on any other dimension where an "ideal–actual" discrepancy is possible, so is self evaluation possible provided the conscious state is objective self awareness. There will be some instances where an ideal or aspiration does not exist, but in such cases a negative evaluation can result from a contradiction; thus, whenever the person behaves in a manner discrepant from his beliefs, engages in two contradictory behaviors, or states a logical contradiction, the result will be a negative evaluation when he is objectively self aware, provided that he focuses on the discrepancy.

In summary, the state of objective self awareness will lead to a negative self evaluation and negative affect whenever the person is aware of a self contradiction or a discrepancy between an ideal and his actual state. The greatest negative affect will be experienced when a substantial discrepancy is salient for the person; thus we would expect the state of objective self awareness to be maximally painful immediately following a failure experience. But even without a prior failure, or loss in self esteem, we would argue that the objective state will be uncomfortable when endured for considerable time intervals. As the individual examines himself on one dimension after another he will inevitably discover ways in which he is inadequate, and at such a point he will prefer to revert to the subjective state.

It is obvious that self evaluation is not characteristic of subjective self awareness. In subjective self awareness, attention is directed away from the self, and because a comparison between standards of correctness and behavior will not take place, there can be no self evaluation. For example, an English teacher might say "ain't" to her class, but fail to engage in self evaluation because the focus of her attention is drawn outward, toward the condition of her class. However, she will experience the bodily feelings and so forth that are concomitant with the subjective state, but it is important to note that these feelings cannot bring about a comparison between standards of correctness and behavior. In addition to the element of self evaluation there is a second difference between objective and subjective self awareness that can be most easily described by reference to the subjective state. We have postulated that the subjectively self aware individual focuses his attention outward — toward other people, toward tasks, toward sources of entertainment, and so forth. This would imply that the person's relationship to the environment will be one carrying a feeling of control and mastery, for it is only in objective awareness that he will think about himself as falling short of the ideal of exerting control over the environment. Provided the person does not enter objective self awareness, he will not evaluate his actions and will feel as if he is forcing his energies onto external events. In short, his experience will be that of subject, rather than object.

Given the distinction between objective and subjective self awareness, there are two ways to proceed to construct a theory. We could assume that people fall with regularity into these two categories, or we can presume that both the objective and subjective states are frequently characteristic of everyone, and are determined primarily by factors in the person's present situation. It is the latter approach which we think will be most fruitful, particularly since we would like to think that knowledge of whether a person is in the objective or subjective state is more likely to be gained by having control of his immediate situation than through historical self reports.

We would suggest that the conditions leading to objective self awareness (or subjective self awareness) are nothing more than stimuli that cause the person to focus attention on himself (or on the environment). More generally, whether attention is directed inward or outward is completely determined. We assume that subjective self awareness is the primary state in that the environment is normally a strong enough stimulus to draw attention toward it, which means the self is totally excluded from attention. In order that the person become objectively self aware, it is necessary to create conditions that remind him of his status as an object in the world. Many of these conditions can be impersonal, and examples are looking into a mirror, hearing one's tape-recorded voice, seeing a photograph of oneself, or any other setting where a manifestation or reflection of the person is external to the individual and can be perceived by him. Given that perception of oneself as an object is the only requirement for creation of objective self awareness, it is not logically necessary that part of himself be "externalized" by a device such as a mirror or tape recorder. If he can focus on himself simply by examining his hand, foot, navel, and so forth, objective self awareness may result. The difficulty with the latter method of arousing the objective state is in control of the awareness: it is relatively hard to determine what causes a person to examine his appendages, but it is easy to arrange for his personal aspects to be reflected outside himself.

Among the stimuli that generate objective self awareness are other human beings and as we will see later in discussion of social comparison, social facilitation, and communication sets, an extension of the theory to other people as sources of the objective state serves as a means of understanding several notable social phenomena. The analysis of inanimate sources of the objective state is considerably more simple than the application to humans, since we think that several dimensions of the others who arouse objective self awareness are theoretically important.

In Mead's psychology of self consciousness, taking the role of the other is the key phrase. Taking the role of the other is as if the person enters another's head and observes himself through the other's characteristic ways of perceiving and evaluating the world.

Thus the content and substance of the self is made up of the perceptions and evaluations of the social other. As far as the incidence of self awareness is concerned, Mead implies that a person cannot be self aware unless he is able to take the role of someone else.

The way in which Mead views the phenomenon of self consciousness implies two qualitatively distinct forms of consciousness; an original type of consciousness, and a consciousness which is the result of a unique interaction and combination of two separate consciousness. The original consciousness is a feeling or rudimentary awareness of the organism and the world whose function is the facilitation of the organism's manipulations of the environment. Self consciousness is impossible in this state. Second, there is the type of consciousness which is not consciousness of the other but, rather, a transmutation of the other person's consciousness of oneself into one's own. From any point of view the rudimentary consciousness and empathic consciousness are qualitatively different; this assumption is completely unnecessary within objective self awareness theory.

Mead's version of self consciousness is a social learning relationship between social entities that comes about only when one person adopts the viewpoint of another toward himself. The theory of objective self awareness takes exception to Mead's thinking and social origin theory in general in postulating that self consciousness occurs because consciousness can focus its attention on the self in the same way that attention is focused on any other object. In no way do we assume that the individual is dependent upon the point of view of the other in the sense that Mead intends. If the object-like nature of the self has been discovered and stimuli in the environment are such that consciousness turns in the direction of the self, the person will become objectively self aware.

The major step that the objective self awareness theory of self consciousness takes beyond Mead's theory is the assumption that the components or elements of self consciousness are indigenous to the person's original psychological structure. To be self conscious a person does not have to borrow either the substance of self or an external point of view from another person. Consciousness of self is not different from consciousness of any other object. Just as a person is innately capable of the awareness of the various things in his environment, he is innately capable of the awareness of the object that is his self.

In order to analyze self consciousness from the standpoint of objective self awareness, several general assumptions must be made about the nature of consciousness and self. As far as consciousness is concerned, objective self awareness theory postulates that the individual has one innate consciousness with directional properties. As indicated earlier, we assume that this consciousness can focus upon any object. Furthermore, we assume that consciousness will focus on a particular element if the stimuli that control the directionality of attention direct consciousness to the region of the environment where this element resides.

The conception of two alternative states of self awareness entails separating the possible foci of consciousness into two separate regions: consciousness focuses on the self in objective self awareness and focuses upon whatever is not the self in subjective self awareness. Thus, there are not two qualitatively different consciousnesses but two distinct foci for one type of consciousness.

Objective self awareness theory also defines "self" in a particular way. The self is not an entity that is developed or comes into being through taking the other's viewpoint, nor is it a subjective phenomenon accessible only to the individual. Instead, self is an object-like entity that exists from the moment that the person is conscious of internal and external stimuli and, as such, is present to any other's awareness.

8

Historical and Cultural Perspectives on Consciousness

READINGS

Andrew Lock
Universals in Human Conception

Julian Jaynes
The Mind of *Iliad*

Richard D. Logan
Historical Change in Prevailing Sense of Self

Paul Heelas
Introduction: Indigenous Psychologies

This chapter further elaborates the distinction between consciousness as sentience and consciousness as self-awareness. Human beings appear to be able to do something unique with their sentience: they can create symbols with shared significance. Viewing self-consciousness as 'something human beings do with consciousness' implies that there might be differences in the form it takes in different cultures and across time. The readings in this chapter support and discuss this view.

We have already encountered the opinion in the previous chapter that it is a mistake to see human consciousness as an abstract field waiting for something to enter it. It is, instead, the process of apprehension of objects. To the extent that these objects, including the self, are socially constructed, consciousness will differ from culture to culture.

The extract by Lock begins by asking what functions any indigenous psychology would have to perform to enable people to maintain a stable, coherent and socially transactable self. These must be universals, as no human culture could exist without such selves. If we are self conscious we cannot do so without the objects we recognize and the explanations we accept affording us an inner self; in turn, that self adjusts towards action in the social and cultural world and allows sociocultural institutions to operate. If Lock is correct, then the historical and cultural variations of consciousness described in the extract will all be fulfilling these criteria.

His first universal is that an indigenous psychology should provide an account of the ways and extent to which the individual is considered responsible for his or her actions. Self-awareness is necessary for performance and individuals in any culture must be given a clear idea of what exactly their powers and consequently their responsibilities are. Next, the indigenous psychology of any culture must make clear the boundary between self and non-self. Lock adds that this marks, simultaneously, the distinction between the subjective and the objective. This is a slightly different emphasis on the meaning of the subjective–objective distinction from that employed by Berger and Luckmann in the previous section. In this extract, 'objective' refers to 'things which are not an intrinsic part of the self' and 'subjective' to 'things which are an intrinsic part of the self'.

Lock gives five basic orientations which enable the individual to make his or her actions seem intelligible to self and others. These are: a distinction between what is self and non-self; the recognition of objects, especially the more socially determined and *symbolic* objects; a framework of space and time in which the recognition of objects and the explanation of their relationships takes place; the interpretation of needs; and a sense of what ought and ought not to be done. Society based on self-conscious, self-monitoring individuals cannot work without the help of each of these conventions. Thus wherever we look in the historical past or in different cultures today we should be able to find the objects recognized in consciousness and the meanings associated with them fulfilling each of these functions.

We can see in general terms that indigenous psychologies are necessary with respect to three functions: sustaining the inner self; sustaining and adjusting that self towards social and cultural action; and enabling sociocultural institutions to operate. Before we attempt to explain cross-cultural variations in indigenous psychological formulations from an anthropological perspective we must pursue a cental issue: what is that that is universal to the indigenous?

The concepts of *self* and *culture* are interdependent: one cannot exist without the other. Thus, while it has become commonplace to regard the self as a cultural product, and enquire as to the "environmental" (cultural) factors that lead to the expression or inhibition of this or that aspect of the self, we must not forget the reverse perspective; that culture itself is a product of the self. Selves are constituted within culture, and culture is maintained by the community of selves. The English language is not a particularly good medium in which to discuss such dialectic relations, reflecting as it does an implicit, straightforward cause-and-effect mode of structure; expressing one thing as a cause and predicating another to it as an effect; an example of an indigenous psychological constraint.

Man is not only a *social* animal but also a *cultural* one; and that means that he is also a *moral* animal: thus his society exhibits both a *social order* and a *moral order*. This moral order comprises norms of conduct and effective social scantions, implicit or explicit, to back them up. Hence, the members of such an order are required to recognize some locus of responsibility for their actions in society. This, in turn, implies a

> self-awareness of one's own conduct, self-appraisal of one's conduct with reference to socially recognised standards of value, some volitional control of one's own behaviour, a possible choice of alternative lines of conduct, etc. . . . without the development of self-awareness as an intrinsic part of the socialisation process, without a concept of self that permits attitudes directed towards the self as an object to emerge and crystallize, we would not have some of the essential conditions necessary for the functioning of a human society.
>
> (Hallowell, 1971:83)

And as Hallowell goes on to note, self-awareness is necessary and basic to the successful performance of the many different roles which the individual has to adopt within society. In order for a culture to maintain itself its individual members *must* have some awareness of their social standing with respect to age, sex, hierarchies of social precedent, etc:

> If [they] were not aware of [their] roles they would not be in a position to appraise their own conduct in terms of traditional values and social sanctions.
> (Hallowell, 1971:83)

Thus, self-awareness is at once both a distinctive and necessary component of human life:

> we must assume that the functioning of any human society is inconceivable without self-awareness, reinforced and constituted by traditional beliefs about the nature of the self.
> (Hallowell, 1971:83)

Three things may be said about self-awareness:

(i) *Self-awareness is a socio-cultural product.* To be self-aware is, by definition, to be able to conceive of one's individual existence in an objective, as opposed to subjective, manner. In G. H. Mead's (1934) terms, one must view oneself from "the perspective of the other". Such a level of psychological functioning is only made possible by the attainment of a symbolic mode of representing the world. Again, this mode of mental life is generally agreed to be dependent upon the existence of a cultural level of social organization. We thus come to a fundamental, though apparently tautologous point: that the existence of culture is predicated upon that of self-awareness; and that the existence of self-awareness is predicated upon that of culture. In the same way as in the course of evolution the structure of the brain is seen as being in a positive-feedback relationship with the nature of the individual's environment, so it is with culture and self-awareness: the self is constituted by culture which itself constitutes the self.

(ii) *Culture defines and constitutes the boundaries of the self: the subjective-objective distinction.* It is an evident consequence of being self-aware that if one has some conception of one's own nature, then one must also have some conception of the nature of things other than oneself, i.e. of the world. Further, this distinction must be encapsulated explicitly in the symbols one uses to mark this polarity. Consequently, a symbolic representation of this divide will have become "an intrinsic part of the cultural heritage of all human societies" (Hallowell, 1971:75). Thus, the very existence of a moral order, self-awareness, and therefore human being, depends on the making of *some* distinction between "objective" (things which *are not* an intrinsic part of the self) and "subjective" (things which *are* an intrinsic part of the self).

> This categorical distinction, and the polarity it implies, becomes one of the fundamental axes along which the psychological field of the human individual is structured for action in every culture. . . . Since the self is also partly a cultural product, the field of behaviour that is appropriate for the activities of *particular* selves in their world of culturally defined objects is not by any means precisely coordinate with any absolute polarity of subjectivity–objectivity that is definable.
> (Hallowell, 1971:84)

(iii) *The behavioural environment of individual selves is constituted by, and encompasses, different objects.* Man, in contrast to other animals, can be afraid of, for example, the dark because he is able to populate it with symbolically constituted objects — ghosts, bogey men, and various other spiritual beings. As MacLeod (1947) points out,

> purely fictitious objects, events and relationships can be just as truly determinants of our behaviour as are those which are anchored in physical reality.

In Hallowell's view (1971:87):

> such objects, in some way experienced, conceptualised and reified, may occupy a high rank in the behavioural environment although from a sophisticated Western point of view they are sharply distinguishable from the natural objects of the physical environment. However, the *nature of such objects is no more fictitious, in a psychological sense, than the concept of the self.*
>
> (My italics)

To the sophisticated mind, such objects are not naturally but only symbolically constituted: but in the psychological sense of their affecting a self which draws a boundary between the objective and subjective at a different point, such objects are necessarily as real as the self they are defined in opposition to, and "must thus be considered as relevant variables because they can be shown to affect actual behaviour" (Hallowell, 1971:87). The environment in which man lives may best be described then as a "culturally constituted behavioural environment" (Hallowell, 1971:87).

Culture not only constitutes man's behavioural environment, but also provides him with basic orientations that enable him to act in an intelligible manner in a world so constituted: all these are orientations for the self, and serve to give it its particular structure.

Culture provides a self-concept through the linguistic marking of self from non-self. All languages must provide deictic marking if they are to be humanly serviceable. Personal pronouns, kinship terms and personal names all function to this end. But, as Hallowell (1971:90) notes:

> while one of the constant functions of all cultures . . . is to provide a concept of self along with other means that promote self-orientation, the individuals of a given society are self-oriented in terms of a provincial content of the self-image.

Thus while all cultures will provide a basic vocabulary for self-orientation, none of those lexical or conceptual items need be directly translatable into those of another culture. In using the word "self" here as a universal marker, we are not implying that *our* provincial concept of self, indicated, unfortunately, by the same word, has any universality in human conception.

If the self is recognized and delineated, then so necessarily is the non-self, that being

> a diversified world of objects . . . discriminated, classified, and conceptualised with respect to attributes which are culturally constituted and symbolically mediated through language. . . . Object-orientation . . . provides the ground for

> an intelligible interpretation of events in the behavioral environment on the basis of traditional assumptions regarding the nature and attributes of the objects involved, and implicit or explicit dogmas regarding the "causes" of events.
>
> (Hallowell, 1971:91)

Such orientating cosmologies supply the conceptual frameword that makes human action possible.

Place-names appear to be among the universally occurring categories of deictic markers, for a culturally constituted orientation to a world of objects other than the self *must* be integrated with a spatial orientation of the self that provides a frame of reference for action. Time and place are intertwined for a self-aware being:

> for self-awareness implies that the individual not only knows where he *is*, but where he *was* at some previous moment in time, or where he expects to be in the *future*.
>
> (Hallowell, 1971:93)

This further implies the existence not only of a self-identity component of self-awareness, but a self-continuity one as well. And this brings us full circle to the concept of culture as a moral order in which a temporal dimension for the self is necessary, for in order to maintain roles and assume moral responsibility, "I not only have to be aware of who I am today, but be able to relate my past actions to both past and future behavior" (Hallowell, 1971:95). There is a further important relation between this temporal orientation of the self and self-continuity: the time-span of recalled experiences that are related to the self. Thus we find, for example, in the Hindu doctrine of rebirth that events "recalled from previous lives" may be regarded as having happened to the self. Hallowell (1971:95) notes that in many cultures

> Self-related experiences are given a retrospective temporal span that far transcends the limits beyond which we know reliable accounts of personal experience can be recalled.

A motivational orientation is as necessary for the continuance of a moral order as it is for the orientation of a self within that order, since motivational factors — needs, desires, goals, attitudes — underpin the functioning of human social orders. The world of objects which the self inhabits is not only discriminated but classified as possessing attributes, positive or negative, that are culturally constituted as relevant to the self. Needs necessitate actions and actions require direction: this is why the self requires motivational orientation — to be able to discriminate the relevant objects towards which it must act.

> Since the human social order is also a moral order, there is always the presumption that an individual is not only aware of his own personal identity and conduct in a spatio-temporal frame of reference, but that he is capable of judging his own conduct by the standards of his culture. Thus normative orientation is a necessary corollary of self-orientation.
>
> (Hallowell, 1971:106)

> From this it follows that an individual possesses volitional control over his own acts. But this does not necessarily imply that acts for which the individual is morally responsible will be attributed by him to himself. For as Hallowell goes on to note (1971:101):
>
> > Just as, in terms of a given self-image, naturalistic time and space may be transcended in self-related experience and the self may interact socially with other-than-human selves, so in the moral world of the self the acts for which the self may feel morally responsible may not all be attributed to waking life, nor to a single mundane existence, nor to interpersonal relations with human beings alone.

We have thus far presented a view of consciousness as something which has evolved, phylogenetically to start with and then culturally, when symbols mediate experience. Most people would assume that, once symbols were employed, human consciousness sprang more or less fully formed into place. Few would describe its development in such a way as to deny that human beings had anything approaching what we now take as consciousness as recently as 3000 years ago. Yet this is precisely the thesis of Julian Jaynes in his book, *The Origin of Consciousness in the Breakdown of the Bicameral Mind*.

Jaynes analyses the language of Homer's *Iliad* and finds no words for consciousness or mental acts; nor does he find volition as part of the explanation of events. His conclusion is that the actions of its characters spring not from conscious plans or motives but from the voices of gods. These voices are, according to Jaynes, the same voices that schizophrenics are said to hear, or what we would call auditory hallucinations. Thus, in our sense of the word, Homeric man was not self-conscious at all.

> As soon as we go back to the first written records of man to seek evidence for the presence or absence of a subjective conscious mind, we are immediately beset with innumerable technical problems. The most profound is that of translating writings that may have issued from a mentality utterly different from our own. And this is particularly problematic in the very first human writings. These are in hieroglyphics, hieratic, and cuneiform, all — interestingly enough — beginning about 3000 B.C.. None of these is entirely understood. When the subjects are concrete, there is little difficulty. But when the symbols are peculiar and undetermined by context, the amount of necessary guesswork turns this fascinating evidence of the past into a Rorschach test in which modern scholars project their own subjectivity with little awareness of the importance of their distortion. The indications here as to whether consciousness was present in the early

Egyptian dynasties and in the Mesopotamian cultures are thus too ambiguous for the kind of concerned analysis which is required.

The first writing in human history in a language of which we have enough certainty of translation to consider it in connection with my hypothesis is the Iliad. Modern scholarship regards this revenge story of blood, sweat, and tears to have been developed by a tradition of bards of *aoidoi* between about 1230 B.C. when, according to inferences from some recently found Hittite tablets, the events of the epic occurred and about 900 or 850 B.C., when it came to be written down. I propose here to regard the poem as a psychological document of immense importance. And the question we are to put to it is: What is mind in the Iliad?

The answer is disturbingly interesting. There is in general no consciousness in the Iliad. I am saying 'in general' because I shall mention some exceptions later. And in general therefore, no words for consciousness or mental acts. The words in the Iliad that in a later age come to mean mental things have different meanings, all of them more concrete. The word *psyche*, which later means soul or conscious mind, is in most instances life-substances, such as blood or breath: a dying warrior bleeds out his *psyche* onto the ground or breathes it out in his last gasp. The *thumos*, which later comes to mean something like emotional soul, is simply motion or agitation. When a man stops moving, the *thumos* leaves his limbs. But it is also somehow like an organ itself, for when Glaucus prays to Apollo to alleviate his pain and to give him strength to help his friend Sarpedon, Apollo hears his prayer and "casts strength in his *thumos*" (Iliad, 16:529). The *thumos* can tell a man to eat, drink, or fight. Diomedes says in one place that Achilles will fight "when the *thumos* in his chest tells him to and a god rouses him" (9:702ff.). But it is not really an organ and not always localized; a raging ocean has *thumos*. A word of somewhat similar use if *phren*, which is always localized anatomically as the midriff, or sensations in the midriff, and is usually used in the plural. It is the *phrenes* of Hector that recognize that his brother is not near him (22:296), this means what we mean by "catching one's breath in surprise". It is only centuries later that it comes to mean mind or 'heart' in its figurative sense.

Perhaps most important is the word *noos* which, spelled as *nous* in later Greek, comes to mean conscious mind. It comes from the word *noeein*, to see. Its proper translation in the Iliad would be something like perception or recognition or field of vision. Zeus "holds Odysseus in his *noos*." He keeps watch over him.

Another important word, which perhaps comes from the doubling of the word *meros* (part), is *mermera*, meaning in two parts. This was made into a verb by adding the ending *-izo*, the common suffix which can turn a noun into a verb, the resulting word being *mermerizein*, to be put into two parts about something. Modern translators, for the sake of a supposed literary quality in their work, often use modern terms and subjective categories which are not true to the original. *Mermerizein* is thus wrongly translated as to ponder, to think, to be of divided mind, to be troubled about, to try to decide. But essentially it means to be in conflict about two actions, not two thoughts. It is always behavioristic. It is said several times of Zeus (20:17, 16:647), as well as of others. The conflict is often said to go on in the *thumos*, or sometimes in the *phrenes*, but never in the *noos*. The eye cannot doubt or be in conflict, as the soon-to-be-invented conscious mind will be able to.

These words are in general, and with certain exceptions, the closest that anyone, authors or characters or gods, usually get to having conscious minds or thoughts. We shall be entering the meaning of these words more carefully in a later chapter.

There is also no concept of will or word for it, the concept developing curiously late in Greek thought. Thus, Iliadic men have no will of their own and certainly no notion of free will. Indeed, the whole problem of volition, so troubling, I think, to modern psychological theory, may have had its difficulties because the words for such phenomena were invented so late.

A similar absence from Iliadic language is a word for body in our sense. The word *soma*, which in the fifth century B.C. comes to mean body, is always in the plural in Homer and means dead limbs or a corpse. It is the opposite of *psyche*. There are several words which are used for various parts of the body, and, in Homer, it is always these parts that are referred to, and never the body as a whole. So, not surprisingly, the early Greek art of Mycenae and its period shows man as an assembly of strangely articulated limbs, the joints underdrawn, and the torso almost separated from the hips. It is graphically what we find again and again in Homer, who speaks of hands, lower arms, upper arms, feet, calves, and thighs as being fleet, sinewy, in speedy motion, etc., with no mention of the body as a whole.

Now this is all very peculiar. If there is no subjective consciousness, no mind, soul, or will, in Iliadic men, what then initiates behavior?

The characters of the Iliad do not sit down and think out what to do. They have no conscious minds such as we say we have, and certainly no introspections. It is impossible for us with our subjectivity to appreciate what it was like. When Agamemnon, king of men, robs Achilles of his mistress, it is a god that grasps Achilles by his yellow hair and warns him not to strike Agamemnon (1:197ff.). It is a god who then rises out of the gray sea and consoles him in his tears of wrath on the beach by his black ships, a god who whispers low to Helen to sweep her heart with homesick longing, a god who hides Paris in a mist in front of the attacking Menelaus, a god who tells Glaucus to take bronze for gold (6:234ff.), a god who leads the armies into battle, who speaks to each soldier at the turning points, who debates and teaches Hector what he must do, who urges the soldiers on or defeats them by casting them in spells or drawing mists over their visual fields. It is the gods who start quarrels among men (4:437ff.) that really cause the war (3:164ff.), and then plan its strategy (2:86ff).). It is one god who makes Achilles promise not to go into battle, another who urges him to go, and another who then clothes him in a golden fire reaching up to heaven and screams through his throat across the bloodied trench at the Trojans, rousing in them ungovernable panic. In fact, the gods take the place of consciousness.

The beginnings of action are not in conscious plans, reasons, and motives; they are in the actions and speeches of gods. To another, a man seems to be the cause of his own behavior. But not to the man himself. When, toward the end of the war, Achilles reminds Agamemnon of how he robbed him of his mistress, the king of men declares, "Not I was the cause of this act, but Zeus, and my portion, and the Erinyes who walk in darkness: they it was in the assembly put wild *ate* upon me on that day when I arbitrarily took Achilles' prize from him, so what could I do? Gods always have their way." (19:86–90). And that this was no particular fiction of Agamemnon's to evade responsibility is clear in that this explanation is fully accepted by Achilles, for Achilles also is obedient to his gods. Scholars who in commenting on this passage say that Agamemnon's behavior has become "alien to his ego," do not go nearly far enough. For the question is indeed, what is the psychology of the Iliadic hero? And I am saying that he did not have any ego whatever.

Who then were these gods that pushed men about like robots and sang epics through their lips? They were voices whose speech and directions could be as distinctly heard by the Iliadic heroes as voices are heard by certain epileptic and schizophrenic patients, or just as Joan of Arc heard her voices. The gods were organizations of the central nervous system and can be regarded as personae in the sense of poignant consistencies through time, amalgams of parental or admonitory images. The god is a part of the man, and quite consistent with this conception is the fact that the gods never step outside of natural laws. Greek gods cannot create anything out of nothing, unlike the Hebrew god of Genesis. In the relationship between the god and the hero in their dialectic, there are the same courtesies, emotions, persuasions as might occur between two people. The Greek god never steps forth in thunder, never begets awe or fear in the hero, and is as far from the outrageously pompous god of Job as it is possible to be. He simply leads, advises, and orders. Nor does the god occasion humility or even love, and little gratitude. Indeed, I suggest that the god-hero relationship was — by being its progenitor — similar to the referent of the ego-superego relationship of Freud or the self-generalized other relationship of Mead. The strongest emotion which the hero feels toward a god is amazement or wonder, the kind of emotion that we feel when the solution of a particularly difficult problem suddenly pops into our heads, or in the cry of eureka! from Archimedes in his bath.

The gods are what we now call hallucinations. Usually they are only seen and heard by the particular heroes they are speaking to. Sometimes they come in mists or out of the gray sea or a river, or from the sky, suggesting visual auras preceding them. But at other times, they simply occur. Usually they come as themselves, commonly as mere voices, but sometimes as other people closely related to the hero.

The picture then is one of strangeness and heartlessness and emptiness.

> We cannot approach these heroes by inventing mind-spaces behind their fierce eyes as we do with each other. Iliadic man did not have subjectivity as do we; he had no awareness of his awareness of the world, no internal mind-space to introspect upon. In distinction to our own subjective conscious minds, we can call the mentality of the Myceneans a *bicameral mind*. Volition, planning, initiative is organized with no consciousness whatever and then 'told' to the individual in his familiar language, sometimes with the visual aura of a familiar friend or authority figure or 'god', or sometimes as a voice alone. The individual obeyed these hallucinated voices because he could not 'see' what to do by himself.

The extract by Logan deals with the development of consciousness within Western culture during the last sixty years. His central proposition is that different historical periods were characterized by changes in the stance of the beholding 'I', the self as subject. Logan begins his discussion in the middle ages which were marked by the emergence of a distinctly *individual* point of view. Before this time writers had been unable to write about individuals because they lacked an individual point of view. This was still seen as a beholding 'I', unable to intervene in events in significant ways. During the Rennaissance, at least among the elite, an increased sense of power to intervene in events became evident. It was, however, a sense of 'I' which affected society and the physical world rather than being affected by them.

Whereas the Rennaissance individual 'asserted individuality but did not reflect on the fact', the seventeenth- and eighteenth-century individual of the age of rationalism and empiricism also realized that there was a necessity to reflect. 'It was now up to the individual mind as never before to be a source of an ordered model of its world,' writes Logan. The 'I' is still orientated towards the outer world and does not recognize an inner world, but it is a competent and construing 'I', an 'I' that would permit the appearance of a Descartes or a Newton.

It was not until the romantic age of the nineteenth century that the self began to be thought of as an object of interest. Writers confidently took themselves to be objects worthy of interest. In Logan's words, 'the most important object of concern to the self *was* the self.'

Finally, in the twentieth century's 'post-modern' culture of the industrialized West, a stable coherent self seems to have become hard to find. The search may be arduous and sometimes expensive, but in the age of the guru and the therapist, help is always at hand.

My proposition generally is that we consider change in the stance of the self as a beholding 'I' as central in historical change, i.e., change in the basic orientation and relationship of the person to the world and in the basic parameters of experience. Further, other cultural developments can be seen as predicated *upon* different prevailing senses of self, as will be briefly suggested.

I recognize the vast range of conceptions and theories of self, and its essential nature, that is extant. By 'self' for the purposes of this contribution, I am relying simply on the dualistic conception of the self-as-subject (Knower) and self-as-object (Known) — G. H. Mead's (1934) 'I' and 'me' — and on the extent to which prevailing sense of self in various (historical) contexts tends toward one or the other of these two orientations. For example, if a person functions in the world as an independent beholding entity but is not beholding itself, that person exists as 'I'; on the other hand, a self ('I') highly concerned *with* itself has a strong sense of 'me'.

While at least four different periods have been put forth as the time of origin of the self, it is apparent from voluminous evidence that the 'self' described is different for each era and also fundamentally unlike the 'self' of people in the modern age of 'Existentialism' and self-conscious concern with 'Identity'. A progression from something closer to a group identity (in very early times) to an emergence of 'individualism' (at some point ranging from the later Middle Ages through the Renaissance to Romanticism, depending on whom one reads) to a relatively intense concern with self (in the contemporary era) can be discerned.

I would hypothesize that the primary (but not exclusive) self-sense of the Middle Ages was the emergence out of the corporate and the ecclesiological of the non-self-conscious sense of 'I' (I will, I choose, I observe). This is supported, for instance, by the painting of Giotto, generally regarded as a major innovative figure in the history of art. The fact that Giotto originated the use of *linear perspective* in painting shows that he was aware that he viewed the world not according to convention and tradition but from his individual *viewpoint*. He had a sense of where 'I (subject) stand' as a (newly) autonomous being in relation to the world. I would put it that Chaucer had a similar stance, seeing his characters more 'in depth' from a newly autonomous viewpoint. In fact, this sense of individual standpoint — oriented to the world rather than reflected back on itself — is virtually a definition of what it means to *be* an 'I'. Little sense of awareness *of* self entered in, however, and Giotto still painted subject matter dictated by group convention and tradition, and Chaucer wrote within conventional story frames and his 'individual' characters also represented corporate categories.

It seems only logical (as well as suggested by the evidence) that the very first experience of a 'self' could only be the individual's awareness of separateness from the larger whole, that is, 'I', with little awareness *of* self as unique individuality. In keeping with this beginning sense of self, philosophers of the time wrestled within the new discovery of their individual wills versus the Will of God, but still sought to use their new autonomous viewpoint and autonomous reasoning to arrive at a strengthened faith in the larger Corporate Being.

The Renaissance seems to have been a time of increased independent action, self-assertion, and 'intrusion' of the individual self into the world. The individual of the elite asserted his/her desire to have effects and make a difference through the exercise of imagination, as revealed in the flowering of the arts, the pursuit of learning for its own sake, ambitious individual enterprise, and in the seeking of power by which to manipulate others.

The Renaissance individualist asserted individuality but did not reflect on that fact; instead one identified with one's effects on the world and society, and autobiographies consisted of a narrative recounting of such activities rather than self-reflection (Weintraub, 1978). Self-awareness appears to have been most saliently a matter of beholding one's *actions* and *effects* on the world, and beholding others' *re*-actions to same. The 'I' was in the stance of *agent*-in-the world. The prevailing sense of self was as an assertive subject ('I') who affected society rather than being shaped *by* society, the opposite of the more contemporary view.

The rise of Rationalism in the seventeenth century (and its cohorts, logic and mathematics) and of Empiricism (and its ally science) so dominate these centuries that one must look to them to find the status of the elite 'I' in this age and its relation to the world. Self appears limited in its self-consciousness by virtue of being still more oriented to the 'outer' world than to its own 'inner' character (cf. Macmurray, 1933). I would propose the following model for self in this age: Upon a cultural foundation of institutionalized individual instrumentality/agency, which was a legacy of the Renaissance's 'adventurous overreaching' and the Protestant Reformation's success in affecting the religious order, and which had 'cleared away' many traditional forms of knowing the world, a typical elite 'I' of the post-Renaissance beheld a 'cleared world space' in need of re-conceptualization. The fact of the historical legacy (passed on through families via socialization, education, and modelled example) of individual autonomy and individual instrumentality meant that (elite) individuals were already prepared to believe that they could 'make a difference'. With that as given, the next step was a further realization of inherent agency potential in the form of having *constructive* effects on the world. It was now up to the individual mind as never before in history to *itself* be the source of an ordered model of its world. Both Rationalism and Empiricism are predicated upon a *Competent* construing, 'I', capable of initiating and conducting well-reasoned thought and systematic observation (and learning from same), respectively. The 'I' now relates to the world via 'constructive' effects on the world. The Newtonian model of the naturally lawful universe, the new practical technologies (e.g., Jefferson, Franklin), and the efforts to construct a new social order are all similarly predicated.

It is the *individual competence* to reason, know, and choose (e.g. one's own government) that is the added element to the idea of individual liberty in this age. Only a *detached* 'I' could conceive of such systems of thought, or of Empiricism as well.

The nineteenth century seems to see the full tilt of the balance toward the self as object, and the kind of awareness of self ('me') that most today take as a simple and obvious given in their phenomenal worlds.

On the heels of an era that had lionized rationalism, reason, systematic empirical observation, and the like (all reflecting the self as subject relating competently to the world) came the Romantics, for whom the self had clearly become the *object* of their interest. At the same time, in philosophy, 'In the field of self, Kant [in the late eighteenth century] was the first to clearly distinguish the two aspects of the self, the *I* [subject] and the *me* [object] which DesCartes had so conveniently merged in his *Cogito*' (Perkins, 1969, p.38). This underlines the fuller emergence of the self as object, 'inner', and in a 'personal space' in the late eighteenth century, and a shift away from the dominance of self as subject. Authors began explicitly to write not just of recognizable individual characters as in previous eras but to personify them-*selves* in their writings. Romantic writers were 'deeply interested in themselves, their *reactions* to the world, the effect of the world on *them*, and their innermost feelings, which they were willing to bare to the world' (Easton, 1966, p.631). Thus a concern with 'personality' as it is understood today began to emerge. The focus had begun to make an historic shift from 'How do I (subject) reason about and observe the world?', to 'How does the world make me (object) feel?', and 'What has experience made of me?'

Following a long period of the gradual emergence of self-awareness, a point began to be reached where the most important object of concern to the self *was* the self. Thus people began to sense themselves as unique individualities in and shaped by the world. This is the virtual opposite of the Renaissance view described earlier. Marx, for instance, wrote on the impact of the economic system on man. He also declared: 'In labor is the creation of the self [as object].' From God as the source, to individual as the worldly source, to individual as the worldly *product*, seems to be the historical progression.

In terms of an increasing awareness of self as prime object of its own interest, the rapid rise of the field of psychology in the twentieth century is itself a major case in point. Structuralist theory in psychology at the turn of the century focused explicitly on the introspective observation of oneself. Psychoanalysis stressed the intensive probing in depth of the psyche as an *object* or study (indeed, it used the term 'object' itself). Erikson's neo-Freudian concept of *identity* made the self even more explicitly the *object* of study. Behaviourism is particularly illuminating in view of the thesis of this paper. In the concept of operant conditioning, behaviourism recognizes that one operates as subject or agent upon the environment, but it focuses explicitly on the *consequences* of one's actions and on the effects of those 'reinforcements' back on the person as a recipient object.

In still more recent times, and perhaps more in the intellectual upper middle-class, a still further change in prevailing sense of self seems to have occurred. It is reflected in the rise of Existentialist philosophy and much of so-called humanistic psychology, especially that of the 'pop' variety. With a more complete removal from group, from the products of one's labour, and from past history, and with a life devoted to the present, the self as subject ceases to have much of a *created* 'me' to behold. ('I' becomes further 'alienated' from 'me'.) The self as 'I' must then undertake an effort to 'find' a self or to 'make' one in the here-and-now of existence with no major help from others or from the past. Thus, in the 'post-affluent' era, the social elite begin to seek their own individual destinies, in their individual ways, by their own efforts. They try to 'get in touch' with their selves — to find the 'real inner me'; that is, the self as object so long sought now becomes a kind of '*lost object*', so much has it become deeply 'inner' and so deeply is it held to be, not a *created* object, but an essence that is existentially 'there'.

The 'real' self had become 'me', not 'I'. The self as subject has by now become merely the *tool* of the search for self ('know your *self*'), whereas in the Middle Ages what mattered was the search itself, i.e., the subjective self was what mattered *because* it searched. The time has long since passed for an individual to feel whole by realizing simply 'I am'. Now one is bound to seek to be able to say 'I am *me*' (or 'I've got to *find* me' . . . or '*be* me'). Modern psychology theory both reflects and must contend with this new orientation. Thus although modern humanistic psychology extols 'subjective experience', the focus is still on attaining and knowing the 'me', expressed for example as my 'gut feelings', my 'body', my 'head', my 'parent/adult/child', also expressed in the feeling that one's self is under constant observation, and in the emphasis on the self as one's 'best friend', self as consumer, self as victim of the world, and self as beneficiary or *object* of contemplation, stimulation, pleasure, and of affection in our narcissistic age. In Existential psychology, the self as subject is important mostly to *create* or discover a 'me'. Rogerian psychology, from the mainstream of Humanistic psychology, reveals that the essential core of our being today (the organismic self) is something we are alienated from and must get in touch with. It is explicitly *not* the self-as-seeker that is central, but the self-as-*sought-object*.

For the final extract in this chapter we return to the concept of indigenous psychologies as a means of proposing and explaining different objects of consciousness, hence qualitatively different manifestations of consciousness between individuals of different cultures. The extract by Heelas starts by illustrating what is meant by an indigenous psychology. It is easy to overlook or take for granted quite profound distinctions concerning the mind which our indigenous psychology provides for us. For example, our notion of mind exists in time in the form of memories but not in space within a particular body.

When attention is turned to cultures further afield the distinctions, meanings and explanations become more illustrative of the essential relativity of conscious experience. The Dinka of the southern Sudan have, apparently, no sense of mind mediating between exterior influence and experience. They consider that bastion of western cognitive psychology, memory, as an external source of agency. Indigenous psychologies are at the heart of the recognition of emotions, the magical and religious institutions which determine the powers of individuals, and the institutions that describe and explain their state of mental health. Thus, if human consciousness depends on sentience supporting symbols, and if symbols, being social, are created within *specific* social contexts, the consciousness created therein will likewise be specific to that context.

We are becoming increasingly fascinated, if not obsessed, with psychological matters. At the heart of this interest lies a distinctive view of the self. An Englishman assumes, without undue reflection, that he is a unique individual, complete with a mind and an unconscious realm, and perhaps a soul or spirit, which are distinct from his physical body. Our indigenous psychology focusses on the inner, private, self: on emotions, states of consciousness, will, memories, the soul (if one is a Christian) and so forth. It also focusses on the self as agency. We regard ourselves as being capable of acting on the world, exercising our will-power, and we feel that we have the ability to alter many of our psychological attributes (as when we "make up our minds to be calm"). A number of expressions, in fact, focus on the powers of the self with respect to itself — e.g. "self-determination", "self-possession", "self-respect" and "self-assurance".

In California — the home of modern "psychological man" — the self, in a manner of speaking, has been left to itself. Because of the relative loss of faith in scientific and technological progress, because of the critical attitude towards traditional ways of organizing and orientating the self (namely those provided by traditional religions and large-scale institutions, such as industry), more and more people are coming to seek salvation in their inner being, in the psychological. As the American Dream turns with a vengeance from the material world, attention is being directed to understanding "how we tick", to grasping how our powers, our experiences, the quality of our lives, can be improved. Psychology has taken over from traditional institutions and political activism.

A considerable number of indigenous psychologies have been developed and institutionalized as new "religions" of the inner self. Negative feelings for Primal Therapists are attributed to "Primal Pain" and are exorcised by "Primaling"; members of Kerista (a group in the Bay area) improve inner life by means of the "Gestalt-O-Rama" process; Rebirthers ritually enact their original traumatic entry into the world using props which include baths at body temperature and "doctors"; and EST makes use of techniques to liberate the self from the constrictions imposed by public role-playing.

Californian obsession with psychology is clearly seen in the language many adopt, the use of "psychobabble" to catalogue the ego's condition, for encouraging authenticity;

> Hear me. I mean, no way I'm about to laya bad trip on you. I'm not coming down, like heavy duty on value judgements and where it's at. If you've got your head together, you'll know where I'm coming from. You into my space? Wow! I'm really into you, you know. I mean how you went with your initial crisis reaction. Aww-right!!

There is nothing particularly new about psychobabble. What is new, though, is the emphasis on psychological understanding and explanation with respect to a whole range of self-orientated activities.

Apart from obvious examples, such as the literature on handling mental illness, indigenous psychologies are being put to the service of the women's movement, interpersonal relationships, obtaining powerful or unusual states of consciousness, and even activities like keeping pets. Concerning the first, a recent entry in *Private Eye*'s "Pseuds Corner" runs,

> Passionate love, both sexual and non-sexual, has been the focus of my major research for the last five years. I have come to think of the Passion Experience as a developmental crisis in women's lives which is a form of regression in the service of the ego. . . . I found that the consequences of a passion experience were a major reason for women seeking therapeutic assistance.

Concerning the second, the popular press is by no means averse to offering advice about how to behave with others. One especially appealing theme is the advice offered to combine the more permanent institution of marriage with the more momentary emotions, like love. Beverly Hayne, writing in *Company*, acknowledges that "we're conditioned to think that, if a romance hasn't ended in marriage, then it hasn't worked out" and goes on to suggest that "a shared passion for bird-watching is a better reason for marrying than an all-consuming passion for each other".

There is in fact a proliferation of books and articles aiming to make more explicit the connections between "how we tick" and our lives as a whole. Social psychologists have made explicit the psychological significance of seating arrangements, gestures, interpersonal rituals (such as greetings) and the like. Familiar institutions — for example, "keeping pets" — have also been brought within the domain of psychology. Books appear with titles like *Psychic Pets, Astrology for Dogs*, and *Understanding Your Dog*. Just as psychology has run riot with the activity of child-raising, so too is it beginning to encompass the pet world.

So far I have been emphasizing the psychological nature of many aspects of our society, the wide-ranging pervasiveness of our indigenous psychology; how it enters into our lives in a number of ways; and how the organization and development of our psychological lives has been taken up by various institutions.

Finally, I want to draw out the point that our indigenous psychology is both rather strange and multi-faceted, if not contradictory.

Our indigenous psychology is to begin with, fundamentally metaphysical. Our basic notion of "mind" exists in time (we have memories) but not in space (our bodies are separate). It is also in part magical. If someone said to a dear friend "I wish you were dead" and the friend died, he would almost certainly experience guilt or disquiet. Standing at an airport wishing a plane to arrive is implicitly magical (for what good can wishing do?). Exercising will-power to persevere in a climb up a mountain is also not without curiosity.

Our indigenous psychology is also cluttered with a most remarkable bric-à-brac of expressions: "it's at the back of my mind", "I'm in two minds about it", "open-minded", "twisted mind", "brainstorm", "thick-skinned", "heartfelt emotions", "gutless", "venting one's spleen", "exploding with anger", "blowing one's top". These metaphors — if this is what they are — in turn indicate deeper-seated models. Thus catharsis metaphors ("venting") imply a volcano model. Other models are provided by temperature ("hot-blooded"), the body (localizing psychological phenomena), or classification of the senses (McLuhan writes that "we are so visually biased that we call our wisest men *vision*aries or *seers*"), and the distinction between the private and the public (the notion of privacy in the home being used to articulate psychological states as when we say "I am going to unwind when I get home").

It seems as though any model can gain currency. Talk of self certainly varies with context and fashion. Footballers have their language ("I'm choked", "I'm sick as a parrot"), those in Britain who are currently using psychobabble are not adverse to saying "I'm feeling really mellow" or "what space are you in?" and the Bright Young Things of the 1920s were fond of expressions like "that was joy-making".

The nature and number of models and expressions make us reflect on what is happening, especially since they often appear contradictory. We believe in free will yet spend much time reading horoscopes. Christian Scientists show more faith in positive thinking or will-power, Primal Therapists in catharsis, than most of us would accept.

The picture which emerges, then, is of an indigenous psychology which is both extremely important in our lives and, on examination, is strange, multi-faceted and apparently incoherent.

Turning further afield, our basic assumptions and distinctions receive more shocks. Consider first the fate of our basic distinction — between the psychological and the non-psychological — in ancient China. In his struggle to understand the "psychological" thought of Mencius, Richards emphasizes,

> we must from the outset realize that psychology and physics are not two separated studies for early Chinese thought . . . ; and that, however metaphysically abhorrent it may be to us, the mind and its objects are not set over against one another for Mencius, or . . . for any of his fellows.
>
> (Richards, 1932:5)

When we are asked in the *Tao Tê Ching*, where much the same kind of "psychological" language is used,

> Can you keep the unquiet physical-soul from straying,
> hold fast to the Unity, and never leave it?
>
> Can your mind penetrate every corner of the land, but
> you yourself never interfere?

we have but the faintest inkling of what is being demanded. The realm and the capacities of the "psychological" do not suit our concepts.

To our way of thinking the psychologically significant concepts and claims of the great Eastern religions generally take an alien form. Thus the *Mandukya Upanishad* treats the ultimate "Self" as far transcending where we take the self to be:

> The Fourth, say the wise, is not subjective experience, nor objective experience, nor experience intermediate between these two, nor is it a negative condition which is neither consciousness nor unconsciousness. . . . It is pure unitary consciousness. . . . It is the self. Know it alone!

To put it crudely, Eastern thought has often veered in a monistic direction, towards the view that self or mind are ultimate. By emphasizing unitary consciousness, distinctions that we make between self and non-self, the subjective and the objective, take secondary significance and become in various ways misleading if not erroneous.

In other societies these distinctions are applied — but in a manner which is the reverse of what we are accustomed to. The Dinka of the southern Sudan provide a good illustration of what happens to the psychological when, as we might put it, the self is taken away from the human individual. Godfrey Leinhardt reports, in a famous passage,

> The Dinka have no conception which at all closely corresponds to our popular modern conception of the "mind" as mediating and, as it were, storing up the experiences of the self. There is for them no such interior entity to appear, on reflection, to stand between the experiencing self at any given moment and what is or had been an exterior influence upon the self.
>
> (Lienhardt, 1961:149)

Not possessing our notion "mind", the Dinka have no firm basis for thinking of themselves as selves. It follows, for example, that whereas we say "*I* recall to *mind* the time that . . ." the Dinka conceive "memories" in terms of their external source or agency: concerning a man who called his child "Khartoum" (because the man had been imprisoned there), Lienhardt writes,

> it is Khartoum which is regarded as an agent, the *subject which acts*, and not as with us the remembering mind which recalls the place. The man is the *object* acted upon.
>
> (Lienhardt, 1961:150, my italics)

What counts as belonging to the human self also depends on whether or not a distinction is drawn between mind and body. The Gahuku-Gama of New Guinea regard the individual as a "complex, biological, physiological and psychic whole" (Read, 1967:206). Accordingly, they treat the body as an integral part of the self. Injuries to the body are bound up with damage to the personality; loss of bodily substances threatens the self; personality judgements focus on public bodily phenomena (doubts about someone's motives are expressed by the phrase "I don't know his skin"); and greetings, together with other features of interpersonal relationships, take what is for us an unseemly biological and intimate nature:

> The most common form of each of these situations is in the verbal expression "let me eat your excreta", or variations such as "your urine", "your semen", etc., accompanied either by a gesture of the up-turned, open hand to the mouth or by grasping the buttocks or genitals of the individual concerned.
>
> (Read, 1967:207)

Postal (1965) draws an interesting contrast between the psychological significance attached to the body in Kwakiutl and Hopi cultures. The former treat the body as a barrier, protecting the individual from external sources of danger by ensuring that his public self is in order (if a man is dressed properly, engages in ritual purification of the body and so on, he has nothing to fear). The latter see danger as originating with loss of self-control. This allows an individual's "bad heart", here a mentalistic notion inidicating evil intention, to take over. The body is no longer a mask or barrier with respect to a muted inner self, it is now something to be used to reveal or unmask the presence of a "good heart".

No one indigenous psychology is the same as another. Together with variations in the application of the basic oppositions already introduced (psychological/non-psychological, self/non-self, mind/body), indigenous psychologies differ with respect to a number of other (psychologically) significant distinctions. These include inner/outer, conscious/unconscious, internal control (e.g. will)/external control (e.g. fate or destiny), hot/cold, reason/emotion, up/down, public/private, unitary self/fragmented self, and, last but not least, the subjective/objective distinction.

These are all variously conceptualized, applied, emphasized, ignored and muted. How much more variation there is when one comes down to detailed indigenous formulations: the classification and handling of emotions; conceptualizations of memory, imagination, motivation and the senses; and responses to aberrant states of mind or disobediant children. There are significant differences even between French and English. The former moralizes consciousness by rendering both "consciousness" and "conscience" as *conscience*.

Then there are differences in how indigenous psychologies inform sociocultural life. Witchcraft is an inherently psychological institution concerning the powers of individuals or agencies to harm others and disrupt social relationships. Instances of death "by suggestion" show the hold that indigenous psychologies can have. Head-hunting, cannibalism (as myth or reality), spirit possession (arguably sometimes used by social inferiors to translate a sense of being under control into social enhancement) are other examples of psychologically significant institutions. Neither can we discount classificatory systems in general. Godfrey Lienhardt writes, "I have often been told in the Sudan that some men turn themselves into lions, indeed *are* lions existing also in the form of men" (Lienhardt, 1967:98). "Identity" statements of this kind are notoriously difficult to interpret, but it seems apparent that together with other ways of classifying the human, the animate and the inanimate, they have psychological significance. They help regulate and articulate personality judgements ("Nixon is a pig"), behaviour, emotional responses and so on.

Indigenous psychologies also play a central role in how suffering, emotional disturbance and illness (both what we call "physical" and "mental") are conceived and handled. Grace Harris's *Casting out Anger* (1978) is interesting because it gives a detailed account of how an indigenous psychology enters into the life of a society, the Taita of Kenya. "Angry hearts" cause misfortune and suffering. By engaging in "purification" rituals the Taita are able to deal with misfortune and distressful emotions, and, as Harris puts it, declare "the good feelings without which neither individuals nor communities could survive" (Harris, 1978:155). A ritual idiom based on the premise that anger *has* bad effects is employed, in Taita thought, to replace anit-social anger with *sere* ("peace"), thereby ensuring that social relationships run smoothly.

Taita interest in the emotions, their emphasis on them as the *operatis munandi* of social and personal relationships, is by no means unique. There are many examples of rituals functioning to present and maintain the experienced reality of what the person should be like. These often take a form which we are inclined to call "psychotherapeutic": rituals apparently aiming to restore suffering or anti-social individuals to health or social conformity. Indigenous psychologies play an important role in providing systems of meaning and techniques to bridge the gap between aberrant individuals and the sociocultural order where, once transformed, they can play their proper role.

‖ 9 ‖

Language and Consciousness

READINGS

L. S. Vygotsky
Thought and Language

Benjamin Lee Whorf
The Relationship of Habitual Thought and Behaviour to Language

Ludwig Wittgenstein
Philosophical Investigations

Erich Fromm
The Nature of Consciousness, Repression and De-Repression

This chapter continues to develop our thesis that the use of symbol is the basis for reflexive consciousness and that reflexiveness is the distinctive aspect of human consciousness. Language is the most refined, organized and distinctively human use of symbols. Language structures consciousness. In the words of Berger and Luckmann, 'Language makes the subjective world objective and the objective world subjective.' In other words, it is the process whereby private, subjective experience can be cast into objects tradeable with others who share their significance and a particularly precise means by which the subjective experience in other people can register in our own experience.

To appreciate the effect of language on consciousness it is revealing to start by looking not at the effect that one's language has on one's own experience but at its effect on the experience of others. If one speaker utters a word (a vocal significant symbol) such as that which we write as 'daffodil' its effect is as predictable as it is unavoidable by anyone within earshot who speaks that language. It evokes representations of one's experience of that yellow flower which blooms in the spring and is worn by the Welsh on March 1st. No two people have exactly the same experience but, and literally, *to all intents and purposes*, the experience is the same. It is the same if they understand each other and they understand each other if it is the same. As Mead has pointed out, telescoped within the meaning of a word is the sum total of our experience with respect to the object and our associated tendencies to respond in particular relevant ways. This view is shared by Vygotsky in the extract below in which he stresses that the crucial difference between the attitude of perception and of thinking is that the latter is a *generalized* attitude. The same effect of words on experience goes for those which have no specific physical referent such as 'hunger' or 'democracy', although the overlapping of common experience or understanding may be less. The same goes for words which describe processes rather than objects, e.g. 'jump' or 'misunderstand' and for the relationship between word order and the structure of the action described and its meaning in terms of that which is *re*presented in experience.

When we use the audio-vocal channel for communication by words each sound is available to everyone close enough to hear and tends to evoke the same responses if they understand each other. It is an important point for the view which emphasizes that we stimulate ourselves with language in the same way that we stimulate others. This has been put forward as the reason why the audio-vocal channel and not the visual-gestural channel has become the means by which we achieve reflexive self-consciousness. As we suggested in Chapter Seven, language can be considered to precede consciousness, or at least self-consciousness, and the infant becomes able to make sense and be accorded adult psychological status through entering into the same world of linguistic meanings as the community around him or her. From the point of view of consciousness there is much to be learned by considering not the development of language within the individual but the development of the individual within language.[1]

Before we proceed to introduce the specific extracts in this chapter we must say something about the relationship between thinking and consciousness. Three of the readings in this chapter are addressed to language and thinking yet we are presenting them as having something to say about consciousness.

Thinking has typically been considered by psychologists as a goal-directed or problem solving activity. In Piaget's work on cognitive development, the child is seen as passing through a series of stages marked by qualitatively different modes of thinking. The modes of thinking are characterized by the way the infant can act towards objects and the logical operations which he or she has internalized to allow internal and external action towards objects to take place. Adult thinking typically involves mental planning, the running through of courses of action in the imagination. This involves not only the logical linking together of strings of objects according to their real and symbolic implications, but also being able to insert into these strings oneself as an object. Thus adult human thinking depends not only on the use of symbolic objects but on the use of some representation of the self which is as generalized as all the other representations.

A fairly general view on the relationship between consciousness and thinking would be that consciousness can be said to comprise thinking when it involves objects and hence the passage of one object to the next. This rather all-embracing view would not rule out thinking in much of the animal kingdom. What would make it more exclusive would be to refine what is meant by 'object' and the manner in which one object might lead to the next. If 'object' refers to objects designated by symbols then just as the objects could be generalized, conventional and imaginary, so could the ways in which they related to each other. The object 'daffodil' could relate to St. David as well as other flowers and associated sensation and thinking would be altogether different. Thinking would then be the passage of one symbolically designated object to the next or simply the passage from symbol to symbol. This view of adult thinking is very much the symbolic interactionist view of thinking outlined in Chapter Seven. It has the distinction of making thinking dependent not just on symbols but on the symbolic representation of the self and proposes that both are dependent on the social process wherein language emerges.

For the first extract we have chosen two exerpts from *Thought and Language* by L. S. Vygotsky, who carried out most of his work in the 1920s, dying in 1934 at the age of 37. Thus the period of his major work coincided with both Mead in America and Freud in Western Europe. All three died in the 1930s and all three in their own way were deeply influenced by the work of Darwin which gave them both an evolutionary and a developmental (i.e. phylogenetic and ontogenetic) approach to human consciousness.[2] Vygotsky's view is that thinking emerges from and therefore depends upon language, in fact, thinking is 'inner speech'.

Vygotsky precedes the first part of the extract with a criticism of the tendency to separate the sound and the meaning of words. The concept of word meaning is central both to the analysis of language and of thought. It is, in his view, the concept which shows language and thinking to be inextricably bound together as parts of one process. Vygotsky proposes that words, which are part of speaking and therefore part of thinking, should not be separated into sound and meaning. Such a separation is a mistake he attributes to the associationist view of psychology where a sound becomes linked to an object or an experience by continued association. Vygotsky makes clear his view that a word does not refer to a single object but to a category of objects — in fact to a categorization of activities towards an object. In this his view closely parallels Mead's that a vocal symbol stands for or represents a generalized attitude. Vygotsky sees a great difference between sensation and thought: they are two different types of experience. The difference is that thought involves a *generalized* reflection of reality while sensation is direct and specific. Generalization is the essence of word meaning and meaning is as much a part of the realm of language as the realm of thought.

The second part of the extract addresses the important phenomenon of private 'egocentric' speech in children, where it comes from, its purpose and what becomes of it. Piaget[3] had observed that young children spoke to themselves and seemed to direct their own activities in a personal language. This speech was not directed at anybody else, indeed it was not understandable by anyone else. He called it 'egocentric speech' in accordance with the general egocentric world view which he attributed to the infant; and he assumed it to be a linguistic stage which children went through on their way to achieving socialized speech. Piaget believed that this egocentric speech became socialized into adult speech. For Vygotsky the reverse was true: it was socialized adult speech which the child was beginning to use to guide his or her own behaviour and which had not yet been completely internalized as silent speech or thought. Thus both Piaget and Vygotsky recognized the phenomenon of egocentric speech in pre-school children and both considered it important developmentally. Piaget saw egocentric speech as the transition of private mental states on their way to becoming shareable public ones. Vygotsky, on the other hand, saw egocentric speech as shared public speech on its way to becoming internalized by the infant as thinking.

These explanations show a fundamental difference in the theoretical stances of the two. For Piaget the objects of consciousness start within the individual as a result of operating on the physical world and subsequently become public and shareable. For Vygotsky the objects of consciousness are already in the public domain and they become internalized as tools for thought during the child's development.

The view that sound and meaning in words are separate elements leading separate lives has done much harm to the study of both the phonetic and the semantic aspects of language. The most thorough study of speech sounds merely as sounds, apart from their connection with thought, has little bearing on their function as human speech since it does not bring out the physical and psychological properties peculiar to speech but only the properties common to all sounds existing in nature. In the same way, meaning divorced from speech sounds can only be studied as a pure act of thought, changing and developing independently of its material vehicle. This separation of sound and meaning is largely responsible for the barrenness of classical phonetics and semantics. In child psychology, likewise, the phonetic and the semantic aspects of speech development have been studied separately. The phonetic development has been studied in great detail, yet all the accumulated data contribute little to our understanding of linguistic development as such and remain essentially unrelated to the findings concerning the development of thinking.

In our opinion the right course to follow is to use the other type of analysis, which may be called *analysis into units*.

By *unit* we mean a product of analysis which, unlike elements, retains all the basic properties of the whole and which cannot be further divided without losing them. Not the chemical composition of water but its molecules and their behavior are the key to the understanding of the properties of water. The true unit of biological analysis is the living cell, possessing the basic properties of the living organism.

What is the unit of verbal thought that meets these requirements? We believe that it can be found in the internal aspect of the word, in *word meaning*. Few investigations of this internal aspect of speech have been undertaken so far, and psychology can tell us little about word meaning that would not apply in equal measure to all other images and acts of thought. The nature of meaning as such is not clear. Yet it is in word meaning that thought and speech unite into verbal thought. In meaning, then, the answers to our questions about the relationship between thought and speech can be found.

Our experimental investigation, as well as theoretical analysis, suggest that both Gestalt and association psychology have been looking for the intrinsic nature of word meaning in the wrong directions. A word does not refer to a single object but to a group or to a class of objects. Each word is therefore already a generalization. Generalization is a verbal act of thought and reflects reality in quite another way than sensation and perception reflect it. Such a qualitative difference is implied in the proposition that there is a dialectic leap not only between total absence of consciousness (in inanimate matter) and sensation but also between sensation and thought. There is every reason to suppose that the qualitative distinction between sensation and thought is the presence in the latter of a *generalized* reflection of reality, which is also the essence of word meaning; and consequently that meaning is an act of thought in the full sense of the term. But at the same time, meaning is an inalienable part of word as such, and thus it belongs in the realm of language as much as in the realm of thought. A word without meaning is an empty sound, no longer a part of human speech. Since word meaning is both thought and speech, we find in it the unit of verbal thought we are looking for. Clearly, then, the method to follow in our exploration of the nature of verbal thought is semantic analysis — the study of the development, the functioning, and the structure of this unit, which contains thought and speech interrelated.

This method combines the advantages of analysis and synthesis, and it permits adequate study of complex wholes. As an illustration, let us take yet another aspect of our subject, also largely neglected in the past. The primary function of speech is communication, social intercourse. When language was studied through analysis into elements, this function, too, was dissociated from the intellectual function of speech. The two were treated as though they were separate, if parallel, functions, without attention to their structural and developmental interrelation. Yet word meaning is a unit of both these functions of speech. That understanding between minds is impossible without some mediating expression is an axiom for scientific psychology. In the absence of a system of signs, linguistic or other, only the most primitive and limited type of communication is possible. Communication by means of expressive movements, observed mainly among animals, is not so much communication as a spread of affect. A frightened goose suddenly aware of danger and rousing the whole flock with its cries does not tell the others what it has seen but rather contaminates them with its fear.

Rational, intentional conveying of experience and thought to others requires a mediating system, the prototype of which is human speech born of the need of intercourse during work. In accordance with the dominant trend, psychology has until recently depicted the matter in an oversimplified way. It was assumed that the means of communication was the sign (the word or sound); that through simultaneous occurrence a sound could become associated with the content of any experience and then serve to convey the same content to other human beings.

Closer study of the development of understanding and communication in childhood, however, has led to the conclusion that real communication requires meaning — i.e., generalization — as much as signs. According to Edward Sapir's penetrating description, the world of experience must be greatly simplified and generalized before it can be translated into symbols. Only in this way does communication become possible, for the individual's experience resides only in his own consciousness and is, strictly speaking, not communicable. To become communicable it must be included in a certain category which, by tacit convention, human society regards as a unit.

Thus, true human communication presupposes a generalizing attitude, which is an advanced stage in the development of word meanings. The higher forms of human intercourse are possible only because man's thought reflects conceptualized actuality. That is why certain thoughts cannot be communicated to children even if they are familiar with the necessary words.

The meaning of a word represents such a close amalgam of thought and language that it is hard to tell whether it is a phenomenon of speech or a phenomenon of thought. A word without meaning is an empty sound; meaning, therefore, is a criterion of 'word,' its indispensable component. It would seem, then, that it may be regarded as a phenomenon of speech. But from the point of view of psychology, the meaning of every word is a generalization or a concept. And since generalizations and concepts are undeniably acts of thought, we may regard meaning as a phenomenon of thinking. It does not follow, however, that meaning formally belongs in two different spheres of psychic life. Word meaning is a phenomenon of thought only in so far as thought is embodied in speech, and of speech only in so far as speech is connected with thought and illumined by it. It is a phenomenon of verbal thought, or meaningful speech — a union of word and thought.

To get a true picture of inner speech, one must start from the assumption that it is a specific formation, with its own laws and complex relations to the other forms of speech activity. Before we can study its relation to thought, on the one hand, and to speech, on the other, we must determine its special characteristics and function.

Inner speech is speech for oneself; external speech is for others. It would indeed be surprising if such a basic difference in function did not affect the structure of the two kinds of speech. Absence of vocalization per se is only a consequence of the specific nature of inner speech, which is neither an antecedent of external speech nor its reproduction in memory but is, in a sense, the opposite of external speech. The latter is the turning of thought into words, its materialization and objectification. With inner speech, the process is reversed: Speech turns into inward thought. Consequently, their structures must differ.

The area of inner speech is one of the most difficult to investigate. It remained almost inaccessible to experiments until ways were found to apply the genetic method of experimentation. Piaget was the first to pay attention to the child's egocentric speech and to see its theoretical significance, but he remained blind to the most important trait of egocentric speech — its genetic connection with inner speech — and this warped his interpretation of its function and structure. We made that relationship the central problem of our study and thus were able to investigate the nature of inner speech with unusual completeness. A number of considerations and observations led us to conclude that egocentric speech is a stage of development preceding inner speech: Both fulfill intellectual functions; their structures are similar; egocentric speech disappears at school age, when inner speech begins to develop. From all this we infer that one changes into the other.

If this transformation does take place, then egocentric speech provides the key to the study of inner speech. One advantage of approaching inner speech through egocentric speech is its accessibility to experimentation and observation. It is still vocalized, audible speech, i.e., external in its mode of expression, but at the same time inner speech in function and structure. To study an internal process it is necessary to externalize it experimentally, by connecting it with some outer activity; only then is objective functional analysis possible. Egocentric speech is, in fact, a natural experiment of this type.

This method has another great advantage: Since egocentric speech can be studied at the time when some of its characteristics are waning and new ones forming, we are able to judge which traits are essential to inner speech and which are only temporary, and thus to determine the goal of this movement from egocentric to inner speech — i.e., the nature of inner speech.

Before we go on to the results obtained by this method, we shall briefly discuss the nature of egocentric speech, stressing the differences between our theory and Piaget's. Piaget contends that the child's egocentric speech is a direct expression of the egocentrism of his thought, which in turn is a compromise between the primary autism of his thinking and its gradual socialization. As the child grows older, autism recedes and socialization progresses, leading to the waning of egocentrism in his thinking and speech.

In Piaget's conception, the child in his egocentric speech does not adapt himself to the thinking of adults. His thought remains entirely egocentric;

this makes his talk incomprehensible to others. Egocentric speech has no function in the child's realistic thinking or activity — it merely accompanies them. And since it is an expression of egocentric thought, it disappears together with the child's egocentrism. From its climax at the beginning of the child's development, egocentric speech drops to zero on the threshold of school age. Its history is one of involution rather than evolution. It has no future.

In our conception, egocentric speech is a phenomenon of the transition from interpsychic to intrapsychic functioning, i.e., from the social, collective activity of the child to his more individualized activity — a pattern of development common to all the higher psychological functions. Speech for oneself originates through differentiation from speech for others. Since the main course of the child's development is one of gradual individualization, this tendency is reflected in the function and structure of his speech.

Our experimental results indicate that the function of egocentric speech is similar to that of inner speech: It does not merely accompany the child's activity; it serves mental orientation, conscious understanding; it helps in overcoming difficulties; it is speech for oneself, intimately and usefully connected with the child's thinking. Its fate is very different from that described by Piaget. Egocentric speech develops along a rising, not a declining, curve; it goes through an evolution, not an involution. In the end, it becomes inner speech.

Our hypothesis has several advantages over Piaget's: It explains the function and development of egocentric speech and, in particular, its sudden increase when the child faces difficulties which demand consciousness and reflection — a fact uncovered by our experiments and which Piaget's theory cannot explain. But the greatest advantage of our theory is that it supplies a satisfying answer to a paradoxical situation described by Piaget himself. To Piaget, the quantitative drop in egocentric speech as the child grows older means the withering of that form of speech. If that were so, its structural peculiarities might also be expected to decline; it is hard to believe that the process would affect only its quantity, and not its inner structure. The child's thought becomes infinitely less egocentric between the ages of three and seven. If the characteristics of egocentric speech that make it incomprehensible to others are indeed rooted in egocentrism, they should become less apparent as that form of speech becomes less frequent; egocentric speech should approach social speech and become more and more intelligible. Yet what are the facts? Is the talk of a three-year-old harder to follow than that of a seven-year-old? Our investigation established that the traits of egocentric speech which make for inscrutability are at their lowest point at three and at their peak at seven. They develop in a reverse direction to the frequency of egocentric speech. While the latter keeps falling and reaches zero at school age, the structural characteristics become more and more pronounced.

This throws a new light on the quantitative decrease in egocentric speech, which is the cornerstone of Piaget's thesis.

What does this decrease mean? The structural peculiarities of speech for oneself and its differentiation from external speech increase with age. What is it then that diminishes? Only one of its aspects: vocalization. Does this mean that egocentric speech as a whole is dying out? We believe that it does not, for how then could we explain the growth of the functional and structural traits of egocentric speech? On the other hand, their growth is perfectly compatible with the decrease of vocalization — indeed, clarifies its meaning. Its rapid dwindling and the equally rapid growth of the other characteristics are contradictory in appearance only.

To explain this, let us start from an undeniable, experimentally established fact. The structural and functional qualities of egocentric speech become more marked as the child develops. At three, the difference between egocentric and social speech equals zero; at seven, we have speech that in structure and function is totally unlike social speech. A differentiation of the two speech functions has taken place. This is a fact—and facts are notoriously hard to refute.

Once we accept this, everything else falls into place. If the developing structural and functional peculiarities of egocentric speech progressively isolate it from external speech, then its vocal aspect must fade away; and this is exactly what happens between three and seven years. With the progressive isolation of speech for oneself, its vocalization becomes unnecessary and meaningless and, because of its growing structural peculiarities, also impossible. Speech for oneself cannot find expression in external speech. The more independent and autonomous egocentric speech becomes, the poorer it grows in its external manifestations. In the end it separates itself entirely from speech for others, ceases to be vocalized, and thus appears to die out.

But this is only an illusion. To interpret the sinking coefficient of egocentric speech as a sign that this kind of speech is dying out is like saying that the child stops counting when he ceases to use his fingers and starts adding in his head. In reality, behind the symptoms of dissolution lies a progressive development, the birth of a new speech form.

The decreasing vocalization of egocentric speech denotes a developing abstraction from sound, the child's new faculty to 'think words' instead of pronouncing them. This is the positive meaning of the sinking coefficient of egocentric speech. The downward curve indicates development toward inner speech.

We can see that all the known facts about the functional, structural, and genetic characteristics of egocentric speech point to one thing: It develops in the direction of inner speech. Its developmental history can be understood only as a gradual unfolding of the traits of inner speech.

We believe that this corroborates our hypothesis about the origin and nature of egocentric speech. To turn our hypothesis into a certainty, we must devise an experiment capable of showing which of the two interpretations is correct. What are the data for this critical experiment?

Let us restate the theories between which we must decide. Piaget believes that egocentric speech stems from the insufficient socialization of

> speech and that its only development is decrease and eventual death. Its culmination lies in the past. Inner speech is something new brought in from the outside along with socialization. We believe that egocentric speech stems from the insufficient individualization of primary social speech. Its culmination lies in the future. It develops into inner speech.

It is our view that consciousness is not an abstract facility or field, but an activity. The implications of language for this view of consciousness must be considered. To speak is to recognize and realize objects. Languages differ in the ways in which they codify experience and to speak one particular language directs attention to one particular aspect of the physical world while speaking another would direct experience to another. This view of linguistic habits affecting habits of thinking and consequently habits of action is due largely to the work of Benjamin Lee Whorf. Whorf reports that he became aware of the fact that the way in which we speak affects what we notice while working for an insurance company assessing fire-insurance claims. He noticed, for example, that in the case of one particular fire people had become casual about the presence of petrol drums because they were 'empty'. When petrol cans are 'empty' they contain petrol vapour which is far more flammable than the fluid, but 'empty' has connotations of being null or void and this was the association which had prevailed.

This eventually led him to study languages other than 'standard average European' languages, particularly that of the North American Hopi Indians, and to the conclusion that languages comprise a set of habitual ways of organizing and interpreting experience. The habits to which we are directed by our own language are not evident to us and it is only when the linguistically conditioned habits of another unrelated culture are examined that the power of language in this respect becomes evident. The relevance of Whorf's ideas for consciousness comes from the role language must be considered to play when we are conditioned to recognize certain objects and associations.

This selection deals only with certain aspects studied by Whorf involving habitual ways of dealing with numbers, temporal forms of verbs, and the expression of duration, intensity and tendency. The general conclusion is that the Hopi and by implication all speakers 'act about situations in ways which are like the ways they talk about them'. There is no doubt that the Hopi talk about things very differently from speakers of 'standard average European'.

In our language, that is SAE (Standard Average European), plurality and cardinal numbers are applied in two ways: to real plurals and imaginary plurals. Or more exactly if less tersely: perceptible spatial aggregates and metaphorical aggregates. We say 'ten men' and also 'ten days.' Ten men either are or could be objectively perceived as ten, ten in one group perception — ten men on a street corner, for instance. But 'ten days' cannot be objectively experienced. We experience only one day, today; the other nine (or even all ten) are something conjured up from memory or imagination. If 'ten days' be regarded as a group it must be as an "imaginary," mentally constructed group. Whence comes this mental pattern? Just as in the case of the fire-causing errors, from the fact that our language confuses the two different situations, has but one pattern for both. When we speak of 'ten steps forward, ten strokes on a bell,' or any similarly described cyclic sequence, "times" of any sort, we are doing the same thing as with 'days.' CYCLICITY brings the response of imaginary plurals. But a likeness of cyclicity to aggregates is not unmistakably given by experience prior to language, or it would be found in all languages, and it is not.

Our AWARENESS of time and cyclicity does contain something immediate and subjective — the basic sense of "becoming later and later." But, in the habitual thought of us SAE people, this is covered under something quite different, which though mental should not be called subjective. I call it OBJECTIFIED, or imaginary, because it is patterned on the OUTER world. It is this that reflects our linguistic usage. Our tongue makes no distinction between numbers counted on discrete entities and numbers that are simply "counting itself." Habitual thought then assumes that in the latter the numbers are just as much counted on "something" as in the former. This is objectification. Concepts of time lose contact with the subjective experience of "becoming later" and are objectified as counted QUANTITIES, especially as lengths, made up of units as a length can be visibly marked off into inches. A 'length of time' is envisioned as a row of similar units, like a row of bottles.

In Hopi there is a different linguistic situation. Plurals and cardinals are used only for entities that form or can form an objective group. There are no imaginary plurals, but instead ordinals used with singulars. Such an expression as 'ten days' is not used. The equivalent statement is an operational one that reaches one day by a suitable count. 'They stayed ten days' becomes 'they stayed until the eleventh day' or 'they left after the tenth day.' 'Ten days is greater than nine days' becomes 'the tenth day is later than the ninth.' Our "length of time" is not regarded as a length but as a relation between two events in lateness. Instead of our linguistically promoted objectification of that datum of consciousness we call 'time,' the Hopi language has not laid down any pattern that would cloak the subjective "becoming later" that is the essence of time.

The three-tense system of SAE verbs colors all our thinking about time. This system is amalgamated with that larger scheme of objectification of the subjective experience of duration already noted in other patterns —

in the binomial formula applicable to nouns in general, in temporal nouns, in plurality and numeration. This objectification enables us in imagination to "stand time units in a row." Imagination of time as like a row harmonizes with a system of THREE tenses; whereas a system of TWO, an earlier and a later, would seem to correspond better to the feeling of duration as it is experienced. For if we inspect consciousness we find no past, present, future, but a unity embracing complexity. EVERYTHING is in consciousness, and everything in consciousness IS, and is together. There is in it a sensuous and a nonsensuous. We may call the sensuous — what we are seeing, hearing, touching — the 'present' while in the nonsensuous the vast image-world of memory is being labeled 'the past' and another realm of belief, intuition, and uncertainty 'the future'; yet sensation, memory, foresight, all are in consciousness together — one is not "yet to be" nor another "once but no more." Where real time comes in is that all this in consciousness is "getting later," changing certain relations in an irreversible manner. In this "latering" or "durating" there seems to me to be a paramount contrast between the newest, latest instant at the focus of attention and the rest — the earlier. Languages by the score get along well with two tenselike forms answering to this paramount relation of "later" to "earlier." We can of course CONSTRUCT AND CONTEMPLATE IN THOUGHT a system of past, present, future, in the objectified configuration of points on a line. This is what our general objectification tendency leads us to do and our tense system confirms.

In English the present tense seems the one least in harmony with the paramount temporal relation. It is as if pressed into various and not wholly congruous duties. One duty is to stand as objectified middle term between objectified past and objectified future, in narration, discussion, argument, logic, philosophy. Another is to denote inclusion in the sensuous field: 'I SEE him.' Another is for nomic, i.e. customarily or generally valid, statements: 'We SEE with our eyes.' These varied uses introduce confusions of thought, of which for the most part we are unaware.

Hopi, as we might expect, is different here too. Verbs have no "tenses" like ours, but have validity-forms ("assertions"), aspects, and clause-linkage forms (modes), that yield even greater precision of speech. The validity-forms denote that the speaker (not the subject) reports the situation (answering to our past and present) or that he expects it (answering to our future) or that he makes a nomic statement (answering to our nomic present). The aspects denote different degrees of duration and different kinds of tendency "during duration." As yet we have noted nothing to indicate whether an event is sooner or later than another when both are REPORTED. But need for this does not arise until we have two verbs: i.e. two clauses. In that case the "modes" denote relations between the clauses, including relations of later to earlier and of simultaneity. Then there are many detached words that express similar relations, supplementing the modes and aspects. The duties of our three-tense system and its tripartite linear objectified "time" are distributed among various verb categories, all different from our tenses; and there is no more basis for an objectified time in Hopi verbs than in other Hopi patterns; although this does not in the least hinder the verb forms and other paterns from being closely adjusted to the pertinent realities of actual situations.

To fit discourse to manifold actual situations, all languages need to express durations, intensities, and tendencies. It is characteristic of SAE and perhaps of many other language types to express them metaphorically. The metaphors are those of spatial extension, i.e. of size, number (plurality), position, shape, and motion. We express duration by 'long, short, great, much, quick, slow,' etc.; intensity by 'large, great, much, heavy, light, high, low, sharp, faint,' etc.; tendency by 'more, increase, grow, turn, get, approach, go, come, rise, fall, stop, smooth, even, rapid, slow'; and so on through an almost inexhaustible list of metaphors that we hardly recognize as such, since they are virtually the only linguistic media available. The nonmetaphorical terms in this field, like 'early, late, soon, lasting, intense, very, tending,' are a mere handful, quite inadequate to the needs.

It is clear how this condition "fits in." It is part of our whole scheme of OBJECTIFYING — imaginatively spatializing qualities and potentials that are quite nonspatial (so far as any spatially perceptive senses can tell us). Noun-meaning (with us) proceeds from physical bodies to referents of far other sort. Since physical bodies and their outlines in PERCEIVED SPACE are denoted by size and shape terms and reckoned by cardinal numbers and plurals, these patterns of denotation and reckoning extend to the symbols of nonspatial meanings, and so suggest an IMAGINARY SPACE. Physical shapes 'move, stop, rise, sink, approach,' etc., in perceived space; why not these other referents in their imaginary space? This has gone so far that we can hardly refer to the simplest nonspatial situation without constant resort to physical metaphors. I "grasp" the "thread" of another's arguments, but if its "level" is "over my head" my attention may "wander" and "lose touch" with the "drift" of it, so that when he "comes" to his "point" we differ "widely," our "views" being indeed so "far apart" that the "things" he says "appear" "much" too arbitrary, or even "a lot" of nonsense!

The absence of such metaphor from Hopi speech is striking. Use of space terms when there is no space involved is NOT THERE — as if on it had been laid the taboo teetotal! The reason is clear when we know that Hopi has abundant conjugational and lexical means of expressing duration, intensity, and tendency directly as such, and that major gramatical patterns do not, as with us, provide analogies for an imaginary space. The many verb "aspects" express duration and tendency of manifestations, while some of the "voices" express intensity, tendency, and duration of causes or forces producing manifestations. Then a special part of speech, the "tensors," a huge class of words, denotes only intensity, tendency, duration, and sequence. The function of the tensors is to express intensities, "strengths," and how they continue or vary, their rate of change; so that the broad concept of intensity, when considered as necessarily always varying and/or continuing, includes also tendency and duration. Tensors convey distinctions of degree, rate, constancy, repetition, increase and decrease of intensity, immediate sequence, interruption or sequence after an interval, etc., also QUALITIES of strengths, such as we should express metaphorically as smooth, even, hard, rough. A striking feature is their lack of resemblance to the terms of real space and movement that to us "mean the same." There is not even more than a trace of apparent derivation from space terms. So, while Hopi in its nouns seems highly concrete, here in the tensors it becomes abstract almost beyond our power to follow.

The comparison now to be made between the habitual thought worlds of SAE and Hopi speakers is of course incomplete. It is possible only to touch upon certain dominant contrasts that appear to stem from the linguistic differences already noted. By "habitual thought" and "thought world" I mean more than simply language, i.e. than the linguistic patterns themselves. I include all the analogical and suggestive value of the patterns (e.g., our "imaginary space" and its distant implications), and all the give-and-take between language and the culture as a whole, wherein is a vast amount that is not linguistic but yet shows the shaping influence of language. In brief, this "thought world" is the microcosm that each man carries about within himself, by which he measures and understands what he can of the macrocosm.

The SAE microcosm has analyzed reality largely in terms of what it calls "things" (bodies and quasibodies) plus modes of extensional but formless existence that it calls "substances" or "matter." It tends to see existence through a binomial formula that expresses any existent as a spatial form plus a spatial formless continuum related to the form, as contents is related to the outlines of its container. Nonspatial existents are imaginatively spatialized and charged with similar implications of form and continuum.

The Hopi microcosm seems to have analyzed reality largely in terms of EVENTS (or better "eventing"), referred to in two ways, objective and subjective. Objectively, and only if perceptible physical experience, events are expressed mainly as outlines, colors, movements, and other perceptive reports. Subjectively, for both the physical and nonphysical, events are considered the expression of invisible intensity factors, on which depend their stability and persistence, or their fugitiveness and proclivities. It implies that existents do not "become later and later" all in the same way; but some do so by growing like plants, some by diffusing and vanishing, some by a procession of metamorphoses, some by enduring in one shape till affected by violent forces. In the nature of each existent able to manifest as a definite whole is the power of its own mode of duration: its growth, decline, stability, cyclicity, or creativeness. Everything is thus already "prepared" for the way it now manifests by earlier phases, and what it will be later, partly has been, and partly is in act of being so "prepared." An emphasis and importance rests on this preparing or being prepared aspect of the world that may to the Hopi correspond to that "quality of reality" that 'matter' or 'stuff' has for us.

Our behavior, and that of Hopi, can be seen to be coordinated in many ways to the linguistically conditioned microcosm. As in my fire casebook, people act about situations in ways which are like the ways they talk about them. A characteristic of Hopi behavior is the emphasis on preparation. This includes announcing and getting ready for events well beforehand, elaborate precautions to insure persistence of desired conditions, and stress on good will as the preparer of right results. Consider the analogies of the day-counting pattern alone. Time is mainly reckoned "by day" (*taLk, -tala*) or "by night" (*tok*), which words are not nouns but tensors, the first formed on a root "light, day," the second on a root "sleep." The count is by ORDINALS. This is not the pattern of counting a number of different men or things, even though they appear successively, for, even then, they COULD gather into an assemblage. It is the pattern of counting successive reappearances of the SAME man or thing, incapable of forming an assemblage. The analogy is not to behave about day-cyclicity as to several men ("several days"), which is what WE tend to do, but to behave as to the successive visits of the SAME MAN. One does not alter several men by working upon just one, but one can prepare and so alter the later visits of the same man by working to affect the visit he is making now. This is the way the Hopi deal with the future — by working within a present situation which is expected to carry impresses, both obvious and occult, forward into the future event of interest. One might say that Hopi society understands our proverb 'Well begun is half done,' but not our 'Tomorrow is another day.' This may explain much in Hopi character.

This Hopi preparing behavior may be roughly divided into announcing, outer preparing, inner preparing, covert participation, and persistence. Announcing, or preparative publicity, is an important function in the hands of a special official, the Crier Chief. Outer preparing is preparation involving much visible activity, not all necessarily directly useful within our understanding. It includes ordinary practicing, rehearsing, getting ready, introductory formalities, preparing of special food, etc. (all of these to a degree that may seem overelaborate to us), intensive sustained muscular activity like running, racing, dancing, which is thought to increase the intensity of development of events (such as growth of crops), mimetic and other magic, preparations based on esoteric theory involving perhaps occult instruments like prayer sticks, prayer feathers, and prayer meal, and finally the great cyclic ceremonies and dances, which have the significance of preparing rain and crops. From one of the verbs meaning "prepare" is derived the noun for "harvest" or "crop": *na'twani* 'the prepared' or the 'in preparation.'

Inner preparing is use of prayer and meditation, and at lesser intensity good wishes and good will, to further desired results. Hopi attitudes stress the power of desire and thought. With their "microcosm" it is utterly natural that they should. Desire and thought are the earliest, and therefore the most important, most critical and crucial, stage of preparing. Moreover, to the Hopi, one's desires and thoughts influence not only his own actions, but all nature as well. This too is wholly natural. Consciousness itself is aware of work, of the feel of effort and energy, in desire and thinking. Experience more basic than language tells us that, if energy is expended, effects are produced. WE tend to believe that our bodies can stop up this energy, prevent it from affecting other things until we will our BODIES to overt action. But this may be so only because we have our own linguistic basis for a theory that formless items like "matter" are things in themselves, malleable only by similar things, by more matter, and hence insulated from the powers of life and thought. It is no more unnatural to think that thought contacts everything and pervades the universe than to think, as we all do, that light kindled outdoors does this. And it is not unnatural to suppose that thought, like any other force, leaves everywhere traces of effect. Now, when WE think of a certain actual rosebush, we do not suppose that our thought goes to that actual bush, and engages with it, like a searchlight turned upon it. What then do we suppose our consciousness is dealing with when we are thinking of that rosebush? Probably we think it is dealing with a "mental image" which is not the rosebush but a mental surrogate of it. But why should it be NATURAL to think that our thought deals with a surrogate and not with the real rosebush? Quite possibly because we are dimly aware that we carry about with us a whole imaginary space, full of mental surrogates. To us, mental surrogates are old familiar fare. Along with the images of imaginary space, which we perhaps secretly know to be only imaginary, we tuck the thought-of actually existing rosebush, which may be quite another story, perhaps just because we have that very convenient "place" for it. The Hopi thought-world has no imaginary space. The corollary to this is that it may not locate thought dealing with real space anywhere but in real space, nor insulate real space from the effects of thought. A Hopi would naturally suppose that his thought (or he himself) traffics with the actual rosebush — or more likely, corn plant — that he is thinking about. The thought then should leave some trace of itself with the plant in the field. If it is a good thought, one about health and growth, it is good for the plant; if a bad thought, the reverse.

The Hopi emphasize the intensity-factor of thought. Thought to be most effective should be vivid in consciousness, definite, steady, sustained, charged with strongly felt good intentions. They render the idea in English as 'concentrating, holding it in your heart, putting your mind on it, earnestly hoping.' Thought power is the force behind ceremonies, prayer sticks, ritual smoking, etc. The prayer pipe is regarded as an aid to "concentrating" (so said my informant). Its name, *na'twanpi*, means 'instrument of preparing.'

That language use not only affects our habitual ways of thinking and acting but also limits what sort of analysis can be achieved has been an issue in twentieth-century philosophy. Linguistic philosophy was developed as a branch of analytical philosophy by Ludwig Wittgenstein during the 1930s as a method of analysis. A chief concern of linguistic philosophy has been to reveal the mistaken assumptions about the world into which our everyday use of language leads us. It is proposed that at least some philosophical problems can be revealed as stemming from mistaken thinking due to misleading assumptions involved in the use of language. The target of the analysis is thus the language in which ideas are expressed rather than the ideas themselves. Accordingly, its goals are not the solutions of problems but their dissolution through being revealed as problems of language.

The extract which we have taken from Wittgenstein's *Philosophical Investigations* deals with the relationship between speaking and thinking. In paragraph 329 he states that 'language is the vehicle for thought'. He comments, in paragraph 342, on an example used by William James to indicate that thinking is possible without language where James cites a deaf-mute's description of, having recovered the power of speech, thoughts he had had while unable to speak. But how, asks Wittgenstein, can we be sure that the words which he now uses reliably express the sensations he had formerly felt? When he asks, in paragraph 344, whether someone could speak a private language without ever speaking a public one Wittgenstein says this is simply a case of doing always what one does sometimes. Although this may be philosophically satisfactory, it would not be satisfactory from a psychological view which says that language can only be used internally if it has been experienced externally.

327. "Can one think without speaking?" — And what is *thinking*? — Well, don't you ever think? Can't you observe yourself and see what is going on? It should be quite simple. You do not have to wait for it as for an astronomical event and then perhaps make your observation in a hurry.

328. Well, what does one include in 'thinking'? What has one learnt to use this word for? — If I say I have thought — need I always be right? — What *kind* of mistake is there room for here? Are there circumstances in which one would ask: "Was what I was doing then really thinking; am I not making a mistake?" Suppose someone takes a measurement in the middle of a train of thought: has he interrupted the thought if he says nothing to himself during the measuring?

329. When I think in language, there aren't 'meanings' going through my mind in addition to the verbal expressions: the language is itself the vehicle of thought.

330. Is thinking a kind of speaking? One would like to say it is what distinguishes speech with thought from talking without thinking. — And

so it seems to be an accompaniment of speech. A process, which may accompany something else, or can go on by itself.

Say, "Yes, this pen is blunt. Oh well, it'll do." First, thinking it; then without thought; then just think the thought without the words. — Well, while doing some writing I might test the point of my pen, make a face — and then go on with a gesture of resignation. — I might also act in such a way while taking various measurements that an on-looker would say I had — without words — thought: If two magnitudes are equal to a third, they are equal to one another. — But what constitutes thought here is not some process which has to accompany the words if they are not to be spoken without thought.

331. Imagine people who could only think aloud. (As there are people who can only read aloud.)

332. While we sometimes call it "thinking" to accompany a sentence by a mental process, that accompaniment is not what we mean by a "thought". — Say a sentence and think it; say it with understanding. — And now do not say it, and just do what you accompanied it with when you said it with understanding! — (Sing this tune with expression. And now don't sing it, but repeat its expression! — And here one actually might repeat something. For example, motions of the body, slower and faster breathing, and so on.)

333. "Only someone who is *convinced* can say that." — How does the conviction help him when he says it? — Is it somewhere at hand by the side of the spoken expression? (Or is it masked by it, as a soft sound by a loud one, so that it can, as it were, no longer be heard when one expresses it out loud?) What if someone were to say "In order to be able to sing a tune from memory one has to hear it in one's mind and sing from that"?

334. "So you really wanted to say . . . ." — We use this phrase in order to lead someone from one form of expression to another. One is tempted to use the following picture: what he really 'wanted to say', what he 'meant' was already *present somewhere* in his mind even before we gave it expression. Various kinds of thing may persuade us to give up one expression and to adopt another in its place. To understand this, it is useful to consider the relation in which the solutions of mathematical problems stand to the context and ground of their formulation. The concept 'trisection of the angle with ruler and compass', and people are trying to do it, and, on the other hand, when it has been proved that there is no such thing.

335. What happens when we make an effort — say in writing a letter — to find the right expression for our thoughts? — This phrase compares the process to one of translating or describing: the thoughts are already there (perhaps were there in advance) and we merely look for their expression. This picture is more or less appropriate in different cases. — But can't all sorts of things happen here? — I surrender to a mood and the expression *comes*. Or a picture occurs to me and I try to describe it.

Or an English expression occurs to me and I try to hit on the corresponding German one. Or I make a gesture, and ask myself: What words correspond to this gesture? And so on.

Now if it were asked: "Do you have the thought before finding the expression?" what would one have to reply? And what, to the question: "What did the thought consist in, as it existed before its expression?"

336. This case is similar to the one in which someone imagines that one could not think a sentence with the remarkable word order of German or Latin just as it stands. One first has to think it, and then one arranges the words in that queer order. (A French politician once wrote that it was a peculiarity of the French language that in it words occur in the order in which one thinks them.)

337. But didn't I already intend the whole construction of the sentence (for example) at its beginning? So surely it already existed in my mind before I said it out loud! — If it was in my mind, still it would not normally be there in some different word order. But here we are constructing a misleading picture of 'intending', that is, of the use of this word. An intention is embedded in its situation, in human customs and institutions. If the technique of the game of chess did not exist, I could not intend to play a game of chess. In so far as I do intend the construction of a sentence in advance, that is made possible by the fact that I can speak the language in question.

338. After all, one can only say something if one has learned to talk. Therefore in order to *want* to say something one must also have mastered a language; and yet it is clear that one can want to speak without speaking. Just as one can want to dance without dancing.

And when we think about this, we grasp at the *image* of dancing, speaking, etc. .

339. Thinking is not an incorporeal process which lends life and sense to speaking, and which it would be possible to detach from speaking, rather as the Devil took the shadow of Schlemiehl from the ground. — But how "not an incorporeal process"? Am I acquainted with incorporeal processes, then, only thinking is not one of them? No; I called the expression "an incorporeal process" to my aid in my embarrassment when I was trying to explain the meaning of the word "thinking" in a primitive way.

One might say "Thinking is an incorporeal process", however, if one were using this to distinguish the grammar of the word "think" from that of, say, the word "eat". Only that makes the difference between the meanings look *too slight*. (It is like saying: numerals are actual, and numbers non-actual, objects.) An unsuitable type of expression is a sure means of remaining in a state of confusion. It as it were bars the way out.

340. One cannot guess how a word functions. One has to *look at* its use and learn from that.

But the difficulty is to remove the prejudice which stands in the way of doing this. It is not a *stupid* prejudice.

341. Speech with and without thought is to be compared with the playing of a piece of music with and without thought.

342. William James, in order to shew that thought is possible without speech, quotes the recollection of a deaf-mute, Mr. Ballard, who wrote that in his early youth, even before he could speak, he had had thoughts about God and the world. — What can he have meant? — Ballard writes: "It was during those delightful rides, some two or three years before my intitiation into the rudiments of written language, that I began to ask myself the question: how came the world into being?" — Are you sure — one would like to ask — that this is the correct translation of your wordless thought into words? And why does this question — which otherwise seems not to exist — raise its head here? Do I want to say that the writer's memory deceives him? — I don't even know if I should say *that*. These recollections are a queer memory phenomenon,— and I do not know what conclusions one can draw from them about the past of the man who recounts them.

343. The words with which I express my memory are my memory-reaction.

344. Would it be imaginable that people should never speak an audible language, but should still say things to themselves in the imagination?

"If people always said things only to themselves, then they would merely be doing *always* what as it is they do *sometimes*." — So it is quite easy to imagine this: one need only make the easy transition from some to all. (Like: "An infinitely long row of trees is simply one that does *not* come to an end.") Our criterion for someone's saying something to himself is what he tells us and the rest of his behaviour; and we only say that someone speaks to himself if, in the ordinary sense of the words he *can speak*. And we do not say it of a parrot; nor of a gramophone.

345. "What sometimes happens might always happen." — What kind of proposition is that? It is like the following: If "$F(a)$" makes sense "$(x).F(x)$" makes sense.

"If it is possible for someone to make a false move in some game, then it might be possible for everybody to make nothing but false moves in every game." — Thus we are under a temptation to misunderstand the logic of our expressions here, to give an incorrect account of the use of our words.

Orders are sometimes not obeyed. But what would it be like if no orders were *ever* obeyed? The concept 'order' would have lost its purpose.

The last extract in this section deals less exclusively with language and consciousness, although it is still an important theme. It deals with the relationship between language and the repression which creates the unconscious, and points towards Chapter Eleven on psychoanalysis and consciousness and particularly the work of Jaques Lacan. Fromm's interest is also, as his title suggests, in Buddhism and his article directs us to the similarities between the idea of releasing the repressed in psychoanalytic theory and enlightement from habitual (often linguistically determined) modes of thought of which Buddhism speaks. Thus this extract, while dealing with language and consciousness, also moves towards the complementary views which are discussed in Chapter Eleven.

Fromm takes a fairly orthodox view on the conflict between biology and society, or the contradiction between 'the wider human aims which are common to all men' and any particular society's 'need to survive in the particular form in which it has developed'. He differs from an orthodox Freudian dichotomy of the conscious and the unconscious by postulating a *continuum* from awareness of the true nature of things to complete delusion or ignorance. The mechanism which places experience at various places on this continuum he describes as a 'socially conditioned filter'. The relevance of language to this filter is discussed and in this respect we are reminded of the discussion in the earlier extract by Whorf.[4]

The idea of a filter between the total sentient capacity of the human organism and that to which we can attend has been a principal feature of cognitive psychology.[5] The body might be thought to have an almost limitless capacity to monitor its surroundings[6] for information at rudimentary levels of analysis. Any of this stimulation may make its way to become an object of attention. Evolution has bestowed upon us mechanisms such as those concerned with pain or the orienting reflex to ensure that important sources of stimulation force their way over all others to attention. There is much experimental and anecdotal evidence that we can be aware of (in the sense of registering with our senses) an enormous amount of information at the same time. However, we seem able to attend only to one thing at a time and this has led to cognitive psychologists' view of the human being as a limited single channel information processor. The implication has been that there is a process which could be represented as a filter and it has frequently been assumed that this is a function of neural architecture and mechanisms. An alternative view would be that what cognitive psychologists call, in this context, attention is the equivalent to what symbolic interactionists would call 'taking an attitude towards an object'. That is, being self-consciously aware of sensation by *re*presenting it. This is how Fromm deals with the process of being aware of sensations which he refers to as *cerebration*. To feel is one thing but to be aware of feeling is to 'think feelings', a presentation of the feeling to the self. The cerebrating person is an alienated person and Fromm links this process to language.

> Having decided to speak of unconscious and conscious as states of awareness and unawareness, respectively, rather than as "parts" of personality and specific contents, we must now consider the question of what prevents an experience from reaching our awareness — that is, from becoming conscious.
>
> But before we begin to discuss this question, another one arises which should be answered first. If we speak in a psychoanalytic context of consciousness and unconsciousness, there is an implication that consciousness is of a higher value than unconsciousness. Why should we be striving to broaden the domain of consciousness, unless this were so? Yet it is quite obvious that consciousness as such has no particular value; in fact, most of what people have in their conscious minds is fiction and delusion; this is the case not so much because people would be *incapable* of seeing the truth as because of the function of society. Most of human history (with the exception of some primitive societies) is characterized by the fact that a small minority has ruled over and exploited the majority of its fellows. In order to do so, the minority has usually used force; but force is not enough. In the long run, the majority has had to accept its own exploitation voluntarily — and this is only possible if its mind has been filled with all sorts of lies and fictions, justifying and explaining its acceptance of the minority's rule. However, it is not the only reason for the fact that most of what people have in their awareness about themselves, others, society, etc., is fiction. In its historical development each society becomes caught in its own need to survive in the particular form in which it has developed, and it usually accomplishes this survival by ignoring the wider human aims which are common to all men. This contradiction between the social and the universal aim leads also to the fabrication (on a social scale) of all sorts of fictions and illusions which have the function to deny and to rationalize the dichotomy between the goals of humanity and those of a given society.
>
> We might say, then, that the content of consciousness is mostly fictional and delusional, and precisely does not represent reality. Consciousness as such, then, is nothing desirable. Only if the hidden reality (that which is unconscious) is revealed, and hence is no longer hidden (i.e., has become conscious) — has something valuable been achieved. Most of what is in our consciousness is "false consciousness" and that it is essentially society that fills us with these fictitious and unreal notions.
>
> The effect of society is not only to funnel fictions into our consciousness, but also to prevent the awareness of reality. The further elaboration of this point leads us straight into the central problem of how repression or unconsciousness occurs.
>
> The animal has a consciousness of the things around it which, to use R. M. Bucke's term, we may call "simple consciousness." Man's brain

structure, being larger and more complex than that of the animal transcends this simple consciousness and is the basis of *self consciousness*, awareness of himself as the subject of his experience. But perhaps because of its enormous complexity human awareness is organized in various possible ways, and for any experience to come into awareness, it must be comprehensible in the categories in which conscious thought is organized. Some of the categories, such as time and space, may be universal, and may constitute categories of perception common to all men. Others, such as causality, may be a valid category for many, but not for all, forms of human conscious perception. Other categories are even less general and differ from culture to culture. However this may be, experience can enter into awareness only under the condition that it can be perceived, related, and ordered in terms of a conceptual system and of its categories. This system is in itself a result of social evolution. Every society, by its own practice of living and by the mode of relatedness, of feeling, and perceiving, develops a system of categories which determines the forms of awareness. This system works, as it were, like a *socially conditioned filter*; experience cannot enter awareness unless it can penetrate this filter.

The question then, is to understand more concretely how this "social filter" operates, and how it happens that it permits certain experiences to be filtered through, while others are stopped from entering awareness.

First of all, we must consider that many experiences do not lend themselves easily to being perceived in awareness. Pain is perhaps the physical experience which best lends itself to being consciously perceived; sexual desire, hunger, etc., also are easily perceived; quite obviously, all sensations which are relevant to individual or group survival have easy access to awareness. But when it comes to a more subtle or complex experience, like *seeing a rosebud in the early morning, a drop of dew on it, while the air is still chilly, the sun coming up, a bird singing* — this is an experience which in some cultures easily lends itself to awareness (for instance, in Japan), while in modern Western culture this same experience will usually not come into awareness because it is not sufficiently "important" or "eventful" to be noticed. Whether or not subtle affective experiences can arrive at awareness depends on the degree to which such experiences are cultivated in a given culture. There are many affective experiences for which a given language has no word, while another language may be rich in words which express these feelings. In English, for instance, we have one word, "love," which covers experiences ranging from liking to erotic passion to brotherly and motherly love. In a language in which different affective experiences are not expressed by different words, it is almost impossible for one's experiences to come to awareness, and vice versa. Generally speaking, it may be said that an experience rarely comes into awareness for which the language has no word.

But this is only one aspect of the filtering function of language. Different languages differ not only by the fact that they vary in the diversity of words they use to denote certain affective experiences, but by their syntax, their grammar, and the root-meaning of their words. The whole language contains an attitude of life, is a frozen expression of experiencing life in a certain way.

Language, by its words, its grammar, its syntax, by the whole spirit which is frozen in it, determines how we experience, and which experiences penetrate to our awareness.

The second aspect of the filter which makes awareness possible is the *logic* which directs the thinking of people in a given culture. Just as most people assume that their language is "natural" and that other languages only use different words for the same things, they assume also that the rules which determine proper thinking, are natural and universal ones; that what is illogical in one cultural system is illogical in any other, because it conflicts with "natural" logic. A good example of this is the difference between Aristotelian and paradoxical logic.

Aristotelian logic is based on the law of identity which states that A is A, the law of contradiction (A is not non-A), and the law of the excluded middle (A cannot be A *and* non-A, neither A *nor* non-A). Aristotle stated it: "It is impossible for the same thing at the same time to belong and not to belong to the same thing and in the same respect. . . . This, then, is the most certain of all principles."

In opposition to Aristotelian logic is what one might call *paradoxical logic*, which assumes that A and non-A do not exclude each other as predicates of X. Paradoxical logic was predominant in Chinese and Indian thinking, in Heraclitus' philosophy, and then again under the name of dialectics in the thought of Hegel and Marx. The general principle of paradoxical logic has been clearly described in general terms by Lao-Tse: "Words that are strictly true seem to be paradoxical." And by Chuang-tzu: "That which is one is one. That which is not-one, is also one."

Inasmuch as a person lives in a culture in which the correctness of Aristotelian logic is not doubted, it is exceedingly difficult, if not impossible, for him to be aware of experiences which contradict Aristotelian logic, hence which from the standpoint of his culture are nonsensical. A good example is Freud's concept of ambivalence, which says that one can experience love and hate for the same person at the same time. This experience, which from the standpoint of paradoxical logic is quite "logical," does not make sense from the standpoint of Aristotelian logic. As a result, it is exceedingly difficult for most people to be aware of feelings of ambivalence. If they are aware of love, they can not be aware of hate — since it would be utterly nonsensical to have two contradictory feelings at the same time towards the same person.

Any society, in order to survive, must mould the character of its members in such a way that *they want to do what they have to do*; their social function must become internalized and transformed into something they feel driven to do, rather than something they are obliged to do. A society cannot permit deviation from this pattern, beause if this "social character" loses its coherence and firmness, many individuals would cease to act as they are expected to do, and the survival of the society in its given form would be endangered. Societies, of course, differ in the rigidity with which they enforce their social character, and the observation of the taboos for protecting this character, but in all societies there are taboos, the violation of which results in ostracism.

Defining consciousness and unconsciousness as we have done, what does it mean if we speak of *making the unconscious conscious, of de-repression*?

What Freud discovered was the fact that we see reality in a distorted way. That we believe to see a person as he is, while actually we see our projection of an image of the person without being aware of it. Freud saw not only the distorting influence of transference, but also the many other distorting influences of repression. Inasmuch as a person is motivated by impulses unknown to him, and in contrast to his conscious thinking (representing the demands of social reality), he may project his own unconscious strivings onto another person, and hence not be aware of them within.

Taking into account what has been said above about the stultifying influence of society, and furthermore considering our wider concept of what constitutes unconsciousness, we arrive at a new concept of unconsciousness — consciousness. We may begin by saying that the average person, while he thinks he is awake, actually is half asleep. By "half asleep" I mean that his contact with reality is a very partial one; most of what he believes to be reality (outside or inside of himself) is a set of fictions which his mind constructs. He is aware of reality only to the degree to which his social functioning makes it necessary. He is aware of his fellowmen inasmuch as he needs to cooperate with them; he is aware of material and social reality inasmuch as he needs to be aware of it in order to manipulate it. *He is aware of reality to the extent to which the goal of survival makes such awareness necessary.* (In contradistinction in the state of sleep the awareness of outer reality is suspended, though easily recovered in case of necessity, and in the case of insanity, full awareness of outer reality is absent and not even recoverable in any kind of emergency.) The average person's consciousness is mainly "false consciousness," consisting of fictions and illusion, while precisely what he is not aware of is reality. We can thus differentiate between what a person *is* conscious of, and what he *becomes* conscious of. He *is* conscious, mostly, of fictions; he can *become* conscious of the realities which lie underneath these fictions.

There is another aspect of unconsciousness which follows from the premises discussed earlier. Inasmuch as consciousness represents only the small sector of socially patterned experience and unconsciousness represents the richness and depth of universal man the state of repressedness results in the fact that I, the accidental, social person, am separated from me the whole human person. I am a stranger to myself, and to the same degree everybody else is a stranger to me. I am cut off from the vast area of experience which is human, and remain a fragment of a man, a cripple who experiences only a small part of what is real in him and what is real in others.

Thus far we have spoken only of the distorting function of repressedness; another aspect remains to be mentioned which does not lead to distortion, but to making an experience unreal by *cerebration*. I refer by this to the fact that I believe I see — but I only *see words*; I believe I feel, but I only *think feelings*. The cerebrating person is the alienated person, the person in the cave who, as in Plato's allegory, sees only shadows and mistakes them for immediate reality.

This process of cerebration is related to the ambiguity of language. As soon as I have expressed something in a word, an alienation takes place, and the full experience has already been substituted for by the word. The full experience actually exists only up to the moment when it is expressed in language. This general process of cerebration is more widespread and intense in modern culture than it probably was at any time before in history. Just because of the increasing emphasis on intellectual knowledge which is a condition for scientific and technical achievements, and in connection with it on literacy and education, words more and more take the place of experience. Yet the person concerned is unaware of this. He thinks he sees something; he thinks he feels something; yet there is no experience except memory and thought. When he thinks *he* grasps reality it is only his brain–self that grasps it, while he, the whole man, his eyes, his hands, his heart, his belly graps nothing — in fact, *he* is not participating in the experience which he believes is *his*.

What happens then in the process in which the unconscious becomes conscious? In answering this question we had better reformulate it. There is no such thing as "the conscious" and no such thing as "the unconscious." There are degrees of consciousness-awareness and unconsciousness-unawareness. Our question then should rather be: what happens when I become aware of what I have not been aware of before? In line with what has been said before, the general answer to this question is that every step in this process is in the direction of understanding the fictitious, unreal character of our "normal" consciousness. To become conscious of what is unconscious and thus to enlarge one's consciousness means to get in touch with reality, and — in this sense — with truth (intellectually and affectively). To enlarge consciousness means to wake up, to lift a veil, to leave the cave, to bring light into the darkness.

Could this be the same experience Zen Buddhists call "enlightenment"?

Notes

1. For a current but quite technical and difficult selection of research based on such a proposition see J. Shotter and K. J. Gergen, *Texts of Identity*, 1989, Sage.
2. For a discussion of the work of Vygotsky in relation to Darwin, and also of Vygotsky's critique of Piaget's explanation of egocentric speech see pp. 92–

98 of C. Sinha, *Language and Representation: A Socio-naturalistic Aproach to Human Development*, 1988, Harvester Wheatsheaf. A 'reader friendly' discussion of the divergence of Vygotsky's and Piaget's appreciation of egocentric speech can be found on pp. 352–356 of A. R. Lindesmith, A. L. Strauss and N. K. Denzin, *Social Psychology*, 1978, Holt, Rinehart and Winston.

3. J. Piaget, *The Language and Thought of the Child*, 1959, Routledge & Kegan Paul.
4. In fact in the original chapter Fromm makes reference to Whorf's work and provides examples. We have omitted these to save space and because to a certain extent they are already stated in the Whorf extract.
5. In a similar vein Aldous Huxley talks of a 'reducing valve' in *The Doors of Perception*. It was his way of explaining his feeling of the release from perceptual constraint brought about by his consumption of mescaline during his researches into the effects of the drug.
6. From a topological point of view we could state that the alimentary canal is 'outside' the organism and hence part of the 'surroundings'. It would then follow that digestion could be considered to be an aspect of this rudimentary information processing.

10

Psychoanalysis and Consciousness

READINGS

Sigmund Freud
The Dissection of the Psychical Personality

Anna Freud
The Ego as the Seat of Observation

Ernest S. Wolf
'Irrationality' in a Psychoanalytic Psychology of the Self

Harry Guntrip
The Turning Point: From Psychobiology to Object-Relations

Ian Craib
The Unconscious and Language

Most people's view of human consciousness includes the idea that there are memories of objects or events of which we are not currently aware but could be if reminded: a latent, potential awareness. Such a view would also be likely to include the belief that there are events and objects of which we are not only currently unaware but which under normal circumstances we never could be: the unconscious. The idea of an unconscious is very much a legacy of the work of Sigmund Freud and the school of psychoanalytic psychology which he founded. Such has been Freud's influence that any view of consciousness without the idea of an unconscious seems incomplete, and it is hard for us now to imagine any other view or indeed how contentious his ideas must first have seemed to philosophers and lay people alike at the end of the last century.

Freud's concept of the unconscious is important because it shows us that there are restraints on our ability to represent objects in conscious experience. We have, so far, considered the role of symbols and objects in human consciousness, and have implied that their role in the *re*presentation of experience as memory is an important aspect of consciousness. Freud would have had this in common with his contemporary, Mead, and the symbolic interactionist and social constructionist approaches to consciousness which were the subject of Chapter Seven. Freud and Mead, writing in the wake of Darwin, also shared a phylogenetic perspective on human psychology, although this was much more exclusive and ultimately shackling for Freud. Of course, these two giants of twentieth-century thought diverged considerably in their principal interests. While Mead was concerned with the role of society in the formation of self-consciousness, Freud's general concern was with the essential conflict between biology (in the form of 'natural instincts') and society (in the form of 'agreements' on moral imperatives) and how the self-conscious subject deals with it.

Freud was led to the idea of the unconscious through his early work on hysteria and his development of dream analysis as a clinical technique. It seemed clear to him that his patients' anxieties and symptoms were related to the experience of childhood traumas of an implicitly sexual nature. His patients could not in the first place recall these traumas and they showed great resistance to accepting this particular version of their past. Freud, however, demonstrated that it could be inferred from what his patients revealed and, more importantly, what they resisted revealing in the course of therapeutic conversation.

Freud's first theoretical writings on the unconscious involved a threefold division of consciousness: the conscious, the pre-conscious and the unconscious. This early view is usually described as the 'topographical' view because it explained psychological functioning in terms of different locations. The three layers of consciousness proposed were distinguishable in terms of their relative depth. The metaphor initially used by Freud to account for his clinical observations in *Studies on Hysteria* and later in *The Interpretation of Dreams* is thus a spatial one. It is as if some unacceptable memories are placed where they cannot reach consciousness. On the storage of memories Freud was in accord with prevailing psychological orthodoxy: the process involved the 'laying down' of a 'memory trace'. His view on recall diverged from this, with some memory traces stored beyond the reach of consciousness but still indirectly effective. It was as if there were an impermeable barrier between this region, the unconscious, and the other two accessible regions, the pre-conscious and the conscious. It should perhaps be remembered that Freud's theorizing and model building were tools to aid his clinical practice rather than contributions to contemporary academic discourse.

The topographical view was later developed into what has become known as the structural view comprising the ego, the super ego and the id. These were to be considered as structures of personality. Their appearance in Freud's writing represented a development rather than a replacement of the topographical view because the structures operated at specific levels of consciousness. In fact not only was their positioning within the three layers of consciousness an essential aspect of their functioning but their functioning helped to explain the conscious and the unconscious. The structural view of personality was developed in *The Ego and the Id*, which was first published in 1923. It was Freud's last major theoretical statement. The ideas contained in *The Ego and the Id* are repeated and developed in Freud's *New Introductory Lectures* which were published in the last month of 1932. The publication date was to have been early 1933 which was to turn out to be an inauspicious year, the one in which Freud's books were first publicly burned in Berlin. The first extract is from No. 31 of the *New Introductory Lectures*.[1] In it Freud outlines first how the ego becomes split against itself, able to represent itself and its actions in the service of the id. The split results in the super-ego or conscience. He then goes on in the second half of the extract to discuss the id, its relation to the ego and the interaction of both in the creation of the unconscious.

The situation in which we find ourselves at the beginning of our enquiry may be expected itself to point the way for us. We wish to make the ego the matter of our inquiry, our very own ego. But is that possible? After all, the ego is in its very essence a subject; how can it be made into an object? Well, there is no doubt that it can be. The ego can take itself as an object, can treat itself like other objects, can observe itself, criticize itself, and do Heaven knows what with itself. In this, one part of the ego is setting itself over against the rest. So the ego can be split; it splits itself during a number of its functions — temporarily at least. Its parts can come together again afterwards. That is not exactly a novelty, though it may perhaps be putting an unusual emphasis on what is generally known. On the other hand, we are familiar with the notion that pathology, by making things larger and coarser, can draw our attention to normal conditions which would otherwise have escaped us. Where it points to a breach or a rent, there may normally be an articulation present. If we throw a crystal to the floor, it breaks; but not into hap-hazard pieces. It comes apart along its lines of cleavage into fragments whose boundaries, though they were invisible, were predetermined by the crystal's structure. Mental patients are split and broken structures of this same kind. Even we cannot withhold from them something of the reverential awe which peoples of the past felt for the insane. They have turned away from external reality, but for that very reason they know more about internal, psychical reality and can reveal a number of things to us that would otherwise be inaccessible to us.

We describe one group of these patients as suffering from delusions of being observed. They complain to us that perpetually, and down to their most intimate actions, they are being molested by the observation of unknown powers — presumably persons — and that in hallucinations they hear these persons reporting the outcome of their observation: 'now he's going to say this, now he's dressing to go out' and so on. Observation of this sort is not yet the same thing as persecution, but it is not far from it; it presupposes that people distrust them, and expect to catch them carrying out forbidden actions for which they would be punished. How would it be if these insane people were right, if in each of us there is present in his ego an agency like this which observes and threatens to punish, and which in them has merely become sharply divided from their ego and mistakenly displaced into external reality?

I cannot tell whether the same thing will happen to you as to me. Ever since, under the powerful impression of this clinical picture, I formed the idea that the separation of the observing agency from the rest of the ego might be a regular feature of the ego's structure, that idea has never left me, and I was driven to investigate the further characteristics and connections of the agency which was thus separated off. The next step is quickly taken. The content of the delusions of being observed already suggests that the observing is only a preparation for judging and punishing, and we accordingly guess that another function of this agency must be what we call our conscience. There is scarcely anything else in us that we so regularly separate from our ego and so easily set over against it as precisely our conscience. I feel an inclination to do something that I think will give me pleasure, but I abandon it on the ground that my conscience does not allow it. Or I have let myself be persuaded by too great an expectation of pleasure into doing something to which the voice of conscience has objected and after the deed my conscience punishes me with distressing reproaches and causes me to feel remorse for the deed. I might simply say that the special agency which I am beginning to distinguish in the ego is conscience. But it is more prudent to keep the agency as something independent and to suppose that conscience is one of its functions and that self-observation, which is an essential preliminary to the judging activity of conscience, is another of them. And since when we recognize that something has a separate existence we give it a name of its own, from this time forward I will describe this agency in the ego as the '*super-ego*'.

Now, however, another problem awaits us — at the opposite end of the ego, as we might put it. It is presented to us by an observation during the work of analysis, an observation which is actually a very old one. As not infrequently happens, it has taken a long time to come to the point of appreciating its importance. The whole theory of psychoanalysis is, as you know, in fact built up on the perception of the resistance offered to us by the patient when we attempt to make his unconscious conscious to him. The objective sign of this resistance is that his associations fail or depart widely from the topic that is being dealt with. He may also recognize the resistance *subjectively* by the fact that he has distressing feelings when he approaches the topic. But this last sign may also be absent. We then say to the patient that we infer from his behaviour that he is now in a state of resistance; and he replies that he knows nothing of that, and is only aware that his associations have become more difficult. It turns out that we were right; but in that case his resistance was unconscious too, just as unconscious at the repressed, at the lifting of which we were working. We should long ago have asked the question: from what part of his mind does an unconscious resistance like this arise? The beginner in psychoanalysis will be ready at once with the answer: it is, of course, the resistance of the unconscious. An ambiguous and unserviceable answer! If it means that the resistance arises from the repressed, we must rejoin: certainly not! We must rather attribute to the repressed a strong upward drive, an impulsion to break through into consciousness. The resistance can only be a manifestation of the ego, which originally put the repression into force and now wishes to maintain it. That, moreover, is the view we always took. Since we have come to assume a special agency in the ego, the super-ego, which represents demands of a restrictive and rejecting character, we may say that repression is the work of this super-ego and that it is carried out either by itself or by the ego in obedience to its orders. If then we are met by the case of the resistance in analysis not being conscious to the patient, this means either that in quite important situations the super-ego and the ego can operate unconsciously, or — and this would be still more important — that portions of both of them, the ego and the super-ego themselves, are unconscious. In both cases we have to reckon with the disagreeable discovery that on the one hand (super-) ego and conscious and on the other hand repressed and unconscious are far from coinciding.

I return now to our topic. In face of the doubt whether the ego and super-ego are themselves unconscious or merely produce unconscious effects, we have, for good reasons, decided in favour of the former possibility. And it is indeed the case that large portions of the ego and super-ego can remain unconscious and are normally unconscious. That is to say, the individual knows nothing of their contents and it requires an expenditure of effort to make them conscious. It is a fact that ego and conscious, repressed and unconscious do not coincide. We feel a need to make a fundamental revision of our attitude to the problem of conscious–unconscious. At first we are inclined greatly to reduce the value of the criterion of being conscious since it has shown itself so untrustworthy. But we should be doing it an injustice. As may be said of our life, it is not worth much, but it is all we have. Without the illumination thrown by the quality of consciousness, we should be lost in the obscurity of depth-psychology; but we must attempt to find our bearings afresh.

There is no need to discuss what is to be called conscious: it is removed from all doubt. The oldest and best meaning of the word 'unconscious' is the descriptive one; we call a psychical process unconscious whose existence we are obliged to assume — for some such reason as that we infer it from its effects —, but of which we know nothing. In that case we have the same relation to it as we have to a psychical process in another person, except that it is in fact one of our own. If we want to be still more correct, we shall modify our assertion by saying that we call a process unconscious if we are obliged to assume that it is being activated *at the moment*, though *at the moment* we know nothing about it. This qualification makes us reflect that the majority of conscious processes are conscious only for a short time; very soon they become *latent*, but can easily become conscious again. We might also say that they had become unconscious, if it were at all certain that in the condition of latency they are still something psychical. So far we should have learnt nothing new; nor should we have acquired the right to introduce the concept of an unconscious into psychology. But then comes the new observation that we were already able to make in parapraxes. In order to explain a slip of the tongue, for instance, we find ourselves obliged to assume that the intention to make a particular remark was present in the subject. We infer it with certainty from the interference with his remark which has occurred; but the intention did not put itself through and was thus unconscious. If, when we subsequently put it before the speaker, he recognizes it as one familiar to him, then it was only temporarily unconscious to him; but if he repudiates it as something foreign to him, then it was permanently unconscious. From this experience we retrospectively obtain the right also to pronounce as something unconscious what had been described as latent. A consideration of these dynamic relations permits us now to distinguish two kinds of unconscious — one which is easily, under frequently occurring circumstances, transformed into something conscious, and another with which this transformation is difficult and takes place only subject to a considerable expenditure of effort or possibly never at all. In order to escape the ambiguity as to whether we mean the one or the other unconscious, whether we are using the word in the desciptive or in the dynamic sense, we make use of a permissible and simple way out. We call the unconscious which is only latent, and thus easily becomes conscious, the 'preconscious' and retain the term 'unconscious' for the other. We now have three terms, 'conscious', 'preconscious' and 'unconscious', with which we can get along in our description of mental phenomena. Once again: the preconscious is also unconscious in the purely descriptive sense, but we do not give it that name, except in talking loosely or when we have to make a defence of the existence in mental life of unconscious processes in general.

We have come to understand the term 'unconscious' in a topographical or systematic sense as well; we have come to speak of a 'system' of the preconscious and a 'system' of the unconscious, of a conflict between the ego and the system *Ucs.*, and have used the word more and more to denote a mental province rather than a quality of what is mental. The discovery, actually an inconvenient one, that portions of the ego and super-ego as well are unconscious in the dynamic sense, operates at this point as a relief — it makes possible the removal of a complication. We perceive that we have no right to name the mental region that is foreign to the ego 'the system *Ucs.*', since the characteristic of being unconscious is not restricted to it. Very well; we will no longer use the term 'unconscious' in the systematic sense and we will give what we have hitherto so described a better name and one no longer open to misunderstanding. Following a verbal usage of Nietzsche's and taking up a suggestion by Georg Groddeck, we will in future call it the 'id'. This impersonal pronoun seems particularly well suited for expressing the main characteristic of this province of the mind — the fact of its being alien to the ego. The super-ego, the ego and the id — these, then, are the three realms, regions, provinces, into which we divide an individual's mental apparatus, and with the mutual relations of which we shall be concerned in what follows.

The id of course knows no judgements of value: no good and evil, no morality. The economic or, if you prefer, the quantitative factor, which is intimately linked to the pleasure principle, dominates all its processes. Instinctual cathexes seeking discharge — that, in our view, is all there is in the id. It even seems that the energy of these instinctual impulses is in a state different from that in the other regions of the mind, far more mobile and capable of discharge; otherwise the displacements and condensations would not occur which are characteristic of the id and which so completely disregard the *quality* of what is cathected — what in the ego we should call an idea. We would give much to understand more about these things! You can see, incidentally, that we are in a position to attribute to the id characteristics other than that of its being unconscious, and you can recognize the possibility of portions of the ego and super-ego being unconscious without possessing the same primitive and irrational characteristics.

We can best arrive at the characteristics of the actual ego, in so far as it can be distinguished from the id and from the super-ego, by examining its relation to the outermost superficial portion of the mental apparatus, which we describe as the system Perceptual-Conscious. This system is turned towards the external world, it is the medium for the perceptions arising thence, and during its functioning the phenomenon of consciousness arises in it. It is the sense-organ of the entire apparatus; moreover it is receptive not only to excitations from outside but also to those arising from the interior of the mind. We need scarcely look for a justification of the view that the ego is that portion of the id which was modified by the proximity and influence of the external world, which is adapted for the reception of stimuli and as a protective shield against stimuli, comparable to the cortical layer by which a small piece of living substance is surrounded. The relation to the external world has become the decisive factor for the ego; it has taken on the task of representing the external world to the id — fortunately for the id, which could not escape destruction if, in its blind efforts for the satisfaction of its instincts, it disregarded that supreme external power. In accomplishing this function, the ego must observe the external world, must lay down an accurate picture of it in the memory-traces of its perceptions, and by its exercise of the function of 'reality-testing' must put aside whatever in this picture of the external world is an addition derived from internal sources of excitation. The ego controls the approaches to motility under the id's orders; but between a need and an action it has interposed a postponement in the form of the activity of thought, during which it makes use of the mnemic residues of experience. In that way it has dethroned the pleasure principle which dominates the course of events in the id without any restriction and has replaced it by the reality principle, which promises more certainty and greater success.

There are, arguably, three main departures from Freud's version of psychoanalysis which have relevance for the study of consciousness. The first of these is 'ego-psychology': the concentration on the ego at the expense of the id. This field of interest was enhanced through the writing of his daughter Anna Freud, especially in her 1936 volume *The Ego and the Mechanisms of Defense*. The concern in psychoanalysis with the ego is important for understanding consciousness because it suggests an ego conceived more and more as synonymous with self. Freud's early conceptualization of the ego was as a development of part of the id. Its purpose was to satisfy the instinctual energies of the id by organizing dealings with the real world; or in Freudian terminology, to operate on the 'reality principle' rather than on the 'pleasure principle', the immediate gratification of instinctive needs which characterizes the id. Initially the ego seems to have been conceived as little more than a metaphor for the perceptual and learning aspects of early development. In the first extract of this chapter Freud refers to the ego as '. . . the sense organ of the entire apparatus . . .'. The term 'self' refers to the totality of the organism with the emphasis on it being something largely socially constructed and understood. We have already discussed the self in Chapter Seven as a social object, the internalization of other people's actions either directly towards us or which otherwise imply the object we are to them. The move to consider the ego as the dominant aspect of personality in the control of behaviour plays down the driving force of sexual and aggressive energies and emphasizes the autonomous, rational, constructive, self-seeking and self-satisfying aspects of life.

In Anna Freud's writing the ego has begun to become something more than the process of observation itself. The ego is no longer the aspect of the id turned towards the immediate outside world. It is a structure through which both the id and superego can be observed. It can become suspicious if the demands of the id become threatening. Its purpose can be to put the instincts of the id out of action by adopting defensive measures such as repression and projection. In this development of psychoanalysis the unconscious, i.e. that which cannot be represented in conscious experience, arises through the activity of the ego in thwarting the natural expression of instincts as behaviour and experience.

There have been periods in the development of psychoanalytic science when the theoretical study of the individual ego was distinctly unpopular. Somehow or other, many analysts had conceived the idea that, in analysis, the value of the scientific and therapeutic work was in direct proportion to the depth of the psychic strata upon which attention was focused. Whenever interest was shifted from the deeper to the more superficial psychic strata — whenever, that is to say, research was deflected from the id to the ego — it was felt that here was a beginning of apostasy from psychoanalysis as a whole. The view held was that the term *psychoanalysis* should be reserved for the new discoveries relating to the unconscious psychic life, i.e., the study of repressed instinctual impulses, affects, and fantasies.

Such a definition of psychoanalysis was not infrequently met with in analytic writings and was perhaps warranted by the current usage, which has always treated psychoanalysis and depth psychology as synonymous terms. Moreover, there was some justification for it in the past, for it may be said that from the earliest years of our science its theory, built up as it was on an empirical basis, was pre-eminently a psychology of the unconscious or, as we should say today, of the id. But the definition immediately loses all claim to accuracy when we apply it to psychoanalytic therapy. From the beginning analysis, as a therapeutic method, was concerned with the ego and its aberrations: the investigation of the id and of its mode of operation was always only a means to an end. And the end was invariably the same: the correction of these abnormalities and the restoration of the ego to its integrity.

When the writings of Freud, beginning with *Group Psychology and the Analysis of the Ego* (1921) and *Beyond the Pleasure Principle* (1920), took a fresh direction, the odium of analytic unorthodoxy no longer attached to the study of the ego and interest was definitely focused on the ego institutions. Since then the term "depth psychology" certainly does not cover the whole field of psychoanalytic research. At the present time we should probably define the task of analysis as follows: to acquire the fullest possible knowledge of all the three institutions of which we believe the psychic personality to be constituted and to learn what are their relations to one another and to the outside world. That is to say: in relation to the ego, to explore its contents, its boundaries, and its functions, and to trace the history of its dependence on the outside world, the id, and the superego; and, in relation to the id, to give an account of the instincts, i.e., of the id contents, and to follow them through the transformations which they undergo.

We all know that the three psychic institutions vary greatly in their accessibility to observation. Our knowledge of the id — which was formerly called the system *Ucs.* — can be acquired only through the derivatives which make their way into the systems *Pcs.* and *Cs.* If within the id a state of calm and satisfaction prevails, so that there is no occasion for any instinctual impulse to invade the ego in search of gratification and there to produce feelings of tension and unpleasure, we can learn nothing of the id contents. It follows, at least theoretically, that the id is not under all conditions open to observation.

The situation is, of course, different in the case of the superego. Its contents are for the most part conscious and so can be directly arrived at by endopsychic perception. Nevertheless, our picture of the superego always tends to become hazy when harmonious relations exist between it and the ego. We then say that the two coincide, i.e., at such moments the superego is not perceptible as a separate institution either to the subject himself or to an outside observer. Its outlines become clear only when it confronts the ego with hostility or at least with criticism. The superego, like the id, becomes perceptible in the state which it produces within the ego: for instance, when its criticism evokes a sense of guilt.

Now this means that the proper field for our observation is always the ego. It is, so to speak, the medium through which we try to get a picture of the other two institutions.

When the relations between the two neighboring powers — ego and id — are peaceful, the former fulfills to admiration its role of observing the latter. Different instinctual impulses are perpetually forcing their way from the id into the ego, where they gain access to the motor apparatus, by means of which they obtain gratification. In favorable cases the ego does not object to the intruder but puts its own energies at the other's disposal and confines itself to perceiving; it notes the onset of the instinctual impulse, the heightening of tension and the feelings of unpleasure by which this is accompanied and, finally, the relief from tension when gratification is experienced. Observation of the whole process gives us a clear and undistorted picture of the instinctual impulse concerned, the quantity of libido with which it is cathected, and the aim which is pursues. The ego, if it assents to the impulse, does not enter into the picture at all.

Unfortunately the passing of instinctual impulses from one institution to the other may be the signal for all manner of conflicts, with the inevitable result that observation of the id is interrrupted. On their way to gratification the id impulses must pass through the territory of the ego and here they are in an alien atmosphere. In the id the so-called "primary process" prevails; there is no synthesis of ideas, affects are liable to displacement, opposites are not mutually exclusive and may even coincide, and condensation occurs as a matter of course. The sovereign principle which governs the psychic processes is that of obtaining pleasure. In the ego, on the contrary, the association of ideas is subject to strict conditions, to which we apply the comprehensive term "secondary process"; further, the instinctual impulses can no longer seek direct gratification — they are required to respect the demands of reality and, more than that, to conform to ethical and moral laws by which the superego seeks to control the behavior of the ego. Hence these impulses run the risk of incurring the displeasure of institutions essentially alien to them. They are exposed to criticism and rejection and have to submit to every kind of modification. Peaceful relations between the neighboring powers are at an end. The instinctual impulses continue to pursue their aims with their own peculiar tenacity and energy, and they make hostile incursions into the ego, in the hope of overthrowing it by a surprise attack. The ego on its side becomes suspicious; it proceeds to counterattack and to invade the territory of the id. Its purpose is to put the instincts permanently out of action by means of appropriate defensive measures, designed to secure its own boundaries.

The picture of these processes transmitted to us by means of the ego's faculty of observation is more confused but at the same time much more valuable. It shows us two psychic institutions in action at one and the same moment. No longer do we see an undistorted id impulse but an id impulse modified by some defensive measure on the part of the ego. The task of the analytic observer is to split up the picture, representing as it

> does a compromise between the separate institutions, into its component parts: the id, the ego, and, it may be, the superego.
>
> In all this we are struck by the fact that the inroads from the one side and from the other are by no means equally valuable from the point of view of observation. All the defensive measures of the ego against the id are carried out silently and invisibly. The most that we can ever do is to reconstruct them in retrospect: we can never really witness them in operation. This statement applies, for instance, to successful repression. The ego knows nothing of it; we are aware of it only subsequently, when it becomes apparent that something is missing. I mean by this that, when we try to form an objective judgment about a particular individual, we realize that certain id impulses are absent which we should expect to make their appearance in the ego in pursuit of gratification. If they never emerge at all, we can only assume that access to the ego is permanently denied to them, i.e., that they have succumbed to repression.

Concern with the organizing features of the ego, conceived as an entity in its own right, has developed a great deal since the publication of Anna Freud's volume in 1936. In the US particularly, there has been a move to integrate the ego of psychoanalysis with ideas and theory concerning the ego in academic psychology.[2] At the same time this has brought about an integration of ideas from what might be called 'self psychology' which includes work inspired by Mead and the symbolic interactionist school of social psychology. These developments may well reflect the day-to-day concerns of psychoanalysts. The people who come to see therapists have become less and less likely to be suffering from hysterical paralysis of arms and more likely to be concerned about the cohesion of the self and its relations with other people. As Logan contends in Chapter Eight, the twentieth century has become the age in which the discovery of the self as an object has become a preoccupation of the age.

The question of the self within the framework of psychoanalysis has been addressed by a number of writers. One of the most important of these has been Heinz Kohut.[3] Freud dealt with the question of the self and its esteem by inventing the term 'narcissism': a state in which the libido was directed towards the self. Kohut notes that he often found himself in the position of contributing to his patients' subjective feeling of well being by confirming and organizing their partial and fragmented self-image. Kohut saw this phenomenon as repeating what the patients' parents would have done by way of confirming their embrionic sense of self. He believed that in the early stages of the infant's development external objects, chiefly in the form of parents, became central to the development of the self. These objects were termed 'self-objects'. The work of Kohut coincides with important themes in the treatment of self in academic developmental psychology. A feature of Kohut's work in any case is that he outlines a developmental schedule to the role of self-objects in the formation of a cohesive and stable sense of self which moves from the first years of life into adolescence. These points are elaborated and discussed in the extract by Ernest Wolf.

The vicissitudes of the drives and their impact on the ego with its defenses have been well worked out by psychoanalysts, beginning with Freud. What is becoming clearer, with increasing knowledge of the psychology of the self, is that it is not drives *per se* that force pathological adaptation upon the ego; it is only when they are wrenched out of their normal matrix within a cohesive self, by the latter's disintegration, that these drives assume pathological intensity which forces the ego into the distortions that we call neurotic symptom formation.

Freud, in his clinical observations, was most impressed by the quality of 'driveness' of his patients.

Thus the core of psychoanalytic theory, as developed by Freud, is a theory of libidinal drives; that theory served well to conceptualize many of the phenomena associated with the classical psychoneuroses. These syndromes could be adequately understood and successfully treated when conceptualized as the consequences of the repression of infantile libidinal drives, particularly those that are involved in the Oedipus complex. The theory assumes these infantile sexual drives, their further development, and the involvements of the Oedipus complex to be universal, and assigns to inborn, quantitative factors, the decisive role in whether the resulting conflict will turn out to be pathological or within normal limits.

But the psychoneuroses represent only a small fraction of the psychological disorders of man. Extending libidinal theory to encompass the major psychoses and depressions resulted in somewhat awkward theoretical constructs and required the postulation of 'narcissism', defined as the libidinal drive invested in the self as an object. From the point of view of psychological development, narcissism was seen to be a phase which had to be traversed on the way from autoerotism to object love. Theory thus assigned to self-love and its derivatives, such as self-expression and creativity, a quasi-pathological status. Of course one can discern the influence of our Judeo-Christian *Weltanschauung*, with its strong emphasis on altruism and its equally strong condemnation of egotism, in this conceptualization. While a number of analysts made important clinical observations and attempted limited revisions of classic psychoanalytic theory Kohut was the first to systematically study the phenomena of narcissism, and to create a new conceptual framework which allows us to encompass drive-phenomena within the new psychology of the self. But Kohut's new insights and conceptualizations are not derived *ab initio* from theoretical considerations; they stem from the discovery within the psychoanalytic clinical situation of two types of transferences that had not been described before.

Kohut noticed that certain analysands experienced a state of well-being and effective functioning when in the presence of the benignly understanding analyst. In the absence of the analyst, or even in his presence if his understanding was defective, the analysand would experience various degrees of discomfort and tension which might lead to varieties of symptomatic behavior. Kohut recognized that these phenomena were the transference repetitions of an early developmental phase, when the child's integrative capacity is as yet insufficient to organize itself into a coherent system unless it is assisted by the various functions performed by the parent, usually the mother. If one conceptualizes the subjective experience of possessing a coherent configuration of smoothly effective psychological functions as a 'cohesive self', then the child's subjective experience of his 'self' includes those aspects of his parents which are necessary for the coherence of the configuration. Similarly the analysand who comes into an analysis with an insufficiently coherent self may begin to utilize aspects of the analyst to achieve a coherent integration of his self. Looked at in a social framework, one can say that aspects of an object are needed by the subject for his efficient functioning; when experienced by the subject as part of his self they are 'self-objects'.

Kohut distinguishes two types of narcissistic transferences which can be schematically outlined as follows. In one type, the analyst lends cohesion to the analysand's self organization by functioning as a confirming psychological mirror ('mirror transference'). In the other narcissistic transference cohesion is brought to the analysand's self by allowing a psychological merger with the analyst's strength, values and ideals ('idealizing transference'). These transferences correspond to early phases in infant development, when a parent's empathic understanding of the child either lends needed support to its sense of self, including its phase appropriate sense of its own grandeur, or, when the parents' apparent greatness infuses the child's feeble sense of self with power, value and cohesion.

Much of the finer details of the psychological development from infancy into mature adulthood remains a matter for future empirical research, but the major trends can be outlined. In the early months various psychological functions of the infant, whether inborn or acquired, are not yet organized into a cohesive configuration even by the mother's mediating ministrations. Therefore there can be as yet neither an experience of self or of non-self. However, by thinking and acting towards the infant from the very beginning as if he were already a self-conscious person, the mother subtly encourages certain functions while discouraging other functions and she facilitates the emergence of an encompassing overall organization that is experienced as a sense of self, roughly by the age of eighteen months. The self that emerges around this time, however, still includes the mother's functions as a necessary part of its self experience (i.e., the mother is still needed as an organizing participant to achieve an encompassing configuration of otherwise disparate psychological functions). At the same time, but in a different context, the mother can also be perceived as an object, e.g. visually; she is both self and object, that is a self-object. The initially fragile self becomes sturdier with further development, while at the same time the functions of the self-object are increasingly internalized with growing freedom from dependence on the caretaking parent. This process is mainly completed around age eight or nine, so that one then can usually speak of a cohesive self, relatively autonomous and resistant to irreversible regressions.

The smooth functioning of a cohesive organizing configuration at the center of one's personality is experienced as a sense of self, with a feeling of wholeness and well-being. A fragmentation of the self, that is loss of its cohesive integration, is experienced with extreme discomfort, such as feelings of depression or even deadness. Impending fragmentation causes great anxiety, even panic, and total fragmentation is equivalent to psychosis. Even transient and relatively minor losses of cohesion are manifest by such symptoms as hypochondriasis and disturbances of self-esteem. These painful subjective states act as powerful motivators for remedial action, and these actions may have an addiction-like intensity which overrides control by the judgmental faculties of the healthy part of the personality. The developmental trauma is usually unconscious and hardly ever understood by the subject, so that its link to the current precipitating circumstance is obscure. Separation from a needed self-object, for example the transient absence of a spouse, is thus usually thought of as trivial and its psychological meaning and implications are not recognized by the narcissistically vulnerable person. The sudden emergence of, let us say, depression and an associated irresistible urge to act out a sexual perversion may strike both the afflicted person and his friends as irrational. Irrationality is here compounded of narcissistic vulnerability and the imperative nature of the urge to escape the pain that this vulnerability may entail. Achieving an irreversible cohesive self is concomitant with achieving freedom from the imperative instrusions of narcissistic tensions into the functioning of 'rational' judgment. (It will be recalled that I am limiting the discussion to one type of irrationality, associated with disorders of self-cohesion, and am not concerned with other conditions of gross irrationality, e.g., the psychoses.)

We can summarize as follows. An infant born into the kind of average expectable environment for which he was genetically pre-adapted will have an average expectable psychological development and manifest an average expectable behavior for that environment. This will be experienced by him, as well as observed by others, as 'rational'. Failure of the environment to be expectably average will likely result in failure of an average expectable psychological development, and be associated with mental states and behaviors which will strike most observers as more or less 'irrational'. The irrationality, however, will begin to make sense with increased understanding based on empathic reconstruction of the pathogenically experienced psychological history.

Since such an account of 'irrationality' is tied to a developmental story, I must give a brief systematic outline of the developmental sequences that lead to the establishment of a cohesive self. Psychology, of course, has little to say about the earliest beginnings of mental functioning as a superordinate organization emerges out of and subsumes the basic neurobiological organizations of the nervous system. Kohut has presented evidence that indicates a facilitating influence of the caretaking person on the contents of the emerging self system. At any rate, a caretaking person who is empathically in tune with the infant responds to its shifting needs in a somewhat selective manner, encouraging the inclusion of certain contents in the self while causing other contents to be discarded. Little is known about the processes involved, but somehow around the age of 18 months, perhaps, the various disparate nuclei of what later becomes the self seem to come together into an integrated configuration which, however, for a long time still lacks stability. Particularly, tension regulation within this self system is deficient and requires the soothing participation of the caretaking self-object. The self-object's functions, though myriad and changing in step with the age-appropriate needs of the child, can usefully be grouped under two headings: availability for mirroring responses and availability for idealization. A properly empathic self-object will be sufficiently in tune to further the relatively undisturbed development of a cohesive self. This self (cohesive by virtue of the self-object's appropriate availability and responsiveness) consists from the beginning of a nuclear grandiose self and a nuclear idealized self-object. The grandiose part contains the self's blissful experience of itself in a state of unlimited power and beauty, a configuration akin to an exhibitionistic drive demanding recognition and confirmation. Everyone is familiar with a mother's extravagant conviction and praise for her baby's unsurpassed grandeur, now and in the future. It is these confirming responses that strengthen the child's nuclear grandiose self in the face of an otherwise unpleasantly depreciating reality, which gradually forces the abandonment, or at least the modification, of the associated grandiose fantasies of the child. *Pari passu* some of the nuclear grandiose self's diminishing grandeur is replaced by increasing idealization of the admired idealized parent whose strength, wisdom and beauty can be experienced as part of the self, i.e. the idealized self-object. Again, a youngster's need to be proud of his family and his feeling dejected, if not shattered, by a parent's traumatically sudden and severe failure, is familiar enough to everyone. Optimal frustration and gradual disappointments facilitate growth and internalization; maximum frustration and severely traumatic disappointments cause the repression of grandiose fantasies in unaltered archaic form and leave the adult with a still vulnerable self which remains dependent on self-objects.

In the course of normal development, grandiose fantasy and exhibition-

> ism are gradually modified and internalized into healthy ambition and pleasure in one's own functioning. The idealized self-object is gradually de-idealized and its values and ideals become part of the self's value system, i.e. the ego ideal. This process seems relatively complete and stable around early latency, that is by about 8 years of age. However, adolescence seems to provide another opening for re-adjustment, when the whole self temporarily undergoes a period of rapid changes and fluid boundaries. The adolescent transformations of the self are as much influenced by the onset of puberty as they are set in motion by the age-appropriate massive disillusionment with the whole adult world, especially the parents. Idealization of peers or of youthful idols temporarily serves to regulate inner tensions, until the consolidation of newly internalized values and goals, (i.e. sexual, social and vocational identities) result in a stable ego-ideal governing a newly restored cohesive self. Transformations of the self probably continue throughout life, especially in conjunction with the major turning points such as marriage, parenthood, middle age crisis and old age. The details are yet to be studied.

The second departure from orthodox Freudian theory which has relevance for consciousness is due to the 'object relations' school of psychoanalysis which can be exemplified in the work of Melanie Klein. Its relevance to consciousness takes us back to the importance of objects in consciousness. We have examined the role of social interaction and language in the recognition and representation of objects in conscious experience. The object relations school of psychoanalysis also focuses on the emergence of objects in the infant's early experience. However, in Kleinian theory neither other people nor even the environment play a part in the first stages of the process of object relations. As the extract by Harry Guntrip explains, the infant has to admit the outside world in order to find somewhere to project its terror in the face of its own death instinct. The instincts are projected onto external objects which are then readmitted or 'introjected' within the child. This startling theory proposes a '. . . wholly internal origin of the active psychic life of the baby'. One less radical implication of object relations theory is that the ego is seen as object seeking rather than pleasure seeking. This means that rather than seeking to dissipate the sexually based energy which Freud referred to as 'libido' or the destructive energy of his later writing, the ego's chief concern is to relate to the objects in its immediate environment.

> I regard Melanie Klein's work as the decisive breakthrough in the development of psychodynamic object-relational thinking.

Freud's structural theory was based on the concepts of the control of instincts (the id) by the ego, under pressure from the external environment which led to the growth of the superego. Melanie Klein's structural theory developed in an entirely different way, eventuating in the concept of an internal psychic world of ego-object relationships.

Klein regarded an infant as an arena for an internal struggle between what at first were conceived of as the life and death instincts, sex and aggression, from the very start, quite apart from environmental influences. This ruthless inner drama then becomes projected onto the outer world, as the infant's brain and sensory organs develop the capacity to discern external objects. This means that the infant is never able to experience real objects in any truly objective way, and the way he does experience them depends more on his own innate make-up than on their real attitude and behavior to him. Basically, what he sees in his environment, is what he reads into it, mainly from his own internal terror of his own threatening death instinct. Segal tells us that "the death instinct is projected into the breast." This is then reintrojected, so that his experience of the outer world simply serves to magnify his impressions and double his anxieties on account of the internal dangers arising out of his permanently split nature. When Melanie Klein finally added an innate biologically determined constitutional envy to the infant's handicaps for any approach to reasonable and friendly objectivity in personal relationships, she seems to have left the environment with no real role to play at all.

If the environment plays such a minor and secondary role, it is little more than a mirror to reflect back to the baby its already existing internal conflicts. It would seem then that such a theory could have little to contribute to object-relational thinking. There could be, one would think, no genuine object-relationships, when the objects-world seems to be of so little primary and intrinsic value. Freud himself did not discount the environment in that way.

Melanie Klein took over Freud's biological mysticism of Eros and Thanatos, and saw human life as an intense hidden dramatic tragedy, a psychodynamic and fearful struggle between the forces of love and death inherent in the baby's constitutional make-up. Quite clearly, in Klein's estimation, the death instinct overshadows the love or life instinct, and is the true and ultimate source of persecutory and all other forms of anxiety.

This fundamental and innate conflict becomes observable, she held, in the infant's fantasy life as soon as it is developed enough to achieve clear expression, and we must remember that in clinical work with very small children, she found this internal fantasy world already well developed in children of between two and three years of age. This is not a matter of theory, but of verifiable, and now already verified, clinical fact, and it must begin to develop much earlier to be so complex by the fourth year of life. It is, moreover, an internal world in which the child is living in fantasied and highly emotion-laden relationships with a great variety of good and bad objects that turn out ultimately to be mental images of parts or aspects of parents. At the most primitive level they are part-objects, breast or penis images, and later on they develop into whole-objects that are in a variety of ways good or bad in the infant's experience. *Life now is viewed, in this internal world of fantasy and feeling, as a matter of ego-object relationships.* This may seem surprising in view of the fact that the Kleinian metapsychology only allows a secondary role to the external world. The infant can never experience the outer world directly, but only through the medium of the projection of its own innate death instinct, and its fear of and struggle against it. These internal bad objects first come into being as an introjection of the projected version of the infant's own innate badness and destructiveness, and they have now become worked up in its experience into parent images. Thus the external object world is forced on us again by the highly personal and psychodynamic nature of the infant's internal fantasy world. The fact is that, whatsoever the tortuous theoretical means, in Melanie Klein we find the term "ego" correlated not now so much with the term "id" as in Hartmann and Freud, but more and more with the term "object."

Klein's use of the term "id" appears to endorse Freud's instinct theory, but Freud's instincts do relate directly to external objects. Hanna Segal states, "Instincts are by definition object-seeking," which had already been explicitly stated in those words by Fairbairn (in order, however, to stress that their aim was not pleasure, but the object that gives pleasure). But in Kleinian metapsychology, instincts are lost in the dim primitive mists of the mystic forces of Eros and Thanatos warring inside the infant, irrespective of what goes on outside. They have, in fact, by making use of the outer world, now become transmuted into internal objects. Kleinian instincts are primitive forces locked in combat inside the infant's nature. The child's first love-object is its own primitive ego, in primary narcissism. Naturally, we have to remember that at birth there is no ego in a conscious sense, but there is a psychic self with ego-potential, out of which the sense of self-hood can gradually grow. For Klein, its entire psychic life is essentially bound up with itself, and out of this internal life consisting essentially of a hostile tension between two contradictory forces, a pattern world is created into which the child's experience of the external world is fitted. What seems to be by far the most important element in this solipsistic theory is that the child's first anxiety concerns its first hate-object. This is its own death instinct, which aims to bring about the organism's return to the inorganic state. The child could have no reason for projecting its love or life instinct, if such a phenomenon is conceivable. But if it is conceivable, it would most certainly have good reason to "project its death instinct," which threatens it with psychic destruction. It is only at this point that the Kleinian scheme finds it necessary to have an external environment into which this dangerous internal component can be extruded by the defensive illusion of projection. And now, the die is cast, the existence of external objects has been admitted and proven to be indispensable. They are indispensable because the infant is supposed to need them to project its death instinct into them, beginning with the mother's breast. But they are also inescapable, for they now constitute a real external threat that the infant has no real means of dealing with physically. It can only try to deal with it inside its own mental life again. The bad breast, now seen as containing a frightful destructive force, is introjected, and this death instinct now turns up inside no longer as an instinct but as an object, literally so perceived and fantasied. Because of her conception of the wholly internal origin of the active psychic life of the baby, Melanie Klein has to use external objects, and external object-relations, as a means of giving concrete expression to these theoretical primary forces and their hypothesized internal relations. What emerges as of first importance in all this is not the more than dubious metapsychology of this biological mysticism but the way in which Klein brings to the front the highly important defensive procedures of projection and introjection that are certainly clinically verifiable facts; and then, of even greater importance, the fact that she has now interpreted the essence of the psychic life of the incipient person in fully ego-object relational terms. It is true that external objects are, apparently, valued not as objects in themselves but as receptacles for projection. However, the result comes to much the same thing in the end, namely the development of an inner world of fantasy that is actually object-relational, and is a counterpart of the ego's relations with the world of real objects that form its physical environment, centered in the mother. This is the real core of Melanie Klein's work. By a very devious and quite unnecessary theoretical route, based on hypotheses that hardly any other analysts but Kleinians accept, she arrived at the fundamental truth that human nature is object-relational in its very essence, at its innermost heart. This goes beyond all biophysiological theories and is pure psychodynamics. Her much greater stress on projection and introjection in therapeutic analysis is a statement of the interaction of the two worlds, internal and external, in which all human beings live, so that finally the external world wins back the reality and importance that was denied it at the start.

Whereas Freud's theory was basically physiological and biological, I do not think that Klein's theory is in any genuine sense biological at all; it is philosophical, and more like a revealed religious belief than a scientific theory in its basic assumptions. Everything in life for Klein is dominated and overshadowed by the mighty and mysterious forces of life and death, creation and destruction, locked in perpetual struggle in the depths of our unconscious psychic experience, and constituting our very nature as persons. Of the two, it is the death instinct that steals the limelight all the time in Kleinian metapsychology.

What she really did was to display the internal psychic life of small children not as a seething cauldron of instincts or id-drives but as a highly personal inner world of ego-object relationships, finding expression in the child's fantasy-life in ways that were *felt* even before thay could be *pictured* or *thought*. These could come to conscious expression in play and dreams, and be disguised in symptoms and in disturbed behavior-relations to real people in everyday living. The study of the person-ego in object relations comes to be the real heart of Melanie Klein's work, however much it may be disguised by theories, many of which I for one find it quite impossible to accept.

The third major departure from strictly Freudian principles of psychodynamics is due to the work of the French psychoanalyst Jaques Lacan.[4] His contribution to the topic of consciousness is a more sophisticated and explicit model of how past experience affects us but cannot be represented in consciousness, or to put it in language less psychoanalytic, cannot be known by the self. Lacan was critical of ego psychology as perpetuating a myth of the 'true self' which can be reached through analysis, and critical of the emphasis on adapting the ego to the demands placed upon it by society. Freud noted, and the ego psychologists developed, the idea that the ego was split into ego and super-ego and therefore continually in conflict. For Lacanians the more important split is between the natural and hence 'true self' (Freud's unconscious) and the ego which develops to make its demands presentable to others.

Not for Lacan, then, the famous exhortation 'where id was so shall ego be'. For the Lacanian version of psychoanalytic psychology there is no knowing of oneself without entering into the tyranny of language, which is seen as a system which orders experience. Language is thus given the prime role in the formation of the unconscious. It orders how one object of experience can lead to another and affects the natural passage from instinctive impulse to expression. Lacan's return to the unconscious with a linguistic interpretation proposes not just that the unconscious is structured in the same way as language but that it *is* structured by language.

This description of Lacan's work obviously provides a link with Chapter Nine on language. Perhaps more surprisingly, it relates as well to the section which follows on complementary approaches to consciousness.

If British psychoanalysis is object-relations theory, and American psychoanalysis has been dominated by ego-psychology, then modern French psychoanalysis has been dominated by one figure: Jacques Lacan.

By way of an introduction, I shall pick out some central ideas from structural linguistics since Lacan models his conception of the psyche on language. Psychoanalysis has long been known as 'the talking cure', and while this manages to be true and not true at the same time, the nature of language is something which faces every practising analyst in both theory and practice.

Structural linguistics originated in the work of Ferdinand de Saussure; there are plenty of good accounts available and here I am only going to discuss a few ideas central to Lacan's work.

Saussure distinguished between speech and language. Speech is what we say, and everything we say is unique in some way. It cannot, therefore, be the object of scientific knowledge, which deals with things that are common to different situations. The language is an underlying structure, consisting of 'signs' and rules which govern the combination of sounds. By using this battery of signs and rules, we can produce our statements. There are several implications to this: the first is that time or history is a surface phenomenon; the underlying structure of a language does change, but very slowly. More rapid changes in speech can be accounted for by the rules that exist, unchanging, on the underlying level. The second is a change in how we think of the speaker. Traditional rationalist conceptions view the person, the ego or the self as a 'subject', an originator of action. *I* decide what to say and when to say it. From Saussure we gain a conception of the 'subject' as something or somebody who only exists in the framework of the language, whose utterances and thoughts are the product of the language to which they are 'subjected'. It was for a while fashionable to talk of people 'being spoken', rather than of people speaking. In psychoanalysis, the unconscious can be seen as the underlying structure, of which our conscious thoughts and actions are the product. Rather like the language, the unconscious for Freud is timeless, and it is not governed by logic.

In describing the language, the concept of a sign is the constant and most important theme in Saussure's work. A sign is seen as a combination of a 'signifier' and a 'signified'. The former, the signifier, is the material element, if we are speaking it is the sound (not the meaning) or if we are writing it is the marks on the paper; the signified is the idea to which the sounds become attached. A language is a structure of signs and rules that governs their combination with each other. Note that the signified is the idea and not the thing that the sign refers to. In common sense we often tend to think of words as somehow inextricably attached to things: we cannot think of a table without seeing the object in front of us. With some thought, however, it becomes clear that, at least on those occasions when we do not have a table in front of us, the word carries the idea of the table rather than a thing in itself. We cannot eat our lunch off the concept of a table. There is no *necessary* connection between a word and a thing: we could call tables tulips and tulips tables and nothing would change in the world; gardens and living rooms remain the same. Another way of saying this is that signs are 'conventional'. A matter of agreement. We agree that on the road a red traffic light means 'stop' and green 'go': it could as well be the other way round, and the same is true of linguistic signs. The meaning of each sign is further defined by its relation to other signs: red, amber and green form a system in which each colour takes it meaning from the others. In another system, say on a snooker table, the colour red has another meaning. Lacan's way of making this point is with a drawing of two identical doors, but one has a man drawn on it, the other a woman. The difference of a few lines in the drawing gives a different meaning to each door.

Lacan emphasises the signifier — it is change in the signifier that produces a change in the signified. The end-result is a conception of the world as governed or determined by language, by signifiers. Lacan, in common with many literary critics influenced by structuralism and post-structuralism, emphasises the importance of two familiar figures of speech: metonymy and metaphor. Metonymy involves using a part to represent a whole: we might say a factory has 500 hands when we mean 500 workers. Metaphor is describing one thing in terms of another, for example, to describe life as a journey. I am not clear about this interpretation of Saussure, but I find the idea of metaphor most useful: we do not have access to something we call reality, only to the sign, or rather, for Lacan, the signifiers: language is a metaphor for reality. We live in a world of metaphors, constantly being referred from one to the other.

One essential part of Freud's work is his analysis of dreams — 'the royal road to the unconscious'. I am introducing it here because the linguistic ideas I have been discussing can act as a metaphor for the workings of the psyche and this is best seen in Freud's analysis of dreams. For Lacan, the unconscious is governed by metaphor and metonymy.

In sleep the protective defences we erect in our waking lives are relaxed and unconscious thoughts come closer to the surface. Some defences remain, and therefore dreams are not the direct expression of unconscious ideas; rather, they are disguised in a process that Freud refers to as dream-work. There are a number of aspects to dream-work. The first is the process of symbolisation: we do not dream ideas, we dream pictures and sometimes words. An idea has to be symbolised, represented somehow. Freud gives numerous examples in *The Interpretation of Dreams*.

The two central aspects of dream-work Freud called 'condensation' and 'displacement', directly comparable to metonymy and metaphor. Condensation is the process of combining several, perhaps many, unconscious ideas into one dream symbol.

Displacement involves moving away from a symbol that is too close to the original idea to one that is safely removed from it.

Displacement is the dream equivalent to metaphor, condensation to metonymy.

Finally, there is the process that Freud called 'secondary revision'. This is a sort of logical gloss we give to our dream images to form them into something like a story, rather than an irrational combination of symbols. All of these processes serve to hide the original unconscious idea. In waking life, the rational is usually dominant, but the psychoanalyst listens to the patient's talk as if it were metaphor or metonymy.

For Lacan it is not that the unconscious works *like* language. It depends upon language and is actually constituted by language, by the child's entry into language. Before language there is no unconscious; hence psychoanalysis is the talking cure. Language literally creates our world for us. Whereas, perhaps, traditional idealism can be typified as maintaining that the world is a product of thought, for Lacan and post-structuralism generally the world and thought are a product of language. This is where the conventional nature of signs, and the lack of any necessary connection between signs and things is relevant: the things in our world are what they are because that is what we decide to call them.

The child's entry into language is, then, for Lacan, the point at which the unconscious appears and the child becomes part of the human world proper. This involves what Freud would call the renunciation of instincts, Lacan the replacement of desire by 'demand', in which the word replaces the desired thing in the infant's psyche, becoming a substitute for a reality that can never be reached. He makes much of Freud's report of a game played by his nephew making objects appear and disappear and at the same time producing the German sounds *fort* and *da* — there it goes and here it is. I imagine that most people who have had regular contact with a young child will have played or witnessed such a game. Freud analysed it on a number of levels. Only one is important for Lacan: it is an instinctual renunciation, where the child gives up its desire for the constant presence of the object (mother) and replaces it with a word that s/he can keep all the time.

The entry into language carries with it a division, both within the child and between the child and the outside world and other people. Here the radical difference between Lacan and ego-psychology appears. The split within the child is between what for Freud is the ego and the id, for Lacan the part of ourselves that we experience as the speaking subject whom we present for others, and the true self, which remains unconscious. From Lacan's point of view, ego-psychology supports and strengthens a fundamental alienation from self, and works solely with the self-for-others; the true purpose of psychoanalysis is to work with the true self. Lacan is notoriously difficult to understand and he does not help matters by referring to this true self as the 'Other', always with a capital O.

There is a further split within the child in which what conventional psychoanalysis calls the 'ideal ego' develops. What Lacan offers is a sort of gloss on the idea, which at one level I find a useful analogy to capture a process, and on the other, a very inadequate summary of a complex and difficult process. The analogy involves the idea of the child looking into a mirror, and Lacan identifies what he calls a 'mirror-phase'. This actually precedes language acquisition, beginning at six to eight months, and is the time when a child learns to recognise its reflection and the reality, and the difference between his or her own image and that of other people. The important thing about it is that as the child recognises the image, s/he gains for the first time a sense of the totality of its body, and as it watches the reflection sees itself apparently in control of itself. Previously, it has experienced itself as a host of conflicting desires and unconnected or uncontrollable parts not properly distinguished from the world. Lacan puns on the French 'omelette' or 'hommelette', the little man who is also a scrambled bundle of desires.

This recognition begins a fundamental alientation, the phantasy attempt to identify with the image and be perfectly together and in control of oneself. Lacan talks of three realms or areas of existence. There is the 'Real' to which, in the last analysis, we do not have access as we never perceive it except through language; there is the 'Imaginary', which has to do with our phantasised identification with the perfect mirror-image, the ideal ego; and the 'Symbolic', the realm of maturity, which we enter into an appropriate language, surrendering the impossible desires that take us into the imaginary. To explain this surrender, Lacan relies, as did Freud, on the oedipal conflict, but he gives it a particular linguistic twist. The father for Lacan is a symbolic figure, the representative of social rules; Lacan talks not about the real father but about 'The name of the father' and there is another French pun here between name, *nomme* and no, *non*. It is the father who as symbolic figure intervenes between the child and its mother. The child is forced to replace his desire for the mother by a symbol. For Lacan, even more than for Freud, the little boy's history is the essential model.

Notes

1. Although the lectures are called *New* Introductory Lectures on Psychoanalysis, they add to rather than replace the original *Introductory Lectures on Psychoanalysis* given between 1915 and 1917. This is indicated by the fact that Freud started the numbering of the new lectures at 29 to run consecutively with the 28 of the *Introductory Lectures*. Furthermore, compared with the earlier *Introductory Lectures*, the *New Introductory Lectures* are quite technically demanding in places. Freud was at his best when adopting the rhetoric of the lecture to give an exposition of new and demanding ideas and this may have attracted him to the lecture format even though they were not to be delivered in public. In fact, as Strachey (1973) points out it was also Freud's idea to write and publish the lectures to aid the ailing finances of the psychoanalytic publishing business so there were at least two reasons for their appearance as lectures. Whatever the reason for their appearance, lecture 31 of the *New Introductory Lectures* gives a clear statement of the structural aspects of personality and the importance of consciousness and the unconscious,

ideas that first emerged ten years earlier in *The Ego and the Id* and in *The Unconscious* which was published in 1915.

2. See G. W. Allport, 'The Ego in Contemporary Psychology', *The Psychological Review*, 50: 451–78.
3. Kohut's work is presented in H. Kohut, *The Analysis of the Self*, 1971, International University Press. An excellent introduction can be found in M. Patton *et al*, 'Kohut's Psychology of the Self' in *J Counseling Psychology*, 1982, 268–282.
4. A selection of Lacan's writing can be found in J. Lacan, *Écrits*, 1977, Tavistock Publications.

11

Complementary Approaches to Consciousness

READINGS

John Hurrell Crook
The Quest for Meaning: Models of Mind and Ego-transcendence

Eugene Taylor
Asian Interpretations: Transcending the Stream of Consciousness

Claudio Naranjo
The Domain of Meditation

Alan Watts
Empty and Marvellous *and* Sitting Quietly, Doing Nothing

The following chapter might conceivably have also been called 'alternative' approaches to consciousness. We emphatically choose 'complementary' because it suggests that the approaches described here can be considered alongside views already outlined. It is our hope and intention that *each* chapter of this book offers an approach to consciousness which complements all of the others. We started by saying that consciousness was going to be a difficult thing about which to talk directly and scientifically. Our approach therefore has been indirect, offering a number of perspectives each inadequate on its own but combining to produce a clearer view of the familiar but elusive phenomenon.

It would be fair to say that the extracts in all the preceding chapters (except perhaps one or two in Chapter Two on 'Consciousness and the Physical World'[1]) can be subsumed under the same general 'view of the world', what Berger and Luckmann[2] refer to as a 'symbolic universe'. If the extracts in this chapter share one characteristic it is that they come from philosophical traditions which do not recognize the same dualistic distinctions between the organism and the environment, the self and the social context, the mind and the body, and the thinker and the thought which are usually implicit in western philosophy.

One might enquire as to what is wrong with this dualism. It has been rather useful in objectifying the world for rational examination. It has been pointed out that the technology of east and west ran much in parallel until the time of Descartes in the first half of the seventeenth century. At this point in western history the view began to emerge, as Logan in Chapter Seven states, that truth was '. . . a knowable single system of natural laws and material things . . .'. Associated with this particular view of the world was the belief that human experience was not part of it. We could think about it and represent it *because* we were outside it. Some of the extracts in Chapter Two propose that our attempts to measure the world affect what we find and that we are therefore very much part of the process of knowing. The personal cost of this dualism is the feeling of separateness which people may feel both from their physical and social surroundings and from the personal worlds of others. Furthermore, science has not been so successful in understanding the 'internal' world we inhabit and our control stops short of self-control.

A look at the titles of the extracts in this last chapter gives the impression that we are presenting matters bordering on the religious rather than scientific. Indeed the first extract, on Buddhism, actually begins with a discussion of religion in the context of history and cultural evolution. Religion and science are both quests for meaning, to use the phrase employed by Crook in the first reading. Religion in a broad sense both indicates and addresses questions which science is unable directly to answer. Since these questions are centrally concerned with consciousness, selfhood and identity, approaches from cultures which do not make the same distinctions between religion and science as we in the West can add a great deal to our understanding of consciousness. They can also add to our understanding of the nature of scientific knowledge itself.

The first extract in this chapter is taken from John Crook's *The Evolution of Human Consciousness*. The book deals with both the biological and cultural evolution of consciousness. The extract begins with his statement that 'death-awareness' is an inevitable consequence of self-awareness. Religion started as the 'quest for meaning' with respect to the end of personal life and to diminish the dread which accompanies its anticipation. The beginnings of religion are identified in chants and incantations wherein the feeling of power over the unknown is achieved by naming it. At this stage there would be no moral or strictly 'religious' aspects to the ritual chants since there was no sense of autonomy and self-control to require a code of approved conduct. The chants, however, might be thought of as the earliest forms of meditative practice where a self-directed change in awareness was produced.

Crook notes the decline of religion in the West and suggests it is due to Christianity's insistence, from the time of the Reformation, on an historical link between a single god and the mortal individual. The demise of religion came about as the growth in philosophy and science tended to weaken that link. Buddhism, it would appear, offers a way of confronting questions of meaning and existence without compromising our contemporary sophisticated scientific view of the universe.

Although Crook discusses Buddhism as a religion, he also points out that it has no theistic pretensions and provides both a subtle understanding of human consciousness and practical advice on how to achieve a more valid understanding of it. It offers a practical methodology, which Crook calls 'subjective empiricism', quite different from the introspectionism of early psychologists such as Wundt and Titchener.

The third part of the extract outlines the model of mind contained within the *Abidhamma*, the great canon written in the Pali language purporting to contain the verbatim teachings of the Buddha — the person who, 2,500 years ago, achieved great insight into the nature of self-consciousness.

An inevitable consequence of the evolutionary development of the self-process and the ability to analyse personal events within the continuum of time was the growing awareness of death, of the fact that life itself had an end. Death-awareness and self-awareness are essentially two sides of the same coin, for the discovery that the body dies implies the ending also of personal experience. The problem that has teased man has centred around the question whether this is really so. The dread of death is the dread of the termination of personal existence. It is clearly anchored in the attachment to the construct of self as an identity. Ever wriggling on this hook, mankind throughout history has sought to make away with the pain of this realization, to rationalize an explanation for death, and to argue for some kind of continuity for the living experiencing principle. Those whose ideological stance on this matter brings comfort and relief are likely to achieve eminence in a fearing society and, when their beliefs are associated with selfless ethics in personal life, such persons receive charisma. Such of course are the great priests. Both the doctrines of life after death and the simpler ideas of reincarnation bring some hope of a personal resurrection or renewed round of existence. The release from grief and the revival of hope in face of despair are both achieved by such comforting thoughts.

In his book *The social reality of religion* Berger (1969) pointed out that the social world arises from the externalization of human consciousness to contruct a world of roles, rules, institutions, and beliefs. Society appears to be a fact just as does the organic world; yet it is actually more a product of pattered consciousness than it is some 'thing' in itself. Born into such a world, the individual internalizes the values and beliefs of the social environment so that the structures of the outer consciousness become the subjectivity of his growing person. Yet from time to time, and especially in the face of death, the fragility of this consciously created world becomes apparent. Then the realm of social convention faces its shadow, a world of comfortless non-humanity before which a person, socially structured as he is, cannot but feel afraid. Religion is concerned with the sacred quality of power which man finds present in nature. He attempts to coerce this power within his conventional world, to socialize the unsocializable. His structured notions of the 'cosmos', in which he traditionally takes central place, still face the dread of 'chaos'; nature is not so easily fitted in and the raw aloneness of the person before the void of death keeps breaking through.

> The sacred cosmos, which transcends and includes man in its ordering of reality, thus provides man's ultimate shield against the terror of anomy. To be in a 'right' relationship with the sacred cosmos is to be protected against the nightmare threats of chaos. To fall out of such a right relationship is to be abandoned on the edge of the abyss of meaninglessness. (Berger 1969, p. 27).

The structure of religious belief, dogma, commonly comprises an exercise in theodicy; that is to say rationalization and ritual that attempt to maintain a safe interpretation of the sacred cosmos despite the intrusions of potentially faith-breaking experiences of suffering, evil, and death. To stay within the 'canopy' of belief is security; to venture forth or to be forced to do so as cultural change devalues a prevalent system of meaning, is an adventure fraught with unavoidable fear and anguish. Yet in most periods of history some individuals of great fortitude have sought more robust responses to the obvious facts of death and the unlikelihood of personal rebirth in any form. Such thinkers move towards a philosophy and a practical psychology of acceptance, which inevitably necessitates a profound personal exploration of the meaning of existence for oneself.

Readily available to all is the knowledge that life consists of birth, reproduction, and death; that health and wealth are hazardous and that even the most carefully conceived plans can be circumvented by arbitrary interventions of fate. While social action operates by rules and through roles moulded by the conventions of a socio-economic system, 'life' itself escapes convention and provides only uncertain criteria for meaning. Making sense of life often entails a cultural projection; man's place in nature is thus seen in the images of seasonal necessities, of agricultural gods, or the fanciful dictates of science fiction. Ultimate meaning escapes practical sense and the awareness of this creates of class of arbiters who seek to understand the whole and relate to it for the communal good.

Making sense of life and death and, for some, the subscription to a culturally prevalent explanation, is a vital activity; for upon it depends an individual's sense of persoanl worth and purpose and his acceptance of an ultimately unavoidable fate. While collective explanations of the ultimate are usually patterned upon those of the proximate necessities of life, it is the former that provides the superordinate meanings that apply to all persons and relate humanity and the cosmos in human consciousness.

It is today not easy to imagine the primitive mental world from which systems of meaning that crystalized into the 'great' religions came. Early man was surrounded by inexplicable forces all more powerful than himself and as capable of his destruction as of the provision of some good. The world he experienced was one in which the enigmatic character of things was barely distinguished from the powerful forces within the psyche itself. The 'objective' and the 'subjective' were not distinguishable: experience was a flux of forces with contentment balanced against fear and unkind fate. To stave off fear and fate became the task of the magician, the shaman who, himself often possessed of strong psychological attributes, could move with unusual confidence in a world of elemental forces.

One of the chief ways of coming to grips with things is through naming them. Name and form are indeed the two first principles of the earliest religious writings of the Indian subcontinent (Govinda 1961). To name a form gives power. Forms are the embodiments of names; power through the name gives power over form. The utterance of a name with the mind appropriately concentrated upon its object was found to affect the extent to which the object had power over experience. In this way power could be transferred from the object to the subject and confident action could be once more restored. The great scriptures of the Vedas consist largely of incantations of this sort, chants expressive of an elementary relationship to a deeply experienced world, a fresh world not yet caged in intricate intellectual analyses that distance the subject from the object of his attention.

These chants or incantations 'were as free from moral or religious considerations as the prescriptions of a physician . . . because religion had not yet become an independent value, had not yet separated itself from nature. A child will satisfy his desires with a natural and innocent egotism — innocent, because not yet being conscious of ego . . . if the self has not yet become conscious of itself there cannot be any concept of morality. The individuality of man had not yet been discovered and for that reason there could not be any idea to preserve it' (Govinda 1961). The idea of self as we know it today was not present in either Rigvedic or Homeric times and the 'hymns' of those periods reveal a different functional consciousness from that of later civilized worlds (Jaynes 1976). The chanting of names or 'ultimate propositions' as Rappaport calls them formed no doubt a means for inducing shifts in consciousness which later became the basis for developed meditation practices.

The idea of gods arose from human powers which were projected onto objects and later turned into abstracted representations or themes. As Govinda puts it, man was universalized and the world anthropomorphized. The rituals involving incantations were now aimed outward for power was placed in the abstract god.

It was not at that stage realized how far this abstraction actually remained within the mind; for in experience it was projected outward, perhaps into an image. There were many rituals designed to bring the power of the project back into the person. Wearing the mask of the God, a man became the god; and initiation rites, as at Eleusis, gave access to the powers of the projected agent. To manage these rituals a class of specialists emerged, shamans, priests, and other intermediaries between 'God' and man.

The great world civilizations developed religious systems with varying degrees of humanistic recognition. Often there is an outward theism, a canopy for the masses, and a secret esoteric tradition (e.g. the Caballah in Judaism, Alchemy within Christendom, Sufism in Mohammedanism) in which the focus is less on the outer projection and more on the inward task of understanding the self. Today, living as most of us do in multicultural societies, where people 'have' a particular religion or none at all, it is often difficult to realize how monolithic and all-pervasive the great religions once were. They were virtually identical with the great cultures of the past, a total source of reference for ultimate meanings for countless inhabitants, embracing orthodoxies defining whole cultures, and to differ from them was heresy, punishable perhaps by death.

Berger (1969) has traced the slow downfall of religion in the West to themes detectable in the historicism and rationalism of Judaic religion which sprang into the forefront of Christian thought at the time of the Reformation. Rather than a human life which in microcosm re-enacted the great macrocosmic cycles of projected naturalistic deities, Jewish and

Christian thinkers insisted on an actual historical link between a single god and the affairs of men. When this link was weakened by philosophical humanism and the rise of science the whole structure became unstable.

Given the contemporary sophistication of science and the emphasis upon the behavioural sciences in higher education, it is clear that in a personal quest for meaning no naïve religious, metaphysical, or ethical system is likely to gain adherents among educated people today. The fact that Buddhism is at present said to be the fastest-growing religion in Britain suggests that here at least is a system of world and life interpretation that is sufficiently sophisticated to appeal to contemporary young people who must comprise one of the most educated generations ever to have lived on Earth. This present-day example of man as a maker of meaning merits scrutiny here for it illustrates particularly clearly how a system of meaning can function in moulding and sustaining individual life, and its potential for social influence.

The merits of Buddhism in the contemporary West seem to lie in the following prime characteristics:

(1) It lacks both metaphysical and ethical dogmatism, and the stance of its most prominent figures is not authoritarian.

(2) It provides a profound and subtle analysis of the human dilemma which never trails off into empty philosophical speculation or a purely intellectual analysis, but both starts and ends in a close examination of human experience.

(3) This analysis is anchored in what I term a 'subjective empiricism'; an 'experimental' study of experience under the closely controlled circumstances of the everyday life of a community and in precisely structured psychophysical disciplines. This tradition stems, indeed, from the Buddha himself, who set out to discover in personal experiment what the limits of human understanding of the self might be.

(4) Buddhism invites those concerned with problems of life to follow a phenomenological pathway which, in its various forms, remains recognizably that of its originator. The exhaustive nature of this path becomes apparent only to a practitioner, and a merely intellectual questioner is given short shrift.

(5) Although contemporary Buddhism is as much afflicted with pluralism as is Christianity — Zen masters, Tibetan lamas, Thai monks, and Indian gurus jostle for space in the advertising sections of Buddhist journals — there is an underlying unity which is no mere matter of doctrinal belief. This unity focuses on the notion of individual search. 'Go work out your salvation with diligence' is a message for individual persons. A path is provided that can yield psychological results which enable the individual to meet 'chaos' on its own ground without fear. The task is to tread the way. Faith in a doctrine is of little importance. An inner atheoretical knowledge of self is acquired by transmission from a teacher or by personal discovery.

(6) The Buddha was directly concerned as much with social issues as he was with individual self-salvation. This fact is little understood but, as Ling (1973) makes clear, the creation of the *Sangha*, or community of practitioners, was intended to lead to the formation of a state governed by rational principles derived from personal and social understanding. Indeed, such a state did in fact emerge for important periods in the histories of India, Ceylon, Burma, Thailand, Tibet, and Japan, and China was at times much influenced by a combination of Taoism and Buddhism.

The Buddhist model of mind provides an embracing set of concepts concerning consciousness that are more inclusive than any model based solely upon modern psychological inquiry.

This defect in contemporary thought arises mainly because of the reductionist tenets of so much of modern psychology. Indeed, without a renewed concept of man as an active agent with powers there has been little hope of a sustained modelling of mind that can treat human experience with any fullness.

What sort of 'meaning' then does a Buddhistic approach provide? The Buddha's system sprang naturally from the story of his life. There seems no reason for doubting the validity of the essential elements of this story. Raised in an influential household he was at first a spoilt child given every luxury and from whom the difficulties of life were hidden. The discovery of the facts of sickness, old age, and death came as a personal shock. The sight of a wandering Sadhu on a quest for 'truth' opened the young man's mind to the possibility of a path or way of life. Finally, unable to resist his questioning as to the meaning of life and death, he bade his sleeping wife and son goodbye and left to become a wandering mendicant. After a period of intense asceticism that provided training for his body and mind, the young Siddhatha, having nearly killed himself with extremes, came to the conviction of their futility. He treated himself kindly, fed well and sat beneath a tree resolved to complete his quest. In profound meditation the inner barriers finally collapsed and he experienced 'enlightenment'. By this is meant that the boundaries of his ego-structured mind dissolved in an inner act of disidentification. He found himself in an unbounded, timeless, non-naming, primordial, and all-inclusive state of awareness which he recognized as the ground of experiencing within which all cognitive structures of named form are created and back to which they can dissolve. In the days that followed he not only systematized his understanding through logical comprehension but, after a period of doubt, decided that this was no mere private event but a solution that he could and should convey to others of his time.

The Four Noble Truths of his first discourse were:

(1) The fact of suffering: all human beings experience a lack of ease, an anxiety if not actual pain in the very fact of being conscious, alive, and self-aware. Self-awareness entails death-awareness also.
(2) This suffering (*dukkha*) is due to a clinging or attachment to phenomena, which being naturally impermanent, cannot give the

satisfaction of security. Above all, attachment to one's own individuality (ego) is the prime source of suffering, for this too is necessarily impermanent, undergoes change, and dies. Indeed, upon analysis it is seen to be no entity at all, but a process. The illusion that one is an entity to be preserved is the prime source of psychological pain.

(3) There is a possibility of relief from suffering which stems from a profound inner realization concerning the nature of the self-process.

(4) The path or way to such a realization entails attention to right views, intentions, speech, action, livelihood, effort, mindfulness, and concentration. The word 'right' implies that the feature in question should be expressive of a stance in life that is primarily directed at comprehending and transcending personal egoism and which involves constant watchfulness for its fulfilment. Effort, mindfulness, and concentration are basically the means whereby right livelihood is maintained.

The essential message is that while personal growth must involve self-definition and a role in life (livelihood) too great an emphasis on individuation produces a defensive ego-consciousness that becomes closed to change through fear concerning its own maintenance. This closure is a sign of ignorance (*avijja*) which only a profound emotional education can dispel.

The Buddha's initial insight was developed by generations of scholars who lived within the order of monks he created. This order, the Sangha, was by no means parasitical on society. In exchange for alms, robes, books, and simple shelter the monks provided education, spiritual advice, insight, and support for those who remained householders. It was a system of mutual exchange. Buddhism was, however, to remain primarily monastic and it was this that led, on the one hand, to the depth of psychological scholarship and, on the other, to a sociological frailty. When the monasteries were destroyed by Islam, the roots of the praxis thus failed to survive in the populace at large. The result was a return in India to the complexities of a village-based and polytheistic Hinduism which, in its subsequent philosophical development by Shankara, owed much to the earlier insights of the Buddhists.

Buddhist scholarship entailed an analysis of the fruits of experience and especially of the effects of meditation. The analysis is described in the *Abidhamma* (also spelt *Abhidharma* in some books), a voluminous body of literature only recently subjected to study by adequately equipped western scholars (Govinda 1961; Guenther 1976). The model of mind in this literature underwent a variety of transformations throughout a long history but the prime ideas in early Buddhist thought were already comprehensive.

It is essential to realize that the *Abidhamma* analysis addresses itself to the nature of experience. It treats a process, and the terms in the argument refer to functions that interrelate so closely in the system that they cannot properly be examined as elements mechanistically separable from one another. Experience is determined by attitude (*citta*) which in the average person is a product of conditionings that are the results of previous actions (*Kamma*, e.g. Karma). The conditioning of the average person comprises a structured striving in relation to acquired wants (superimposed upon basic biological needs) and an anxiety about both the achievements and the existence, status, or significance of the person himself. In short, greed, anger, and illusion are the drives that maintain the attitude of a conditioned mind (*citta*) and produce the whole panorama of experiences which, as a result of actions entailed by them, yield a yet further set of similar events. The karmic process is cyclic and self-maintaining. Ignorance (*avijja*) of the cause of the cycle maintains the striving, desiring, or wanting associated with expectations of permanence and security. The full cycle is represented in Fig. 13.1. Ignorance and precipitated karmic consequences from the past (*saṅkhāras*) produce a consciousness (*viññāna*) involving the naming of particulars (*nām-rūpa* in the world of forms, *rūpadhātu*) and the patterned cravings for ego-satisfaction (*tanhā*). If these cravings are temporarily satisfied the result is attachment to the object (*upādāna*). Failure to achieve satisfaction leads to longing in disappointment (*lobha*) or aversion to the apparent obstacles (*dosa*). The patterns of attachment result in what we may call character-formation of identity, the 'becoming a person' (*bhava*), while sustained disappointment and aversion ends in disintegration (death — *jarāmarana*).

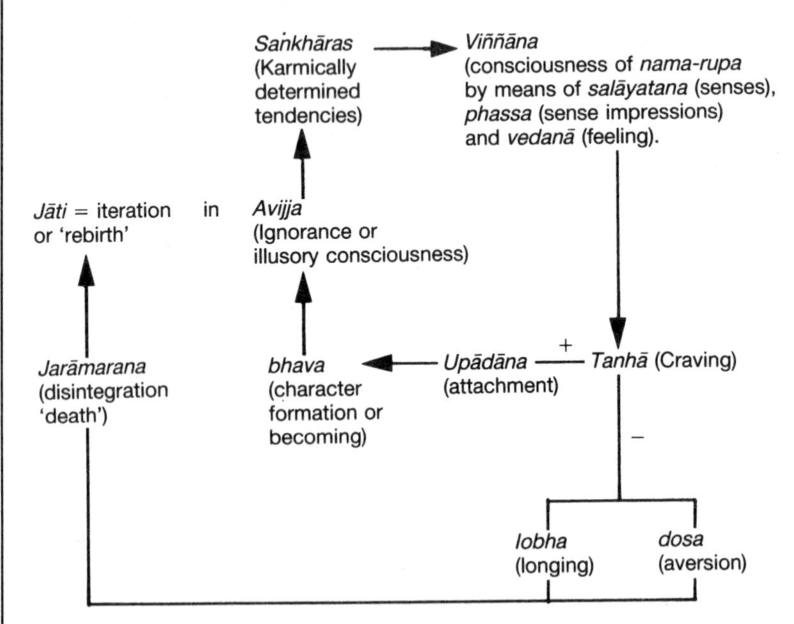

FIG. 13.1. The cycle of dependent origination in the Pali *Abidhamma*. (After Govinda 1961).)

Character-formation initiates through *jāti* — the principle of 'rebirth' — further karmically conditioned actions under illusion (*avijja*), and the cycle, in the form of successive spirals in time, recurs.

The analysis of perception is detailed. *Nāma-rūpa* includes body awareness involving the activation of the senses (*salāyatana*). This results in sensory experience (*phassa*), which includes the recall of earlier images and emotional feelings (*vedanā*) in relation to particular objects.

The growth of identity (*bhava*) through emotional attachment to ideas (*upādāna*) implies a clinging to the perceived; yet since life is in flux this involves an inevitable contradiction and disappointment. It is important to realize that ignorance of the nature of the whole conceptual and emotional cycle is not seen as causal to the cycle in any simple sense nor as some cosmogenic principle, but rather as the *attitudinal condition* that assembles the whole iterating cycle of consciousness in this mode. The whole system of the conditional arising of the co-related functions is known as the dependent origination (*paṭiccasamuppāda*) of mind.

The cycle of dependent origination begins to be broken through a change in attitude that arises once the possible effects of following the Eightfold path have been glimpsed either intellectually or in deeper experience. The eight steps in the path do not really form a sequence: they operate at differing levels. Right views and intentions arise from intellectual appreciation and create a favourable attitude of aspiration for right speech, action, and livelihood which involve ethically determined activity in daily life. These are sustained by right effort, mindfulness, and concentration (the last two being esentially components of the meditative praxis in everyday life) which yield a heightened awareness of being that tends towards a harmony of mind and equanimity. The Eightfold path components are seen as opposed to the functions descriptive of the cycle of dependent origination. Thus

Right views oppose illusion
Right intentions oppose karmic tendencies
Right speech/thought opposes false comprehension of name and form.
Right livelihood opposes uncontrolled emotional expression (*vedanā, tanhā, upādāna*)

Right effort Right mindfulness Right concentration	oppose	illusory craving (*tanhā*) and its consequent effects in becoming (*bhava*).

In his *Principles of Psychology* William James typified human consciousness as the stream of thought. This captures the sense of one sensation or object of thought leading to the next which leads to the next and so on in an apparently seamless and continuous manner. Eugene Taylor shows that this concept of flow underlies the great Chinese philosophy of Taoism. Consequently it is to be found in Zen Buddhism which is the result of a fusion between Taoism from China and Buddhism from

India. Buddhism deals with the dissatisfaction and suffering brought about by our tendency to cling to what are impermanent ever-changing phenomena. Impermanence is also a central feature in Taoism which, like James, characterizes human life and experience as a stream. Ignorance of the nature of this stream and actions predicated on such ignorance interfere with it to produce all manner of ill effects. The acceptance and non-interference with the stream is referred to as *te*. The flow continuously reverses and every action, perception and thought implies its opposite. The Taoist attitude does not recognize 'things' other than as conventions important only for objective communicable thought. Thus Taoism and consequently Zen are very alert to the idea of 'objects' and their role in consciousness.

The theme of non-attachment to the stream and avoidance of trying to control it is at the heart of Zen Buddhism. The idea of a stream is also revealed in yoga, where there is an attempt to control the disparate stream of mental images and bodily sensations by joining them in a single pointed concentration. Of all these eastern complementary approaches to consciousness, it is perhaps yoga which most explicitly reveals the fallacy and consequences of the mind-body dualism which underpins the Western world view.

An examination of the historical roots of American psychology reveals that we have a variety of different frames of reference within which to regard the phenomena of consciousness. There is, first of all, cognitive psychology — what we would call the mainstream of the present-day scientific, academic orientation. Secondly, there is the interpretive framework of the psychoanalytic tradition, which emphasizes dynamics of the unconscious, and most currently, its relation to adaptive functioning of the ego. Thirdly, we have the experiential orientation of the Humanistic movement, advocates of which tend to focus on the creative relationship between consciousness and the unconscious, chiefly through the visualization of preconscious mental processes. Finally, there is the relatively new expression called *Trans-personal Psychology*, which focuses on inner exploration, and the actualization of "ultimate states" of consciousness (Sutich, 1976) achieved through the practice of personal disciplines.

Each of these historical traditions in its own way contributes a unique perspective to understanding the phenomena of consciousness. Yet, we must raise the question as to whether or not any of these are in and of themselves sufficient interpretive frames of reference for an adequate understanding of Asian notions concerning the same phenomena. The answer must in every case be an emphatic, No. Cognitive psychologists, and their behavioristic predecessors, have tended to reduce the rich variety of definitions for the notion of consciousness to specified, measurable constructs. Asian definitions are thus only admissible when they have been cast into the framework of the scientific method, the presuppositions of which some analysts claim are surprisingly protestant in character (Bakan, 1967). In other words, Asian concepts of consciousness are dealt with only when they have been "made Western." The psychoanalytic tradition, on the other hand, has attempted to interpret Asian notions by translating them into Freudian terms (Alexander, 1931), while humanistic and transpersonal psychologists have relied more on their own phenomenological definitions, or the definition of one single Asian spiritual text or teacher only. In all cases, no one has sought to bring together in a very broad way knowledge of how the various classical psychologies of Asia have, themselves, comparatively defined the term.

How then do we as Westerners even presume to have a sensible comprehension of consciousness according to the way the Asian traditions understand it, especially when attempting to comprehend such a thing is at best a hazardous proposition, full of innumerable pitfalls arising from our mutual cultural biases, readily liable to misinterpretation because of radically different philosophical presuppositions, not to mention linguistic barriers, incomparability of basic sources, and the sheer absence of documented material? One means however, may be suggested that has been relatively unexplored by contemporary psychologists, although none other than William James himself was aware of it through his relationship with the famous Harvard Orientalist, Charles Rockwell Lanman (Lanman, in Lyall, 1899). This is the wealth of psychological constructs objectively documented by a tradition of Western scholarship in the scientific study of comparative religion and philosophy.

The basic data of this effort deals with material covering more than three thousand years, including primary translations of significant texts from the original languages, as well as literally thousands of secondary references in the scholarly periodical literature.

While William James suggests that because of the current state of our psychology we must content ourselves with an examination of only the personal aspects of the stream of consciousness (1958, p. 28), classical Chinese Taoism has described it more broadly as the unrealized dimension of all of life. This personal aspect of the stream, James called the "me," or empirical ego, which appears to be made up of different subcomponents, such as the biological, the material, the social, and the spiritual. In Taoism this is expressed as *tao* (pronounced "d" as in "dow," which rhymes with "how"), transliterated into English with a smaller case letter "t." It is their designation for the individual flow the stream of consciousness. The pure sense of "I," on the other hand — that enigmatic illuminating character of consciousness in James's terms — is called *Tao*, transliterated with a capital "T." It refers simply to that flow of all of life — the passage of time, the change of the seasons, the entire milieu of history and evolution; in short, everything that we in the West define as what the individual flow of consciousness is *not*.

In Taoism the individual flow of consciousness and the flow of all of life are continually, at every moment, seen to mutually interpenetrate one another. The reason, according to Lao-tzu, legendary author of the *Tao te Ching*, that we have inner conflicts, problems in relationship, and evil in society, is because of the inability of the normal personality to understand that certain definite laws of change govern both the flow of consciousness and life in the same way. People thus act *as if* the stream of their thoughts and feelings did not have the far-reaching consequences that inevitably they must produce. Such ignorance artificially creates interference with the stream, and the result is neuroticism, psychosomatic conversion reactions, unfulfilled relationships, and the like.

Te, however, is the Taoist notion that expresses the power or energy of things to interpenetrate and change into each other. It is the very principle of transformation itself. Specifically, it means "to plant" in the sense of "potential," a latent power or virtue inherent in something. Psychologically, it means "a power over the outside world undreamt of by those who pit themselves against matter while still in its thralls" (Waley, 1958, p. 46). *Te* is the power not seen by those who are so caught up in the context of the stream of their thoughts and feelings that they fail to see the influences of the *process* of cognitive thought on their current state of physical, emotional, and psychological well-being.

An individual develops *te*, or the power to change the inner conditions of mental ill-health, by comprehending the paradoxical law that governs the flow of the stream, that the Tao in motion is constantly reversing itself. In other words, consciousness is cyclic, that everything is at some point always changing into its opposite. Thus, it is poetically said of *Tao*:

> The coldest day of winter
> Is the first day of spring.

And so of *te*:

> To beget
> To nourish
> To beget but not to cherish
> To achieve but not to claim
> This is called the Mystic Virtue(*te*).

But how, might we ask, does one acquire *te*? Two answers that the Taoists give are behavioral, through the practice of Chinese forms of Yoga and meditation; and mental, by developing the attitude of *wu*, literally "no-thing," which means transcending the unconscious influences of the stream through spontaneous noninterference with it.

More correctly, however, the technical term is not just *wu*, but *wu-wei*, or the "magical passivity of inward quietness." *Wu* should be taken to mean "no-thing" in the sense of nothing obscuring the clarity or lucidity of the stream of awareness. On the other hand, *wei* means "obscure because of being so small or dark" (Waley, 1958, p. 187). It refers to the subtlety of things that are normally unconscious, particularly our inability to see that the stream of both consciousness and the unconscious are perpetually changing into one another. *Wu-wei* thus means that by inward quietness, or by not interfering with the natural course of things, these fundamental changes of life are revealed.

Wu-wei has been variously translated to mean noninterference, non-doing, the secret of action without deeds, or actionless activity. The Taoists, however, did not mean that one should never act, but that one should be fluid and changing enough to always know how to adjust one's self to circumstances. But how to make circumstances work for us? Through *non*interference with the natural course of events. It is a negatively stated expression about an essentially positive condition. It does not mean negative in the sense of bad or pessimistic, but rather, is used in terms of "opposite," as in "looking in the opposite place from where you would expect to find the answer." This is why it is said:

> What is of all things most yielding (water)
> Can overwhelm that which is of all things most hard(rock).
> Being substanceless it can enter even where there is no space.
> That is how I know the value of action that is actionless.
> But that there can be teaching without words,
> Value in action that is actionless,
> Few indeed can understand. (Waley, 1958, p. 197)

In adopting the attitude of wu, the realized man thus harmonizes with the stream, sees into its inner meaning, transcending it through noninterference. He blends with it by becoming completely still and so paradoxically transformed consciousness through his inaction.

While the Zen tradition in Japan is essentially Buddhist in origin, it may be considered as distinctly different from its Indian counterparts (which we will discuss shortly), principally because of its historical antecedents in Chinese soil. Zen in many respects must therefore claim a closer affinity with the general spirit of Taoism. This we see in the Zen expression of *wu-nien*. As with Taoist *wu-wei, wu-nien* means not repression, nor forcefully stopping thought, but rather, nonattachment and consequently nonreinforcement of the stream of thought and feelings.

In the work of Hui-neng, legendary founder of the Zen tradition in China as it later developed in Japan, *wu-nien* had its greatest expression as "no-thought-ness." It was Hui-neng who said, "No-thought is not to think, even when involved in thought." Paraphrasing his own translation of Hui-neng, the scholar Yampolski writes:

> Thoughts are conceived as advancing in progression from past to present to future, in an unending chain of successive thoughts. Attachment to one instant of thought leads to attachment to a succession of thoughts, and thus to bondage. But by cutting off attachment to one instant of thought, one may . . . cut off attachment to a succession of thoughts and thus attain to no-thought, which is a state of enlightenment. (Yampolski, 1967, p. 116)

No-thought thus means watching a thought go by without identifying with it emotionally or cognitively with more thoughts and feelings. One then sees a thought arise, burst forth into the field of conscious awareness, reach its zenith, and then begin to degenerate, decay, and finally disappear. Thoughts come and go without any consequence. They have not produced any associated thoughts. No words have been spoken as a result of attaching one's self to them, nor have any behaviors been initiated.

Hui-neng further taught that as soon as one instant of thought is cast off the person awakens into a completely different state. This is because when one becomes aware of the field of consciousness after quieting the flow of thoughts and feelings, consciousness itself is seen to change. In other words, consciousness of what is normally unconscious causes both consciousness and the unconscious to become something else. The two no longer are the same that they were before because both were defined in relation to each other. What they both become is just *wu* — awareness with no conceptual thought. Under the proper instruction of a Zen master, this leads to what is called in Zen the experience of *satori*, or "a breakthrough into the reality of one's own true nature," a deep inner state where one is a witness to that which is "pure, genuine, vast, and illuminating" about one's self (Chang, 1959, p. 80).

Yoga, being a Sanskrit word derived from the root *yuj* (discipline) means to join or yoke, as in "yoking" the wandering train of our thoughts by concentrating on them, collecting the normally scattered rays of the mind and joining them in one-pointed concentration with the body, integrating mind and body so that both are transcended. This "yoking" or taming of the mind is generally described as the purification of consciousness by promoting inward calmness. This is a process of transcending the stream that is technically called *nirodha*, a "burning out" of undesirable characteristics of mental life through the practice of inner examination and self-control, both of which are essential to produce the conditions necessary for insight.

Nirodha first refers to a quieting of the flow of waking conscious impressions. This is expressed in the famous phrase, "yoga is the restriction of the fluctuations of mental activity." Restriction in this sense is a condition brought about through extended practice in detaching the senses from their attachment to objects in the external material world. Thus, like Taoist *wu*, or *wu-nien* in Zen, it is not a form of repression, the automatic blocking out of contents from consciousness, nor is it to be termed supression, the willful, active, and directed disregard of conscious contents, therefore, forcing them into the unconscious. Rather, restriction refers to a relaxed, natural cutting off of the immediate influx of sense impressions, and so also to their lack of influence in stimulating cognitive perceptual thought processes. More subtly, however, *nirodha* refers to a "burning out" of the unconscious determinants of conscious thoughts, a "burning up" of the seeds of waking conscious impressions.

To understand this it is important to examine the meaning of consciousness *per se* in Yoga. The technical term for consciousness is *citta* (pronounced "ch" as in "church") which appears in the normal field of awareness in the form of changing cognitive thoughts (*vritti*) and simultaneously in the unconscious as "latent impressions" (*samskaras*). Every waking cognitive impression that we have, according to Yoga, psychology, is accompanied by a corresponding unconscious impression or "seed" that lies dormant in the unconscious, ripening like a piece of fruit until it is then ready to sprout forth into the field of awareness to produce more cognitive thoughts under the appropriate future conditions. Hence, we have memories, visual pictures, emotions, and the like, associated with a thought in the field of awareness or accompanying it as a "penumbra" in James's terms, that peripheral tinge to the passing of each cognitive impression. Note here the similarity of this notion of "unconscious seed" to Freud's description of the spontaneous eruption of unconscious repressions into consciousness (Freud, 1951) and to Tolman's famous experiments demonstrating the phenomena of latent learning (Tolman, 1957).

All thought, whether conscious or unconscious, in Yoga is classified as to whether it hinders or helps in the process of self-realization. Hindered mental activities include such things as what we in the West would call factual knowledge, all types of verbal communication, misconceptions, delusions, sleep, and memory, especially that aspect of memory called forgetting. Those that are unhindered, or liberating, include good thoughts, insights, and inner realizations. Unhindered kinds of thoughts tend to balance or neutralize the influence of hindered thoughts, and so need to be actively cultivated in the beginning. Eventually, however, Yoga means restricting both hindered and unhindered thought, good thoughts as well as bad, since even the "thought of enlightenment" can become another limiting preconceived idea, and so be for the unwary just another diabolical mental game.

Transcending the stream thus means not only quieting the flow of thoughts on the surface of the mind, but also eliminating the possibility that any thought will sprout forth from the unconscious. How to accomplish this is perhaps one of the most interesting aspects of Yoga psychology and involves the active cultivation of insight into one's own inner thought processes through the practice of personal discipline.

We have already noted that Buddhism (and some other Eastern philosophies) offers a practical way of discovering the value of their teaching for oneself. In fact it is part of their philosophy that there are severe limits on what may be communicated through the symbols of language and that it is necessary for everyone to embark on their own programme of self-examination and instruction. Thus they insist that the practices which foster 'subjective empiricism' are essential to realizing insight. Clearly language is going to be severely limited as a tool for enlightenment. As Wittgenstein made clear, we cannot expect language to help us realize the effects of language.

Meditation is the generic term to describe the practice deemed necessary by these Eastern approaches to consciousness. Claudio Naranjo points to the diversity of practices which are given the name 'meditation'.

He proposes that these practices are unified by the idea of seeking a *presence*. The presence involves stillness: stillness of physical activity and stillness in mental activity. It is only from direct observation within that state of stillness that we can begin to see how little control we can bring to bear on the passage from one object of thought to the next. It is only in such stillness that we can hope to begin to experience the world as it is rather than how we have become conditioned to see it.

The word "meditation" has been used to designate a variety of practices that differ enough from one another so that we may find trouble in defining what *meditation* is.

Is there a commonality among the diverse disciplines alluded to by this same word? Something that makes them only different forms of a common endeavor? Or are these various practices only superficially related by their being individual spiritual exercises? The latter, apparently, is the point of view of those who have chosen to equate meditation with only a certain type of practice, ignoring all the others that do not fit their description or definition. It is thus that in the Christian tradition meditation is most often understood as a dwelling upon certain *ideas*, or engaging in a directed intellectual course of activity; while some of those who are more familiar with Eastern methods of meditation equate the matter with a dwelling on anything *but* ideas, and with the attainment of an aconceptual state of mind that excludes intellectual activity.

The distinction between ideational versus non-ideational is only one of the many contrasting interpretations of the practices called meditation. Thus, while certain techniques (like those in the Tibetan Tantra) emphasize mental images, others discourage paying attention to any imagery; some involve sense organs and use visual forms (mandalas) or music, and others emphasize a complete withdrawal from the senses; some call for complete inaction, and others involve action (mantra), gestures (mudra), walking, or other activities. Again, some forms of meditation require the summoning up of specific feeling states, while others encourage an indifference beyond the identification with any particular illusion.

The very diversity of practices given the name of "meditation" by the followers of this or that particular approach is an invitation to search for the answer of what meditation is *beyond its forms*. And if we are not content just to trace the boundaries of a particular group of related techniques, but instead search for a unity within the diversity, we may indeed recognize such a unity in an *attitude*. We may find that, *regardless of the medium* in which meditation is carried out — whether images, physical experiences, verbal utterances, etc. — the task of the meditator is essentially the same, as if the many forms of practice were nothing more than different occasions for the same basic exercise.

If we take this step beyond a behavioral definition of meditation in terms of a *procedure*, external or even internal, we may be able to see that meditation cannot be equated with thinking or non-thinking, with sitting still or dancing, with withdrawing from the senses or waking up the senses: meditation is concerned with the development of a *presence*, a modality of being, which may be expressed or developed in whatever situation the individual may be involved.

Just as the spirit of our times is technique-oriented in its dealings with the external world, it is technique-oriented in its approach to psychological or spiritual reality. Yet, while numerous schools propound this or that method as a solution of human problems, we know that it is not merely the method but *the way in which it is employed* that determines its effectiveness, whether in psychotherapy, art, or education. The application of techniques or tools in an interpersonal situation depends upon an almost intangible "human factor" in the teacher, guide, or psychotherapist. When the case is that of the intrapersonal method of meditation, the human factor beyond the method becomes even more elusive. Still, as with other techniques, it is the *how* that counts more than the *what*. The question of the right attitude on the part of the meditator is the hardest for meditation teachers to transmit, and though it is the object of most supervision, may be apprehended only through practice.

It might be said that the attitude, or "inner posture," of the meditator is both his path and his goal. For the subtle, invisible *how* is not merely a *how to meditate* but a *how to be*, which in meditation is exercised in a simplified situation. And precisely because of its elusive quality beyond the domain of an instrumentality that may be described, the attitude that is the heart of meditation is generally sought after in the most simple external or "technical" situations: in stillness, silence, monotony, "just sitting." Just as we do not see the stars in daylight, but only in the absence of the sun, we may never taste the subtle essence of meditation in the daylight of ordinary activity in all its complexity. That essence may be revealed when we have suspended everything else but *us*, our presence, our attitude, beyond any activity or the lack of it. Whatever the outer situation, the inner task is simplified, so that nothing remains to do but gaze at a candle, listen to the hum in our own ears, or "do nothing." We may then discover that there are innumerable ways of gazing, listening, doing nothing; or, conversely, innumerable ways of *not* just gazing, not just listening, not just sitting. Against the background of the simplicity required by the exercise, we may become aware of ourselves and all that we bring to the situation, and we may begin to grasp experientially the question of attitude.

While practice in most activities implies the development of habits and the establishment of conditioning, the practice of meditation can be better understood as quite the opposite: a persistent effort to detect and become free from all conditioning, compulsive functioning of mind and body, habitual emotional responses that may contaminate the utterly simple situation required by the participant. This is why it may be said that the attitude of the meditator is both his path and his goal: the unconditioned state is the freedom of attainment and also the target of every single effort. What the meditator realizes in his practice is to a large extent how he is failing to meditate properly, and by becoming aware of his failings he gains understanding and the ability to let go of his wrong way. The right way, the desired attitude, is what remains when we have, so to say, stepped out of the way.

If meditation is above all the pursuit of a certain state of mind, the practice of a certain attitude toward experience that transcends the qualities of this or that particular experience, a mental process rather than a mental content, let us then attempt to say what cannot be said, and speak of what this common core of meditation is.

A trait that all types of meditation have in common, even at the procedural level, gives us a clue to the attitude we are trying to describe: all meditation is a *dwelling upon* something.

While in most of one's daily life the mind flits from one subject or thought to another, and the body moves from one posture to another, meditation practices generally involve an effort to stop this merry-go-round of mental or other activity and to set our attention upon a single object, sensation, utterance, issue, mental state, or activity.

"Yoga," says Patanjali in his second aphorism, "is the inhibition of the modifications of the mind." As you may gather from this statement, the importance of dwelling upon something is not so much in the *something* but in the *dwelling upon*. It is this concentrated attitude that is being cultivated, and, with it, attention itself. Though all meditation leads to a stilling of the mind as described by Patanjali, it does not always consist in a voluntary attempt to stop all thinking or other mental activity. As an alternative, the very interruptions to meditation may be taken as a temporary meditation object, by dwelling upon them. There is, for example, a Theravadan practice that consists in watching the rising and falling of the abdomen during the breathing cycle. While acknowledging these movements, the meditator also acknowledges anything else that may enter his field of consciousness, whether sensations, emotions, or thoughts. He does it by mentally naming three times that of which he has become aware ("noise, noise, noise," "itching, itching, itching") and returning to the rising and falling. As one meditation instructor put it: "There is no disturbance because any disturbance can be taken as a meditation object. Anger, worry, anxiety, fear, etc., when appearing should not be suppressed but should be accepted and acknowledged with awareness and comprehension. This meditation is for dwelling in clarity of consciousness and full awareness."

The practice described above is a compromise of freedom and constraint in the direction of attention, in that the meditator periodically returns to the "fixation point" of visual awareness of his respiratory movements. If we should take one further step toward freedom from a pre-established structure, we would have a form of meditation in which the task would be merely to be aware of the contents of consciousness at the moment. Though this openness to the present might appear to be the opposite of the concentrated type of attention required by gazing at a candle flame, it is not so. Even the flame as an object of concentration is an ever-changing object that requires, because of its very changeability, that the meditator be in touch with it moment after moment, in sustained openness to the present. But closer still is a comparison between the observation of the stream of consciousness and concentration on music. In the latter instance, we can clearly recognize that a focusing of attention is not only compatible with, but indispensable to, a full grasp of the inflections of sound.

Our normal state of mind is one that might be compared to an inattentive exposure to music. The mind is active, but only intermittently are we aware of the present. A real awakening to the unfolding of our psychic activity requires an effort of attention greater and not lesser than that demanded by attending to a fixed "object" like an image, verbal repetition, or a region of the body. In fact, it is because attention to the spontaneous flow of psychological events is so difficult that concentrative meditation *sensu stricto* is necessary either as an alternative or a preliminary.

Attending to one's breath, for instance, by counting and remaining undistracted by the sensations caused by the air in one's nose, is a much more "tangible" object of consciousness than feeling-states and thoughts, and by persisting we may discover the difference between true awareness and the fragmentary awareness that we ordinarily take to be complete. After acquiring a taste of "concentrated state" in this situation and some insight into the difficulties that it entails, we may be more prepared for the observation of "inner states."

Such a "taste" can be regarded as a foretaste, or, rather, a diluted form of the taste the knowledge of which might be the end result of meditation. In the terminology of Yoga, that ultimate state is called *samadhi*, and it is regarded as the natural development of *dhyana*, the meditative state, itself the result of an enhancement or development of *dharana*, concentration. Dharana, in turn, is regarded as a step following *pranayama*, the technique of breathing control particular to Yoga, which entails just such a concentrative effort as the spontaneous breathing of Buddhist meditation.

The process leading from simple concentration to the goal of meditation (*samadhi, kensho*, or whatever we may want to call it) is thus one of progressive refinement. By practicing attention we understand better and better what attention is; by concentrating or condensing the taste of meditation known to us we come closer and closer to its essence. Through this process of enhancing that *attitude* which is the gist of the practice, we enter states of mind that we may regard as unusual and, at the same time, as the very ground or core of what we consider our ordinary experience. We would have no such "ordinary" experience without awareness, for instance, but the intensification of awareness leads us to a perspective as unfamiliar as that of the world which intensified scientific knowledge reveals to us — a world without any of the properties evident to our senses, materiality itself included.

Awareness, though, is only a facet of that meditative state into whose nature we are inquiring. Or, at least, it is only a facet if we understand the term as we usually do. The meditator who sets out to sharpen his awareness of awareness soon realizes that awareness is inseparable from other aspects of experience for which we have altogether different words, and so intertwined with them that it could be regarded as only conceptually independent from them.

Let us take the classical triad *sat-chit-ananda* according to the formulations of *Vedanta*, for instance. On the basis of the experiential realizations in which we are interested here, these three are our true nature and that of everything else, and the three are inseparable aspects of a unity: *sat* means being; *chit*, consciousness of mind; *ananda*, bliss.

From our ordinary point of view, these three seem quite distinct: we can conceive of being without bliss or awareness, of awareness without bliss. From the point of view of what to us is an unusual or "altered" state of consciousness, on the other hand, the individual sees his very identity in another light, so that he *is* consciousness. His very being is his act of awareness, and this act of awareness is not bliss-ful but consists *in* bliss. While we ordinarily speak of pleasure as a reaction in *us* to *things*, the meditator in samadhi experiences no distinction between himself, the world, and the quality of his experience because he *is* his experience, and experience is of the nature of bliss. From his point of view, the ordinary state of consciousness is one of not truly experiencing, of not being in contact with the world or self, and, to that extent, not only deprived of bliss but comparable to a non-being.

Special states of consciousness are not more expressible than states of consciousness in general, and are bound to the same limitation that we can only understand what we have already experienced. Since the goal of meditation is precisely something beyond the bounds of our customary experience, anything that we might understand would probably be something that it is not, and an attachment to the understanding could only prevent our progress.

One writer who is able both to be scholarly and to communicate convincingly in a language accessible to many is Alan Watts. We have chosen an extract from two chapters in his book, *The Way of Zen*. The first examines our tendency to explain natural phenomena in terms of cause and effect. This is an attitude which obscures our ability to see that '. . . human experience is determined as much by the nature of the mind and the structure of the senses as by the external objects which it reveals.' This is a sentiment echoed in the extract by Leslie White to which we have already referred. There follows an analysis of the role of symbols in the creation of an object with which we identify, which G. H. Mead himself might have provided.

The second part of the extract briefly follows up the consequences of identifying with an 'ego image' of ourselves. One such consequence is that we can never decide whether actions are spontaneous or initiated by 'acts of will'. It suggests that we have little control over what objects of experience we can represent. Our control of deliberate remembering and deliberate forgetting seems limited. What we cannot stop ourselves doing we must be doing spontaneously, so we are in the strange position of spontaneously doing what we do not want to do. It begins to look as though things will just happen whether we will them or not. This need not be a cause for alarm, however, and Watts quotes a Zen Master: 'Nothing is left to you at this moment but to have a good laugh.'

> When everyone recognizes beauty as beautiful,
> there is already ugliness;
> When everyone recognizes goodness as good,
> there is already evil.
> 'To be' and 'not to be' arise mutually;
> Difficult and easy are mutually realized;
> Long and short are mutually contrasted;
> High and low are mutually posited; . . .
> Before and after are in mutual sequence.
>
> Tao Te Ching, 2.

To see this is to see that good without evil is like up without down, and that to make an ideal of pursuing the good is like trying to get rid of the left by turning constantly to the right. One is therefore compelled to go round in circles.

The logic of this is so simple that one is tempted to think it over-simple. The temptation is all the stronger because it upsets the fondest illusion of the human mind, which is that in the course of time everything may be made better and better. For it is the general opinion that were this not possible the life of man would lack all meaning and incentive. The only alternative to a life of constant progress is felt to be a mere existence, static and dead, so joyless and inane that one might as well commit suicide. The very notion of this 'only alternative' shows how firmly the mind is bound in a dualistic pattern, how hard it is to think in any other terms than good or bad, or a muddy mixture of the two.

Yet Zen is a liberation from this pattern, and its apparently dismal starting point is to understand the absurdity of choosing, of the whole feeling that life may be significantly improved by a constant selection of the 'good'. One must start by 'getting the feel' of relativity, and by knowing that life is not a situation from which there is anything to be grasped or gained — as if it were something which one approaches from outside, like a pie or a barrel of beer. To succeeds is always to fail — in the sense that the more one succeeds in anything, the greater is the need to go on succeeding. To eat is to survive to be hungry.

But the viewpoint is not fatalistic. It is not simply submission to the inevitability of sweating when it is hot, shivering when it is cold, eating when hungry, and sleeping when tired. Submission to fate implies someone who submits, someone who is the helpless puppet of circumstances, and for Zen there is no such person. The duality of subject and object, of the knower and the known, is seen to be just as relative, as mutual, as inseparable as every other. We do not sweat *because* it is hot; the sweating is the heat. It is just as true to say that the sun is light because of the eyes as to say that the eyes see light because of the sun. The viewpoint is unfamiliar because it is our settled convention to think that heat comes first and then, by causality, the body sweats. To put it the other way round is startling, like saying 'cheese and bread' instead of 'bread and cheese'. Thus the *Zenrin Kushu* says:

> Fire does not wait for the sun to be hot,
> Nor the wind for the moon, to be cool.

This shocking and seemingly illogical reversal of common sense may perhaps be clarified by the favourite Zen image of 'the moon in the water'. The phenomenon moon-in-the-water is likened to human experience. The water is the subject, and the moon the object. When there is no water, there is no moon-in-the-water, and likewise when there is no moon. But when the moon rises the water does not wait to receive its image, and when even the tiniest drop of water is poured out the moon does not wait to cast its reflection. For the moon does not intend to cast its reflection, and the water does not receive its image on purpose. The event is caused as much by the water as by the moon, and as the water manifests the brightness of the moon, the moon manifests the clarity of the water. Another poem in the *Zenrin Kushu* says:

> Trees show the bodily form of the wind;
> Waves give vital energy to the moon.

To put it less poetically — human experience is determined as much by the nature of the mind and the structure of its senses as by the external objects whose presence the mind reveals. Men feel themselves to be victims or puppets of their experience because they separate 'themselves' from their minds, thinking that the nature of the mind-body is something involuntarily thrust upon 'them'. They think that they did not ask to be born, did not ask to be 'given' a sensitive organism to be frustrated by alternating pleasure and pain. But Zen asks us to find out 'who' it is that 'has' this mind, and 'who' it was that did not ask to be born before father and mother conceived us. Thence it appears that the entire sense of subjective isolation, of being the one who was 'given' a mind and to whom experience happens, is an illusion of bad semantics — the hypnotic suggestion of repeated wrong thinking. For there is no 'myself' apart from the mind-body which gives structure to my experience. It is likewise ridiculous to talk of this mind-body as something which was passively and involuntarily 'given' a certain structure. It *is* that structure, and before the structure arose there was no mind-body.

Our problem is that the power of thought enables us to construct symbols of things apart from the things themselves. This includes the ability to make a symbol, an idea of ourselves apart from ourselves. Because the idea is so much more comprehensible than the reality, the symbol so much more stable than the fact, we learn to identify ourselves with our idea of ourselves. Hence the subjective feeling of a 'self' which 'has' a mind, of an inwardly isolated subject to whom experiences involuntarily happen. With its characteristic emphasis on the concrete, Zen points out that our precious 'self' is just an idea, useful and legitimate enough if seen for what it is, but disastrous if identified with our real nature. The unnatural awkwardness of a certain type of self-consciousness comes into being when we are aware of conflict or contrast between the idea of ourselves, on the one hand, and the immediate, concrete feeling of ourselves, on the other.

When we are no longer identified with the idea of ourselves, the entire relationship between subject and object, knower and known, undergoes a sudden and revolutionary change. It becomes a real relationship, a mutuality in which the subject creates the object just as much as the object creates the subject. The knower no longer feels himself to be independent of the known; the experiencer no longer feels himself to stand apart from the experience. Consequently the whole notion of getting something 'out' of life, of seeking something 'from' experience, becomes absurd. To put it in another way, it becomes vividly clear that in concrete fact I have no other self than the totality of things of which I am aware. This is the Hua-yen (Kegon) doctrine of the net of jewels, of *shih shih wu ai* (Japanese, *ji ji mu ge*), in which every jewel contains the reflection of all the others.

The sense of subjective isolation is also based on a failure to see the relativity of voluntary and involuntary events. This relativity is easily felt by watching one's breath, for by a slight change of viewpoint it is as easy to feel that 'I breathe' as that 'It breathes me'. We feel that our actions are voluntary when they follow a decision, and involuntary when they happen without decision. But if decision itself were voluntary, every decision would have to be preceded by a decision to decide — an infinite regression which fortunately does not occur. Oddly enough, if we had to decide to decide, we would not be free to decide. We are free to decide because decision 'happens'. We just decide without having the faintest understanding of how we do it. In fact, it is neither voluntary nor involuntary. To 'get the feel' of this relativity is to find another extraordinary transformation of our experience as a whole, which may be described in either of two ways. I feel that I am deciding everything that happens, or, I feel that everything, including my decisions, is just happening spontaneously. For a decision — the freest of my actions — just happens like hiccups inside me or like a bird singing outside me.

Man's identification with his idea of himself gives him a specious and precarious sense of permanence. For this idea is relatively fixed, being based upon carefully selected memories of his past, memories which have a preserved and fixed character. Social convention encourages the fixity of the idea because the very usefulness of symbols depends upon their stability. Convention therefore encourages him to associate his idea of himself with equally abstract and symbolic roles and stereotypes, since these will help him to form an idea of himself which will be definite and intelligible. But to the degree that he identifies himself with the fixed idea, he becomes aware of 'life' as something which flows past him — faster and faster as he grows older, as his idea becomes more rigid, more bolstered with memories. The more he attempts to clutch the world, the more he feels it as a process in motion.

Social conditioning fosters the identification of the mind with a fixed idea of itself as the means of self-control, and as a result man thinks of himself as 'I' — the ego. Thereupon the mental centre of gravity shifts from the spontaneous or original mind to the ego image. Once this has happened, the very centre of our psychic life is identified with the self-controlling mechanism. It then becomes almost impossible to see how 'I' can let go of 'myself', for I am precisely my habitual effort to hold on to myself. I find myself totally incapable of any mental action which is not intentional, affected, and insincere. Therefore anything I do to give myself up, to let go, will be a disguised form of the habitual effort to hold on. I cannot be intentionally unintentional or purposely spontaneous. As soon as it becomes important for me to be spontaneous, the intention to be so is strengthened; I cannot get rid of it, and yet it is the one thing that stands in the way of its own fulfilment. It is as if someone had given me some medicine with the warning that it will not work if I think of a monkey while taking it.

While I am remembering to forget the monkey, I am in a 'double-bind' situation where 'to do' is 'not to do', and vice versa. 'Yes' implies 'no', and 'go' implies 'stop'. At this point Zen comes to me and asks, 'If you cannot help remembering the monkey, are you doing it on purpose?' In other words, do I have an intention for being intentional, a purpose for being purposive? Suddenly I realize that my very intending is spontaneous, or that my controlling self — the ego — arises from my uncontrolled or natural self. At this moment all the machinations of the ego come to nought; it is annihilated in its own trap. I see that it is actually impossible not to be spontaneous. For what I cannot help doing I am doing spontaneously, but if I am at the same time trying to control it, I interpret it as a compulsion. As a Zen master said, 'Nothing is left to you at this moment but to have a good laugh.'

In this moment the whole quality of consciousness is changed, and I feel myself in a new world in which, however, it is obvious that I have always been living. As soon as I recognize that my voluntary and purposeful action happens spontaneously 'by itself', just like breathing, hearing, and feeling, I am no longer caught in the contradiction of trying to be spontaneous. There is no real contradiction, since 'trying' is 'spontaneity'. Seeing this, the compulsive, blocked, and 'tied-up' feeling vanishes. It is just as if I had been absorbed in a tug-of-war between my two hands, and had forgotten that both were mine. No block to spontaneity remains when the trying is seen to be needless. As we saw, the discovery that both the voluntary and involuntary aspects of the mind are alike spontaneous makes an immediate end of the fixed dualism between the mind and the world, the knower and the known. The new world in which I find myself has an extraordinary transparency or freedom from barriers, making it seem that I have somehow become the empty space in which everything is happening.

Notes

1. These are, of course, the very extracts which point to the relevance of this particular 'complementary view' to modern physics.
2. Berger, P. L. and T. Luckmann, *The Social Construction of Reality*.

References to Readings

Alexander, F., 'Buddhistic Training as an Artificial Catetonia', *Psychological Review*, 18: 129–145.
Bakan, D., *On Method*, 1967, San Francisco: Jossey-Bass.
Berger, P.L., *The Social Reality of Religion*, 1969, London: Faber and Faber.
Chang, G.C.C., *The Practice of Zen*, 1959, New York: Harper & Row.
Duval, S. and Wicklund, R.A., *A Theory of Objective Self-awareness*, 1972, New York: Academic Press.
Easton, S., *The Western Heritage*, 1966, New York: Holt, Rinehart & Winston.
Eccles, J.C., *The Human Psyche*, 1980, New York: Springer.
Eddington, A.S., *The Nature of the Physical World*, 1928, Cambridge: Cambridge University Press.
Fouts, R.S., 'Capacities for Language in the Great Apes' in *Proceedings of the IXth International Congress of Anthropological and Ethnological Sciences*, 1973, The Hague: Mouton & Co.
Freud, S., *The Psychopathology of Everyday Life*, 1951, New York: New American Library.
Gallup, G.G., 'Self-recognition in Primates: A Comparative Approach to the Bi-directional Properties of Consciousness', *American Psychologist*, 32: 329–338.
Govinda, Lama A., *The Psychological Attitude of Early Buddhist Philosophy*, 1961, London: Rider.
Griffin, D.R., *The Question of Animal Awareness: Evolutionary Continuity of Mental Experience*, 1976, New York: Rockefeller University Press.
Guenther, H.V., *Philosophy and Psychology in the Abidhamma*, 1976, Berkeley: Shambhala.
Hallowell, A.I., *Culture and Experience*, 2nd Ed., 1971, Philadelphia: University of Pennsylvania Press.
Harris, G.G., *Casting out Anger*, 1978, Cambridge: Cambridge University Press.
Hayes, C., *The Ape in Our House*, 1951, New York: Harper.
Hayes, K. and Hayes, C., 'The Intellectual Development of a Home-raised Chimpanzee', *Proceedings of the American Philosophical Society*, 6: 705–19.
Holloway, R.L., Review of Jerison, H.J., *Evolution of the Brain and Intelligence, Science*, 184: 677–679.
Ince, L.P., Brucker, B.S., and Alba, A., 'Reflex Condition in Spinal Man', *Journal of Comparative and Physiological Psychology*, 92: 796–802.
James, W., *The Principles of Psychology*, 1890, New York: Henry Hold & Co.
James, W., *Talks to Teachers on Psychology*, 1958, New York: Norton.

Jaynes, J., *The Origins of Consciousness in the Breakdown of the Bicameral Mind*, 1976, New York: Houghton Mifflin.

Jerison, H.J., *Evolution of the Brain and Intelligence*, 1973, New York: Academic Press.

Jerison, H.J., 'Paleoneurology and the Evolution of Mind', *Scientific American*, 234: 90–101.

Kenny, A.J.P., Longuet-Higgins, H.C., Lucas, J.R., & Waddinton, C.H., *The Nature of Mind*, 1972, Edinburgh: Edinburgh University Press.

Kolb, B., Sutherland, R.J. & Whishaw, I.Q., 'A Comparison of the Contributions of the Frontal and Parietal Association Cortex to Spatial Localization in Rats', *Behavioral Neuroscience*, 97: 13–27.

Lanman, C.R., Handwritten facsimile, in A.C. Lyall, *Asiatic Studies: Religious and Social,* 1899, London: John Murray.

Lashley, K.S., 'The Behaviouristic Interpretation of Consciousness', *Psychological Review*, 30: 237–272, 329–353.

Lashley, K.S., 'Cerebral Organization and Behavior', *Proceedings of the Association for Research on Nervous Mental Disorder*, 36: 1–18. (Reprinted in Beach, F.A., Hebb, D.O., Morgan, C.T. & Nissen, H.W., *The Neuropsychology of Lashley*, 1960, New York: McGraw-Hill.)

Lienhard, G., 'Modes of Thought' in E. Evans-Pritchard et al., *The Institutions of Primitive Society*, 1967, Oxford: Basil Blackwell.

Lienhardt, G., *Divinity and Experience*, 1961, Oxford: Clarendon Press.

Ling, T., *The Buddha*, 1973, Harmondsworth: Penguin Books.

MacLeod, R.B., 'The Phenomenological Approach to Social Psychology', *Psychological Review*, 54: 193–210.

Macmurray, J. (Ed.), *Some Makers of the Modern Spirit*, 1933, London: Methuen.

Mead, G.H., *Mind, Self, and Society*, 1934, Chicago: University of Chicago Press.

Mowrer, O.H., *Learning Theory and Symbolic Processes*, 1960, New York: Wiley.

O'Keefe, J. and Nadel, L., *The Hippocampus as a Cognitive Map*, 1978, Oxford: Clarendon Press.

Oakley, D.A. and Rusell, I.S., 'Subcortical Storage of Pavlovian Conditioning in the Rabbit', *Physiology and Behavior*, 18: 931–937.

Oakley, D.A. 'Cerebral Cortex and Adaptive Behaviour', in D.A. Oakley and H.C. Plotkin (Eds.), *Brain, Behaviour and Evolution*, 1979a, London: Methuen.

Oakley, D.A., 'Learning with Food Reward and Shock Avoidance in Neodocorticate Rats', *Experimental Neurology*, 63: 627–42.

Oakley, D.A., 'Improved Instrumental Learning in Neodecorticate Rats', *Physiology and Behavior*, 24: 357–366.

Oakley, D.A., Eames, L.C., Jacobs, J.L., Davey, G.C.L. and Cleland, G.C., 'Signal-centred Action Patterns in Rats without Neocortex in a Pavlovian Conditionin Situation', *Physiological Psychology*, 9: 135–44.

Pauli, W., in C.G. Jung and W. Pauli, Eds, *The Interpretation of Nature and the Psyche*, 1955, New York: Bollingen.

Perkins, J., *The Concept of Self in the French Enlightenment*, 1969, Geneva: Librairie Droz.

Paiget, J., *Judgement and Reasoning in the Child*, 1966, Totawa, New Jersey: Littlefield and Adams.

Polyanyi, M., *Personal Knowledge: Towards a Post-critical Philosophy*, 1958, New York: Harper & Row.

Popper, K.R. and Eccles, J.C., *The Self and its Brain*, 1977, New York: Springer.

Postal, S., 'Body Image and Identity: A Comparison of Kwakiutl and Hopi, *American Anthropologist*, 67: 455–462.

Premack, D., *Intelligence in Ape and Man*, 1976, Hillsdale, N.J.: Earlbaum.

Read, K.E., 'Reality and the Concept of the Person among the Gahuku-Gama', in J. Middleton, Ed., *Myth and Cosmos*, 1967, New York: Natural History Press.

Richards, I.A., *Mencius on the Mind*, 1932, London: Routledge & Kegan Paul.

Sayfarth, R.M., Cheney, D.L., and Marler, P., 'Vervet Monkey Alarm Calls: Semantic Communication in a Free-ranging Primate', *Animal Behavior*, 28: 1070–1094.

Schaffer, J.A., 'Philosophy of Mind', *Encyclopedia Brittanica Macropedia*, 12: 224–233, Chicago: Enclycopedia Brittanica.

Sherman, B.S., Hoeler, F.K. and Buerger, A.A., 'Instrumental Avoidance Conditioning of Increased Leg Lowering in the Spinal Rat', *Physiology and Behavior*, 25: 123–128.

'Smith, Adam', *Supermoney*, 1972, New York: Popular Library.

Sperry, R.W., 'A Modified Concept of Consciousness', *Psychological Review*, 76: 532–536.

Strachey, J., *The Standard Edition of the Complete Works of Sigmund Freud*, 1973, London: Hogarth Press.

Strusaker, T., *Behaviour and Ecology of Red Columbus Monkeys*, 1967, Chicago: Chicago University Press.

Suarez, S.D. and Gallup, G.G., 'Self-recognition in Chimpanzees and Orangutans, but not Gorillas', *Journal of Human Evolution*, 10: 175–188.

Sutherland, R.J., Kolb, B. and Whislaw, I.Q., 'Spatial Mapping: Definitive Disruption by Hippocampal or Medial Frontal Cortical Damage in the Rat', *Neuroscience Letters*, 31: 271–276.

Sutich, A.J., *The Founding of Humanistic and Transpersonal Psychology: A Personal Account*, 1976, Unpublished Doctoral Dissertation, San Franscisco: Humanistic Psychology Institute.

Thatcher, R.W. and John, E.R., *Foundations of Cognitive Processes*, 1977, Hillsdale, N.J.: Erlbaum.

Tolman, E.C., *Purposive Behaviour in Animals and Man*, 1957, New York: Century Co.

Waley, A., *The Way and its Power*, 1958, New York: Grove Press.

Walker, S., *Animal Thought*, 1983, London: Routledge & Kegan Paul.

Weintraub, K.J., *The Value of the Individual*, 1978, Chicago: University of Chicago Press.

Weitzsacker, C.F., *The Unity of Nature*, 1980, (trans. Zucker, F.J.), New York: Farrar, Strauss & Giroux.

Yampolski, P.B., *The Platform Sutra of the Sixth Patriarch*, 1967, New York: Columbia University Press.

Yeo, A.G. and Oakley, D.A., 'Habituation of Distraction to a Tone in the Absence of Neocortex in Rats', *Behavioral Brain Research*, 8: 403–409.

Index